New Trends in HERA Physics

Proceedings of the Ringberg Workshop

New Trends in HERA Physics

Tegernsee, Germany 25–30 May 1997

edited by **B A Kniehl**
Max Planck Institute for Physics

G Kramer
Hamburg University

A Wagner
DESY, Hamburg

World Scientific
Singapore • New Jersey • London • Hong Kong

Published by

World Scientific Publishing Co. Pte. Ltd.

P O Box 128, Farrer Road, Singapore 912805

USA office: Suite 1B, 1060 Main Street, River Edge, NJ 07661

UK office: 57 Shelton Street, Covent Garden, London WC2H 9HE

British Library Cataloguing-in-Publication Data
A catalogue record for this book is available from the British Library.

NEW TRENDS IN HERA PHYSICS
Proceedings of the Ringberg Workshop

ISBN 981-02-3398-1

Printed in Singapore.

Foreword

The international workshop entitled *New Trends in HERA Physics* took place on 25–30 May 1997 at Ringberg Castle, which overlooks Lake Tegernsee in the foothills of the Bavarian Alps, one of the most picturesque locations to be found in the whole of Germany. The castle was built during the first half of this century by Duke Luitpold in Bavaria (Herzog Luitpold in Bayern), a member of the Wittelsbach family which ruled Bavaria over 800 years, and his friend Friedrich Attenhuber, an all-round artist, architect, and interior decorator. The castle is entirely their creation, from the massive Renaissance-inspired exterior right down to the fittings and furniture, which, in every detail, were designed by Attenhuber himself and executed by native craftsmen. Attenhuber also painted every single picture hanging in the castle. He found his models in the farmhouses around Lake Tegernsee. The castle embodies all trends of art and styles which dominated the first half of this century, combined with local Alpine originality and the individual creative power of its constructors. According the the Duke's last will, the castle passed into the hands of the Max Planck Society after his death, in 1973. The castle was then transformed into a conference venue, where scientists can exchange their latest ideas and discuss problems with their colleagues from all over the world in beautiful surroundings and in a relaxed mountain atmosphere, high above the daily business activities.

This event has continued a series of Ringberg workshops on electroweak and strong interactions in high-energy reactions, which include: *Hadronic Matrix Elements and Weak Decays* on 17–22 April 1988; *Radiative Corrections for e^+e^- Collisions* on 3–7 April 1989; *High Precision versus High Energies in e^+e^- Collisions* on 15–19 April 1991; *Perspectives of the Electroweak Interactions in e^+e^- Collisions* on 4–8 April 1993; *Advances in Non-Perturbative and Perturbative Techniques* on 13–19 November 1994; *Perspectives for Electroweak Interactions in e^+e^- Collisions* on 5–8 February 1995; and *The Higgs Puzzle—What Can we Learn from LEP2, LHC, NLC, and FMC?* on 8–13 December 1996. On the occasion of the 1997 Ringberg workshop, forty-one experts of elementary-particle physics, both theoreticians and experimentalists, from twenty-nine universities and research institutions in ten countries congregated to present their latest results on the various aspects of HERA physics.

The purpose of the workshop was to gather the latest experimental results from HERA and to capture new trends in the phenomenology of high-energy electron-proton collisions. The presentations were by experts for experts, but were to be suitable for a mixed audience of both theoreticians and experimentalists. H1 members also covered ZEUS results and vice versa. This was the right place to point out existing discrepancies between experimental data and theoretical predictions and to identify projects to be attacked in the future.

There was a total of 35 presentations of 45 minutes each. This proceedings volume is to give a comprehensive account of the novel ideas and results discussed by the speakers. The topics include: proton structure function; polarized *ep* scattering; low-x phenomena in deep inelastic scattering (DIS); high-Q^2 phenomena in DIS; searches for leptoquarks and R-parity violation at HERA and LEP2; jet production in DIS; α_s determination; hadronic final states in DIS; parton-hadron duality; QCD instantons in DIS; photon structure function; $\gamma\gamma$ physics at LEP2; photoproduction of jets; photoproduction of light mesons; open charm production; inelastic J/ψ production; elastic vector-meson production; and diffractive *ep* scattering. We hope that the high-energy-physics community at large will find this document useful in their ongoing efforts in understanding the nature of the strong interactions.

At this point, we should like to thank our friends and colleagues who have contributed their shares to these proceedings. We are indebted to the workshop secretary, Mrs. Renate Saffert, for her assistance before, during, and after the workshop. The costs of catering and lodging at Ringberg Castle and the charges for the publication and dissemination of these proceedings were covered in equal parts by the Deutsches Elektronen-Synchrotron at Hamburg and the Max-Planck-Institut für Physik at Munich.

Munich, September 1997

BERND A. KNIEHL
GUSTAV KRAMER
ALBRECHT WAGNER

Contents

Chapter 1

Proton Structure

PROTON STRUCTURE FUNCTION MEASUREMENT AT HERA

LAURENT FAVART
LAL, Université de Paris-Sud, IN2P3-CNRS
91405 Orsay, France
E-mail: favart@mail.desy.de

On behalf of the H1 and ZEUS collaborations

ABSTRACT

The status of HERA structure function measurements is reviewed and a focus is made on the future improvements

1. Introduction

Deep inelastic lepton-nucleon scattering experiments have provided a remarkable insight into the structure of the matter at small distances. The total deep inelastic scattering cross-section may be written in terms of three, a priori, unknown functions, $F_2(x, Q^2)$, $F_L(x, Q^2)$ and $F_3(x, Q^2)$, called structure functions, related to the nucleon gluon and quark structure.

In the single photon exchange approximation, the $F_3(x, Q^2)$ term does not contribute and for the NC the differential ep cross-section is given by:

$$\frac{d^2\sigma}{dx dQ^2} = \frac{2\pi\alpha^2}{xQ^4}\left[(2(1-y)+y^2)F_2(x,Q^2) - y^2 F_L(x,Q^2)\right] \quad (1)$$

$$= \Gamma\, \sigma^{eff}_{\gamma^* p}(x, y, Q^2) \quad (2)$$

where F_L is related to the cross-section of transverse and the longitudinal polarised photons, σ_T and σ_L: $F_L = F_2 - 2xF_1, R = \sigma_L/\sigma_T = F_L/(F_2 - F_L)$ and has a non negligible effect on the cross-section only in the high y region. This cross-section is related by Eq. 2 to the effective virtual photon-proton cross-section $\sigma^{eff}_{\gamma^* p}$ using:

$$\Gamma = \alpha(2 - 2y + y^2)/2\pi Q^2 x. \quad (3)$$

As Eq. 1 shows, the F_2 and F_L structure functions can not be measured at once for given (x, Q^2) values unless y varies keeping (x, Q^2) constant, which corresponds to a beam energy variation. But at y values not too close to 1, the contribution of the F_L term to the cross-section is very small. NC deep inelastic events at small Q^2 (compared to M_Z^2) are therefore used to measure directly F_2 making an assumption on F_L values. The F_3 structure function measurement implies large Q^2 cross-section measurement with significant statistics.

The rise of F_2 with decreasing x has already been established from the previous measurements with the data taken in 1992, 1993 and 1994 by H1 [1,3,5] and ZEUS [2,4,6], extending with each new year of data taking both the kinematic range of the F_2 measurement (up to $1.5 \leq Q^2 \leq 5000$ GeV^2 and $0.78 \leq y \leq 0.01$) and in the precision achieved.

Last results on structure functions from the H1 and ZEUS experiments at HERA are presented in this report. Five topics are covered:

- the extension of the proton structure function F_2 measurement in a continuous way down to $Q^2 = 0.11$ GeV^2 at low x;
- the gluon density extraction from the proton structure function F_2 QCD fit on a wide Q^2 range;
- a determination of the longitudinal proton structure function F_L in a region of $8.5 < Q^2 \leq 35$ GeV^2 and small x between $1.3 \cdot 10^{-4}$ and $5.5 \cdot 10^{-4}$;
- a measurement of the charm contribution $F_2^{c\bar{c}}$ to the inclusive proton structure function F_2;
- NC and CC cross-section measurements at high Q^2, including a remark on the observed excess of events at very high Q^2 and large y.

2. F_2 Measurement at low Q^2 and low x

In 1995, both H1 and ZEUS collaborations upgraded the backward detectors of their experiment to access a smaller Q^2 region. H1 has introduced a more precise drift chamber (BDC) to measure the polar angle of the scattered electron, and a new calorimeter (SpaCal) with better hadronic containment, better granularity and better angular acceptance. The ZEUS detector was upgraded installing a Beam Pipe Calorimeter (BPC) and modules of the rear detector (SRTD) were moved closer to the beam.
Using the 1995 data which includes the upgraded detectors, H1 [7] and ZEUS [8] performed F_2 measurements covering the Q^2 region as given in table 1. The F_2 measurements are shown in function of x in Fig. 1 and the effective virtual photon-proton cross-section in function of W^2 is shown in Fig. 2. They give new experimental information for the transition region between the perturbative and non-perturbative QCD.

In the region of perturbative QCD ($Q^2 \geq 1$ GeV^2), the evolution of F_2 is nicely described by the model of Glück, Reya, Vogt (GRV) [9], which evolves valence like parton distribution from a low starting scale of 0.34 GeV^2 according to NLO DGLAP equations. At $Q^2 \simeq 0$ (see Fig. 2), in the photoproduction region, the measured cross-sections are in agreement with the Donnachie-Landshoff model (DL) [10], which

Table 1. Kinematic range covered by the different analysis

	Luminosity (pb^{-1})	vertex type	Kinematic range
H1-SpaCal	0.11	shifted	$Q^2 \geq 0.35$ GeV^2
ZEUS-SRTD	0.23	shifted	$Q^2 \geq 0.6$ GeV^2
ZEUS-BPC	1.65	nominal	$0.11 \leq Q^2 \leq 0.65$ GeV^2

assumes a Q^2-independent Regge behaviour up to a few GeV^2, due to the exchange of a "soft" pomeron.

One can notice on Fig. 1 the surprising persistent rise of F_2 with decreasing x, even at very low Q^2, although this rise gets weaker with decreasing Q^2. Therefore the Donnachie-Landshoff model seems not to describe the data, even in the lowest Q^2 bin. The weakening of the rise of F_2 is however too strong in the model of GRV for the data below 1 GeV^2.

Several attempts (as CKMT [11], BK [12] or ABY [13]) were made to connect the very low Q^2 behaviour with perturbative QCD regime but non of them is totally successful and leave the underlying dynamics still to be understood (note that the ABY parametrisation, including preliminary F_2 measurement at Q^2 down to 0.11 GeV^{2}from ZEUS in their fit, has less predictive power in the kinematic region in the figures presented here).

3. The Gluon extraction

The high precision obtained on F_2 measurement (about 5% already using the data collected in 1994) in a wide Q^2 range, allows to perform a NLO-DGLAP fit on the data (restricted to $Q^2 > 5$ GeV^2) together with BCDMS [14] and NMC [15] data, as shown in Fig. 3. A good description of F_2 is observed over the whole kinematic range, surprisingly even when the fit is extrapolated at the lowest Q^2 of 1.5 GeV^2 and shows clearly the scaling violation. The gluon density found by the fitting procedure of both collaborations (Fig. 4a) shows a sharp rise at low x.

No new fit has been performed including the F_2 measurement obtained with 1995 data since the fit uses only $Q^2 > 5$ GeV^2 to escape possible higher twist effects, while the improvement of the new measurement is mainly at smaller Q^2. In future, this fit will be completed by other measurements as the charmed contribution to the inclusive F_2, the jet rate measurement, the longitudinal structure function measurement,... and should provide a competitive way to extract the strong coupling constant, α_s.

4. The Longitudinal Structure Function F_L determination

As we have seen, the F_2 and F_L structure functions can not be measured independently in a (x, Q^2) point unless y varies keeping (x, Q^2) constant. The conventional

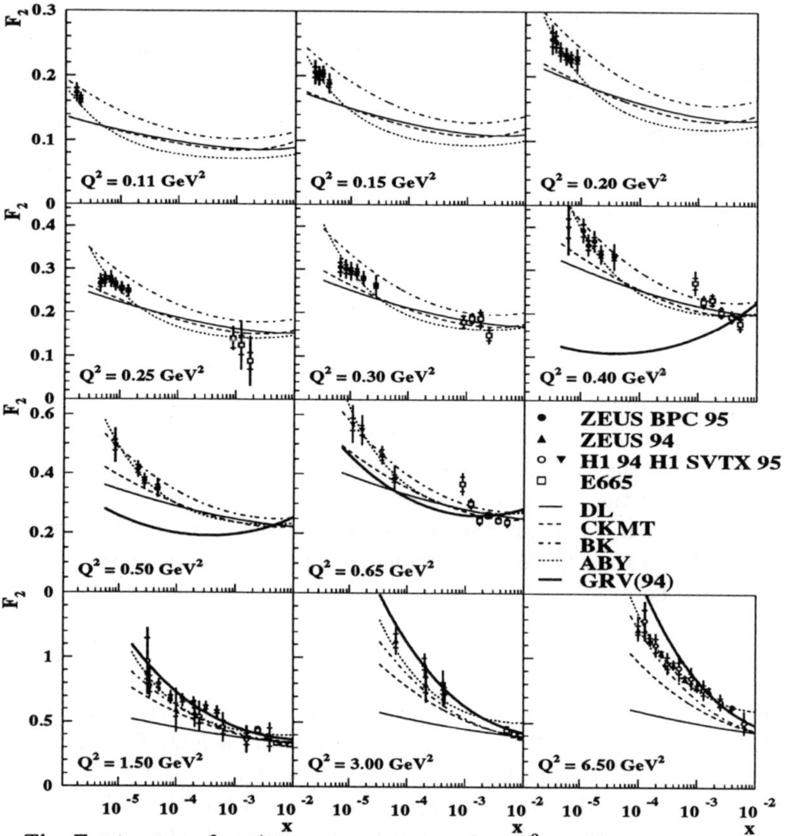

Fig. 1. *The F_2 structure function measurement at low Q^2 and low x compared to fixed target data and to the DL, CKMT, BK, ABY and GRV models.*

method therefore consists in unfolding the DIS cross-section measured using different beam energies as made by various fixed target experiments[16,17]. Another method [18] proposed is to use events containing an initial state radiation with a significant energy, which corresponds to a decrease of the centre of mass energy.

The H1 collaboration has published a determination of F_L at $8.5 < Q^2 < 35$ GeV^2 and at very small x values between $1.3 \cdot 10^{-4}$ and $5.5 \cdot 10^{-4}$, using the so called "subtraction method" [19]: as the behaviour of F_2 can be described with a good precision by a NLO DGLAP fit over several orders of magnitude in Q^2, such a fit is performed on the F_2 data at $y \leq 0.35$, where the influence of F_L is negligible. After an extrapolation up to $y = 0.78$ (keeping x constant) this contribution of F_2 to the cross-section

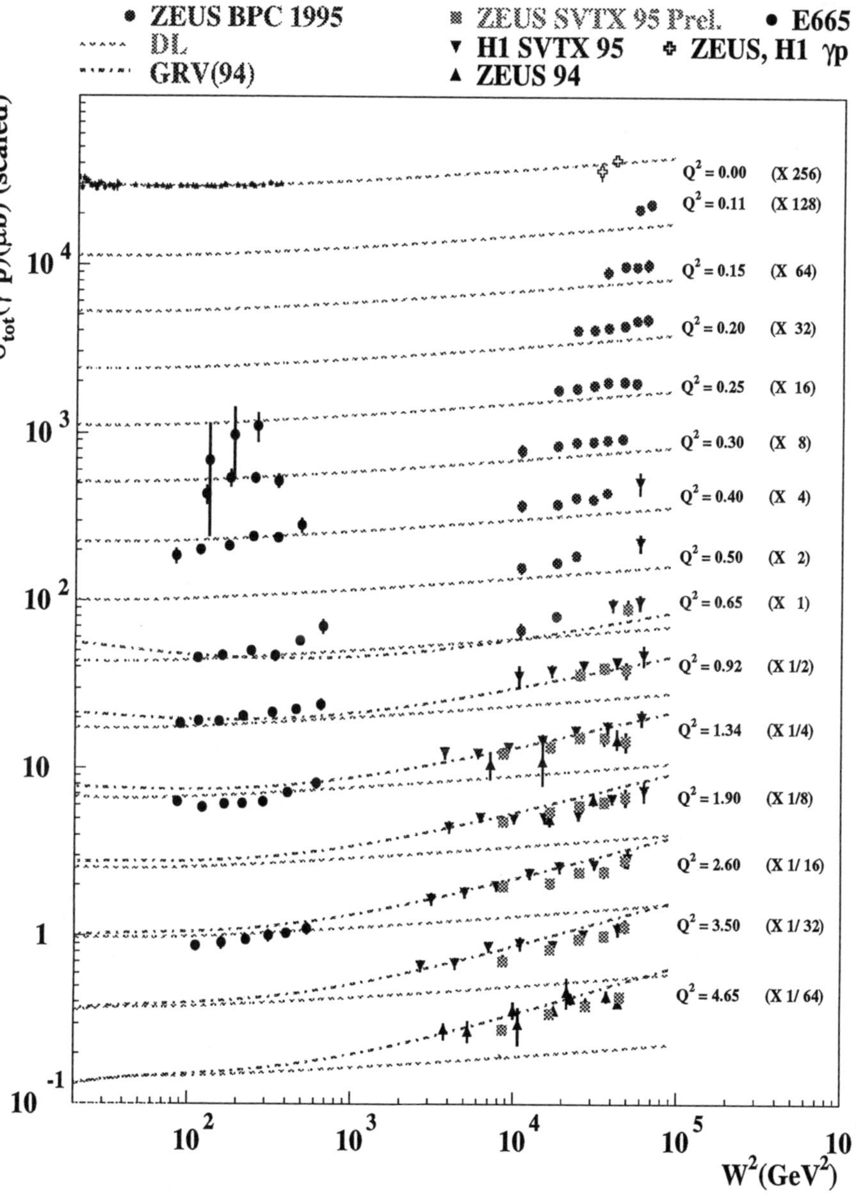

Fig. 2. *The effective virtual photon-proton cross-section measurement in function of W^2; compared to E665 measurement and DL and GRV parametrisations.*

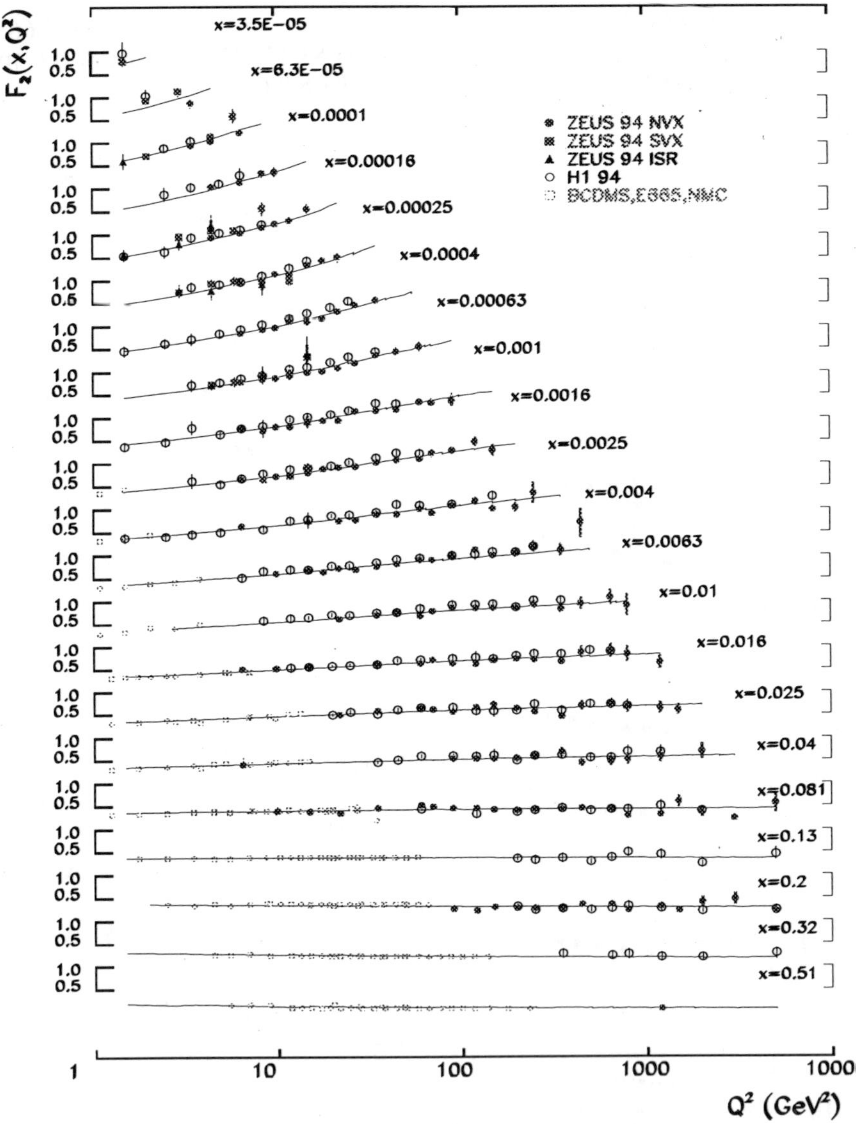

Fig. 3. F_2 *measurement from 1994 data compared to fixed target results (BCDMS, E665, NMC). The full line shows a NLO-DGLAP fit is shown using* $Q^2 > 5\ \mathrm{GeV}^2$

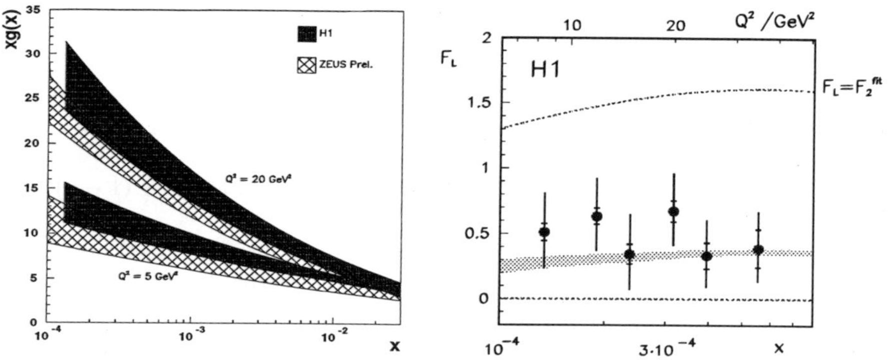

Fig. 4. a) *The gluon density.* b) *Longitudinal structure function F_L extraction.*

is subtracted and the difference attributed to F_L. The results of this procedure is displayed in fig. 4b for six points in x and Q^2.

The F_L values obtained is consistent with the NLO QCD caluculations [21] shown as the band in fig. 4b. It excludes the extreme limits of $F_L = 0$ and $F_L = F_2$ by 2.3 and 4.0 times the total error on F_L.

Concerning the future of the F_L measurement, the possibility to use different beam energies is still in discussion. If such data are taken, they would provide a direct F_L measurement in a large Q^2 range and in a x range depending on the set of centre of mass energies (lowering the beam energies would limit the measurement to a higher x domain). In the other hand, the use of initial state radiative events should provide [20] with 1996 and 1997 data the first direct measurement of F_L at low x and $Q^2 \sim 1$ GeV^2 (which would be complementary to the different beam energies method).

5. Charmed Structure Function Measurement

The extraction of the charmed structure function, $F_2^{c\bar{c}}$, relies on the observation of open charm production in deep inelastic scattering. Tagging of charm is performed by reconstructing D^0 and D^{*+} mesons via their decays into $D^0 \to K^-\pi^+$ and $D^{*+} \to D^0\pi^+$). Taking into account the fragmentation function of $c \to D$ the meson D cross-section has been determined and $F_2^{c\bar{c}}$ derived by H1 [22] and ZEUS [23]. The details of these measurements are given in the heavy-flavour dedicated contribution [24] of this Workshop.

A steep rise of $F_2^{c\bar{c}}$ is observed with decreasing x and is in agreement with the prediction extrapolated from the gluon density of Fig. 4a. An averaged over the kinematic range gives a relative contribution of the charm to the totally inclusive F_2

close to 25% both by H1 and ZEUS.

6. High Q^2 Cross Sections

At high Q^2 the exchange of the W and of the Z^o cannot be neglected anymore, and the differential cross-sections can be rewritten for the NC events as:

$$\frac{d^2\sigma^{NC}(e^{\pm}p)}{dxdQ^2} = \frac{2\pi\alpha^2}{xQ^4}\left(Y_+\mathcal{F}_2(x,Q^2)\mp Y_- x\mathcal{F}_3(x,Q^2)\right) \quad ; \quad Y_{\pm} \equiv (1 \pm (1-y)^2) \quad (4)$$

The $\mathcal{F}_2$, $\mathcal{F}_3$ structure functions of the proton include Z^o-propagator effects and ew-couplings: $\mathcal{F}_2$ contributes symmetrically for e^+ and e^- scattering, while $\mathcal{F}_3$ is parity-violating, so its contribution changes sign when e^+ and e^- beams are exchanged. For the CC events the cross-sections can be expressed directly as a function of the quark densities ($q = q(x, Q^2)$, with $q = u, d, s, c$), G_μ being the Fermi–Coupling constant and M_W is the mass of the W boson in the propagator term.

$$\frac{d^2\sigma^{e^+p}}{dxdQ^2} = \frac{G_\mu^2}{\pi}\frac{M_W^2}{(M_W^2+Q^2)^2}((\bar{u}+\bar{c})+(1-y)^2(d+s)) \quad (5)$$

$$\frac{d^2\sigma^{e^-p}}{dxdQ^2} = \frac{G_\mu^2}{\pi}\frac{M_W^2}{(M_W^2+Q^2)^2}((u+c)+(1-y)^2(\bar{d}+\bar{s})) \quad (6)$$

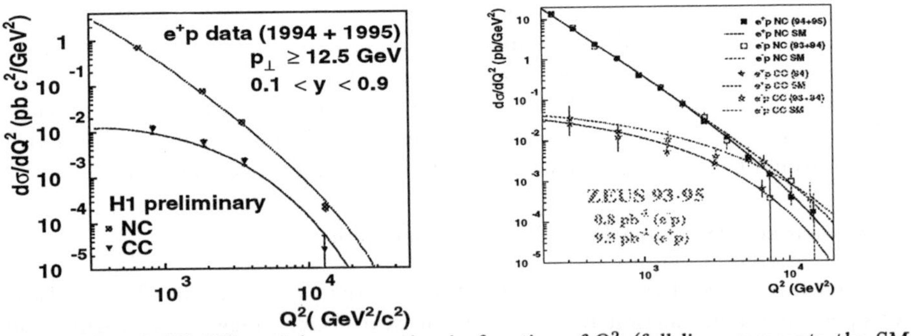

Fig. 5. *NC and CC differential cross-section in function of* Q^2 *(full line represents the SM prediction): left: H1 measurement (1994-1995 data: 6.6* pb^{-1}*), right: ZEUS measurement (1993-1995 data: 20.1* pb^{-1}*).*

The NC and CC differential cross-sections have been measured at high Q^2 ($Q^2 >$ 2500 GeV^2) by both collaborations (see Fig. 5), using the hadrons only to reconstruct the kinematic in order to compare NC and CC with minimal systematic bias. As expected the NC and the CC cross-sections have a comparable size at high Q^2 ($Q^2 \approx M_Z^2$). A good agreement is observed with the Standard Model prediction (full line in Fig. 5) on the full kinematic range available with that statistics.

It is foreseen in 1998 and 1999 to take data with an e^- beam again. Comparing data with colliding positrons and electrons, the sign change in front of the $\mathcal{F}_3$ term should allow a first direct measurement of the F_3 structure function. The y-differential CC cross-section is expected, in future, to provide separate information on the valence and sea quark densities, due to the different y dependence of these two terms. The future e^- beam data combined with previous e^+ data, could then give an access to separated light flavours quark densities.

A focus at the very high Q^2 ($Q^2 > 15000$ GeV^2) and large y ($0.1 \le y \le 0.9$) region shows a significant excess of events observed by both H1 [25] and ZEUS [26]. All details on these analysis and updates are given in dedicated report of this workshop [27] and will not be developed again here. But it seems important to emphasised in the frame of this structure function report, that this excess cannot be explain in a structure function approach in the frame of QCD. In other words, the freedom given by the parametrisations of the quark and gluon densities, by the unknowing of the longitudinal structure function and by the uncertainty on the radiative correction is largely not sufficient to explain the observed excess. The future data will precise the measurement and give a precise significance to the observed signal.

7. Acknowledgements

I would like to thank the organisers for the very exciting workshop and its perfect organisation. I am also grateful to all H1 and ZEUS members who made by their work these results presented at Ringberg, and particular thanks to Gregorio Bernardi and Bernd Surrow for the closer help they offered me.

8. References

1. H1 Collab., I. Abt et al., *Nucl. Phys.* **B407** (1993) 515.
2. ZEUS Collab., M. Derrick et al., *Phys. Lett.* **B316** (1993) 412.
3. H1 Collab., T.Ahmed et al., *Nucl. Phys.* **B439** (1995) 471.
4. ZEUS Collab., M. Derrick et al., *Z. Phys.* **C65** (1995), 379.
5. H1 Collab., S. Aid et al., *Nucl. Phys.* **B470** (1996) 3.
6. ZEUS Collab., M.Derrick et al., *Z. Phys.* **C72** (1996) 399.
7. H1 Collab., C.Adloff et al., DESY 97-042 (1997) subm. to *Nucl. Phys. B*.
8. ZEUS Collab., J.Breitweg et al., DESY 97-135 (1997) subm. to *Phys. Lett. B*.
9. M. Glück, E. Reya and A. Vogt, *Z. Phys.* **C67**. (1995) 433;
10. A. Donnachie and P.V. Landshoff, *Z. Phys.* **C61** (1994) 139.

11. A. Capella et al., *Phys. Lett.* **B337** (1994) 358.
12. B. Badelek, J. Kwiecinski and A. Stasto, Durham preprint DTP/96/16 (1996).
13. K. Adel, F. Barreiro and F.J. Yndutáin, FTUAM 96-39, hep-ph/9610380, (1996).
14. BCDMS Collab., A. C. Benvenuti et al., *Phys. Lett.* **B237** (1990)
15. NMC Collab., M. Arneodo et al., *Phys. Lett.* **B364** (1995) 107.
    NMC Collab., M. Arneodo et al., hep-ph/9610231, (1996).
16. L. Whitlow et al., *Phys. Lett.* **B250** (1990) 93.
17. NMC Collab., M. Arneodo et al, hep-ph/9610231, *subm. to Nucl. Phys.*
18. M.W. Krasny et al., *Z. Phys.* **C53** (1992) 687.
19. H1 Collab., C. Adloff et al., *Phys. Lett.* **B393** (1997) 452.
20. L. Favart et al., *Z. Phys.* **C 72** (1996) 425.
21. E.B. Zijlstra and W.L. van Neeven, *Nucl. Phys.* **B383** (1992) 525.
22. H1 Collab., C. Adloff et al., *Z. Phys.* **C72** (1996) 593.
23. ZEUS Collab., J. Breitweg et al., DESY 97-089, *subm. to Z. Phys.*.
24. K. Daum, *This proceeding.*
25. H1 Collab., C. Adloff et al., *Z. Phys.* **C74** (1997) 191.
26. ZEUS Collab., J. Breitweg et al., *Z. Phys.* **C74** (1997) 207.
27. Y. Sirois, *This proceeding.*

RECENT DEVELOPMENTS IN THE DESCRIPTION OF F_2 AT HERA

A.D. MARTIN
Physics Department, Durham University
Durham, DH1 3LE, UK
E-mail: A.D.Martin@durham.ac.uk

Abstract

We discuss the partonic description of the HERA data for F_2, paying particular attention to a recent k_T factorization approach which unifies DGLAP and BFKL effects. We outline the possible descriptions of the charm component of F_2.

1. Introduction

One of the most striking features of the measurements of deep inelastic scattering at HERA is the strong rise of the proton structure function F_2 as x decreases from 10^{-2} to below 10^{-4}. At first sight it appeared that the rise was due to the (BFKL) resummation of leading $\ln(1/x)$ contributions. In the small x domain the basic dynamical quantity is the gluon distribution $f(x, k_T^2)$ unintegrated over its transverse momentum k_T. Observables are computed in terms of f via the k_T factorization theorem. For example

$$F_2(x, Q^2) = \int^{Q^2} \frac{dk_T^2}{k_T^2} \int_x^1 \frac{dx'}{x'} F_2^{\gamma g}(\frac{x}{x'}, k_T^2) f(x', k_T^2) \qquad (1)$$

where $F_2^{\gamma g}$ is the off-shell gluon structure function which at lowest order is given by the "quark box and crossed-box" contributions, $\gamma g \to q\overline{q}$, see Fig. 1(a). In principle the $\ln(1/x)$ resummation predicts the small x behaviour of f and hence, via (1), the behaviour of F_2. At leading order, for fixed α_S, the BFKL equation gives $f \sim x^{-\lambda}$ as $1/x \to \infty$ with $\lambda = \overline{\alpha}_S\, 4 \ln 2$ (with $\overline{\alpha}_S = 3\alpha_S/\pi$). However, even allowing for the running of α_S, the leading order prediction is found to be too steep to describe the recent, much more precise, measurements of F_2. Of course, at subasymptotic values of $1/x$ the prediction is subject to considerable correction. We return to this in a moment.

On the other hand the conventional DGLAP approach is able to describe the F_2 data at the smallest x values observed (even for $Q^2 \sim 1\text{GeV}^2$) with an appropriate choice of parton distributions at some starting scale Q_0^2 for the evolution. DGLAP evolution effectively resums the leading $\ln Q^2$ contributions, which arise from strongly ordered transverse momenta $Q^2 \gg k_T^2 \gg k_T'^2 \gg \ldots$ along the gluon chain in Fig. 1(a).

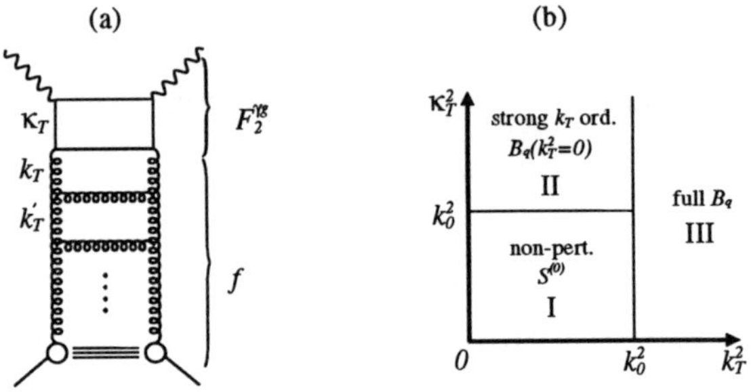

FIGURE 1: (a) The diagrammatic representation of the k_T factorization formula $F_2 = F_2^{\gamma g} \otimes f$ of equation (1). At lowest order $F_2^{\gamma g}$ is given by the quark box shown (together with the crossed box) summed over all types of quark, $F_2^{\gamma g} = \Sigma e_q^2 B_q$. (b) The different regions of integration used to evaluate the sea quark distribution. In regions II and III the sea is driven by the gluon, $S_q = B_q \otimes f$.

Thus we can set $k_T^2 = 0$ in $F_2^{\gamma g}$ in (1) and work in terms of the familiar integrated gluon distribution,

$$g(x, Q^2) = \int^{Q^2} \frac{dk_T^2}{k_T^2} f(x, k_T^2), \tag{2}$$

and (1) then reduces to the DGLAP collinear factorization formula. The free choice of non-perturbative input shapes in x for the parton distributions $g(x, Q_0^2)$, $\overline{q}(x, Q_0^2)\ldots$ at the starting scale for DGLAP evolution means that there is more freedom than in the BFKL approach. In practice this distinction is not so clear cut since non-perturbative and subleading $\ln(1/x)$ effects modify the BFKL prediction. Of course, the issue is not whether DGLAP or BFKL give a better description of F_2 at small x, since both $\ln Q^2$ and $\ln(1/x)$ resummations are necessary. Rather we seek a unified description of F_2 which embodies both of these perturbative QCD effects.

2. Unified BFKL/DGLAP descriptions of F_2

2.1. Collinear factorization approach

A unified BFKL/DGLAP description can be obtained from a DGLAP analysis, but with the $\ln(1/x)$ summations included in the anomalous dimensions γ_{ij} and coefficient functions [1]. Then the splitting functions P_{ij} take the form

$$xP_{gg} = \sum_{n=1} A_n \frac{[\alpha_S \ln(1/x)]^{n-1}}{(n-1)!} \tag{3}$$

$$xP_{qg} = \frac{\alpha_S}{2\pi}\, xP_{qg}^{(0)}(x) + \alpha_S^2 \sum_{n=1} B_n \frac{[\alpha_S \ln(1/x)]^{n-1}}{(n-1)!}. \tag{4}$$

We see that P_{qg} is formally non-leading at small x in contrast to P_{gg}. However, for moderately small values of x, which are relevant for the HERA data, it is found [2] that the resummation effects in P_{qg} have a stronger impact on F_2 than those in P_{gg}. This is due to the fact that the coefficients $A_2 = A_3 = A_5 = 0$ while all the B_n coefficients are non-zero. Moreover, the BFKL resummation effects in P_{qg} can significantly affect the extraction of the gluon distribution from the observed scaling violations of F_2, since $\partial F_2/\partial \ln Q^2 \sim P_{qg}\overline{\otimes} g$, where $\overline{\otimes}$ denotes a convolution in longitudinal momentum only.

We are interested in the case when the input parton distributions are flat in x at small x, so that the rise of F_2 is due to perturbative QCD. Unfortunately, for flat input distributions the small x behaviour of F_2 is found to be sensitive to the renormalization scheme that is adopted.

An interesting development is the possibility of writing evolution equations for observables, for example

$$\frac{\partial F_2}{\partial \ln Q^2} = \Gamma_{22}\overline{\otimes} F_2 + \Gamma_{2L}\overline{\otimes} F_L, \tag{5}$$

where Γ_{ij} are independent of the scheme [3]. It is encouraging that an analysis [4] based on such an approach finds that the $\ln(1/x)$ resummations are not excluded by the data, but rather that their presence improves the description of F_2.

2.2. k_T factorization approach

As we mentioned above, the unintegrated gluon and the k_T factorization theorem provide the natural framework for describing small x physics (see, for example, (1)). The $\ln(1/x)$ resummations, which have to be performed explicitly in the collinear approach, are automatically implicitly included in the integration over the entire k_T^2 phase space of the gluon ladder and in the k_T factorization integrals.

The early attempts [5] to obtain the unintegrated gluon distribution $f(x, k_T^2)$ by numerically solving the BFKL equation were plagued by the dependence on the treatment of the infrared region, $k_T^2 < k_0^2$. Here we describe an improved treatment [6] in which the BFKL equation is arranged so that we need only solve it in the perturbative domain $k_T^2 > k_0^2$ and in which the residual DGLAP contributions are now included. To be precise we solve the unified BFKL/DGLAP equation

$$f(x, k_T^2) = \frac{\alpha_S}{2\pi} \int_x^1 dz P_{gg}(z) \frac{x}{z}\, g(\frac{x}{z}, k_0^2) +$$

$$+\,\overline{\alpha}_S\,k_T^2\int_x^1\frac{dz}{z}\int_{k_0^2}\frac{dk_T'^2}{k_T'^2}\left[\frac{f(x/z,k_T'^2)\,\theta\left(k_T^2/z-k_T'^2\right)-f(x/z,k_T^2)}{|k_T'^2-k_T^2|}+\frac{f(x/z,k_T^2)}{(4k_T'^4+k_T^4)^{\frac{1}{2}}}\right]$$
$$+\,\overline{\alpha}_S\int_x^1\frac{dz}{z}\left(\frac{z}{6}\,P_{gg}(z)-1\right)\int_{k_0^2}^{k_T^2}\frac{dk_T'^2}{k_T'^2}\,f\left(\frac{x}{z},k_T'^2\right)+\frac{\alpha_S}{2\pi}\int_x^1 dz P_{gq}(z)\,\Sigma\left(\frac{x}{z},k_T^2\right).\tag{6}$$

Symbolically this equation for the gluon has the structure

$$f\;=\;\text{non-pert.input}\;+\;f\leftarrow f(\text{BFKL})\;+\;f\leftarrow f(\text{DGLAP}-1)\;+\;f\leftarrow q\text{ singlet}$$

where -1 is taken from DGLAP because it is already included in BFKL. The input term comes from the $k_T^2 < k_0^2$ parts of BFKL and of (DGLAP – 1). We specify the input in terms of a simple two parameter form

$$g(x,k_0^2)\;=\;N(1-x)^{\beta}.\tag{7}$$

In addition to restricting the solution of the BFKL equation to the perturbative region $k_T^2 > k_0^2$ and to including the DGLAP terms, we have also introduced a θ function which imposes the constraint $k_T'^2 < k_T^2/z$ on the real gluon emissions. The origin of this constraint[1] is the requirement that the virtuality of the exchanged gluon is dominated by its transverse momentum $|k'|^2 \simeq k_T'^2$. We take a running coupling $\alpha_S(k_T^2)$, which is consistent with the results of the next-to-leading order $\ln(1/x)$ analyses of Fadin, Lipatov, Camici and Ciafaloni [9].

The final term in (6) depends on the quark singlet momentum distribution Σ. At small x the sea quark components S_q of Σ dominate. They are driven by the gluon via the $g \to q\overline{q}$ transitions, that is

$$S_q\;=\;B_q\otimes f\tag{8}$$

where at lowest order B_q is the box (and crossed box) contribution indicated in Fig. 1(a). Besides the z and k_T^2 integrations symbolically denoted by $\otimes$ the box contribution implicitly includes an integration over the transverse momentum κ_T of the exchanged quark. The evolution equation for Σ may be written in the form

$$\Sigma\;=\;S^{(0)}\;+\;\sum_q B_q(k_T^2=0)\,\overline{\otimes}\,zg(z,k_0^2)\;+\;\sum_q B_q\otimes f\;+\;P_{qq}\otimes S_q\;+\;V\tag{9}$$

where $\overline{\otimes}$ is simply a convolution over longitudinal momentum. The first three terms on the right hand side are the "$B_q \otimes f$" contributions of (8) coming from the regions

[1] A more general treatment of the gluon ladder which incorporates both the BFKL equation and DGLAP evolution is given by the CCFM equation [7], which is based on angular ordering of the gluon emissions. The angular ordered and kinematic constraints lead to similar subleading $\ln(1/x)$ effects, but the kinematic constraint overrides the angular ordered constraint, except when $Q^2 < k_T^2$ in the large x domain [8].

I, II, III of the k_T^2 and κ_T^2 integrations that are shown in Fig. 1(b). First, in the non-perturbative domain, region I, the u, d, s sea quark contribution is parametrized in the form

$$S^{(0)} = C_{I\!P}\, x^{-0.08}(1-x)^8 \qquad (10)$$

consistent with soft pomeron and counting rule expectations, where $C_{I\!P}$ is independent of Q^2. The constant $C_{I\!P}$ is fixed in terms of the two parameters, N and β, of (7) by the momentum sum rule. In region II we apply the strong k_T ordering approximation with $B_q \approx B_q(k_T^2 = 0)$ so that the k_T^2 integration can be carried out to give a contribution proportional to $g(x/z, k_0^2)$ [10]. Finally in region III we evaluate the full box contribution; this gives the main contribution and is responsible for the rise of F_2 with decreasing x. The last two terms in (9) give the sea $\to$ sea evolution contribution, and the valence contribution $V(x, Q^2)$ which is taken direct from a recent parton set. The charm quark component of the sea is given totally by *perturbative* QCD, since for $k_T^2 < k_0^2$ the box $B(k_T^2 = 0)$ is finite as $\kappa_T^2 \to 0$ due to $m_c \neq 0$.

The coupled integral equations (6) and (9) for the gluon f and the quark singlet Σ are solved in the perturbative domain, $k_T^2 > k_0^2$. The only input is the gluon $g(x, k_0^2)$ of (7), where $k_0^2 = 1\text{GeV}^2$. The values of the two input parameters are determined by fitting to the available data for F_2 with $x < 0.05$ and $Q^2 > 1.5\text{GeV}^2$. The continuous curves in Fig. 2(a) show the description of a sample of the data. Overall the fit is excellent; at least as good as that achieved in the recent global analyses. When we repeat the analysis with the kinematic constraint omitted we see that the description (given by the dashed curves in Fig. 2(a)) is not so good and, moreover, the extrapolation of the gluon to $x \approx 0.4$ no longer describes the WA70 prompt photon data.

How important are the $\ln(1/x)$ effects? First we replace the BFKL kernel in (6) by the standard DGLAP splitting function, P_{gg}. We find that the gluon is not changed much in the HERA domain, compare the dashed and dotted curves in Fig. 2(b), as anticipated from the earlier observation that the coefficients $A_2 = A_3 = A_5 = 0$ in (3). On the other hand, when we use pure DGLAP evolution for the quark singlet, as well as the gluon, the difference is pronounced; compare the dashed and dot-dashed curves in Fig. 2(b), indicating the importance of $\ln(1/x)$ effects in P_{qg}.

We argue that it is more advantageous to describe small x observables in terms of a universal unintegrated gluon together with the k_T factorization theorem, rather than to reduce the BFKL effects to collinear form. We list some points to consider. In the unintegrated approach it is straightforward to identify the perturbative contributions which contribute at all Q^2. We can therefore avoid subsuming them in the input distributions. Second, there is a natural way to introduce running α_S, which for sufficiently small x goes beyond the Renormalization Group behaviour. Thirdly the kinematic constraint along the BFKL ladder is easy to implement in the k_T factorization approach – the constraint is a subleading $\ln(1/x)$ effect which appears to embody

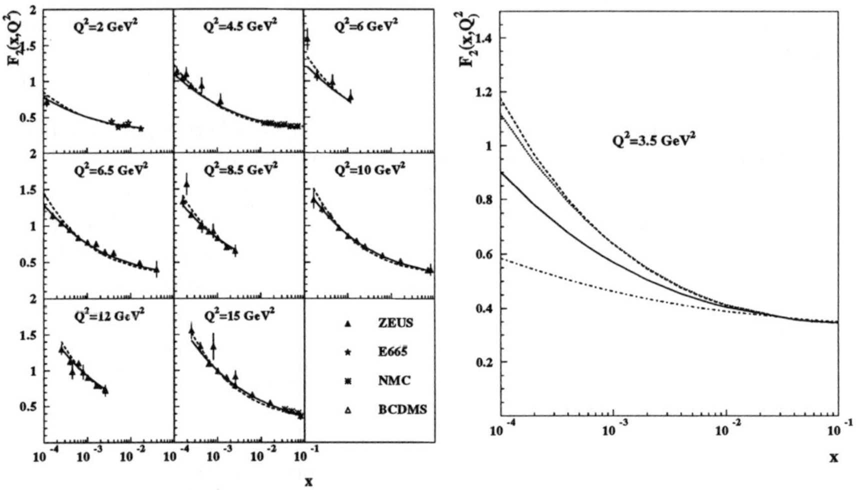

FIGURE 2(a). The two-parameter description of some of the F_2 data at small x using $f(x, k_T^2)$ evaluated with (continuous curves) and without (dashed curves) the kinematic constraint $k_T'^2 < k_T^2/z$. (b) The continuous and dashed curves come from solving (6) and (9) with and without the kinematic constraint. The dotted and dot-dashed curves are obtained using DGLAP in the gluon sector and in both the gluon and quark sectors respectively. The same input is used for all curves.

a major part of the next-to-leading contribution (compare the α_S^2 term in the exponents of [8] and [9]). Another point is that BFKL contains all twists, whereas only the leading twist is retained in the reduction to collinear form. Finally the BFKL kernel and the off-shell gluon structure function $F_2^{\gamma g}$ are calculable perturbatively. We simply use expressions to leading order in α_S. The $\alpha_S \ln(1/x)$ summations are implicit in the integration over the *entire* k_T^2 phase space of the gluon ladder and in the k_T factorization integrals.

In conclusion, the fact that we achieve an excellent two-parameter fit of the F_2 data at small x is not, in itself, remarkable. Other equally good phenomenological fits have been obtained. What is encouraging is that we have a theoretically well-grounded and consistent formalism which, with the minimum of non-perturbative input, is able to give a good *perturbative* description of the observed structure of F_2. Moreover the BFKL/DGLAP components of F_2 are decided by dynamics. In this way we have made a determination of the *universal* gluon distribution $f(x, k_T^2)$ which can be used, via the k_T factorization theorem, to predict the behaviour of other observables at small x. The predictions for F_2 (charm) and F_L can be found in [6].

3. Descriptions of F_2 (charm)

Both H1 and ZEUS have used $c \to D^*$ to measure charm production in the HERA domain [11]. They find $F_2^c/F_2 \sim 25\%$, as compared to the EMC measurement of $F_2^c/F_c \sim 1\%$ at lower fixed-target energies. Moreover, the precision of the measurements of $F_2^c(x,Q^2)$ will improve, particularly when vertex detectors become available. Such measurements should reveal important information on the gluon, which enters at LO through the photon gluon fusion (PGF) process $\gamma g \to c\bar{c}$. Clearly it is important to see how best to include the charm mass effects in the analysis. At NLO we have

$$F_2^c(x,Q^2) \;=\; \frac{8}{9}\int_x^1 dz\frac{x}{z}\,\left[C_{q=c}(z,Q^2,\mu^2)\,c(\frac{x}{z},\mu^2) \,+\, C_g(z,Q^2,\mu^2)\,g(\frac{x}{z},\mu^2)\right] \quad (11)$$

where the coefficient functions $C_c = C_c^{(0)} + \alpha_S C_c^{(1)}$ and $C_g = \alpha_S\, C_g^{(1)}$.

3.1. Massless charm evolution

The simplest approach is to assume that the charm distribution $c(x,Q^2) = 0$ for $Q^2 < \mu^2$, where $\mu^2 \sim m_c^2$, and then to evolve assuming c is a massless parton both in the splitting and coefficient functions. This is the approach used in the MRS and CTEQ global analyses. MRS take $\mu^2 = 2.7\mathrm{GeV}^2$ so as to obtain a satisfactory description of F_2^c of EMC [12], while CTEQ set $\mu = m_c$.

Although phenomenologically successful, the massless model clearly is inadequate in the charm threshold region. For instance an on-shell $c\bar{c}$ pair can be created by photon-gluon (PGF) provided

$$W^2 \;=\; Q^2(1-x)/x \geq 4m_c^2$$

where W is the $\gamma^* g \to c\bar{c}$ c.m. energy. Thus at small x, $c\bar{c}$ production is not forbidden even for $Q^2 < \mu^2$ where the massless approach gives zero. There is a subtle point here. The $c\bar{c}$ production threshold $W^2 = 4m_c^2$ is not the threshold for the onset of the charm distribution. Only if $Q^2 \gtrsim m_c^2$ will the probing virtual photon be able to resolve a c or $\bar{c}$ in the $g \leftrightarrow c\bar{c}$ fluctuations. For smaller Q^2 values $c\bar{c}$ production is entirely a component of the gluon distribution.

3.2 Photon-gluon fusion

In this approach charm is treated as a heavy quark and not a parton. That is we put $c = 0$ for all Q^2 in (1) and use the known $m_c \neq 0$ gluon coefficient function C_g^{PGF} for C_g. This is called a fixed flavour number scheme (FFNS) with $n_f = 3$. In contrast to the situation for massless quarks, there is no collinear divergence in $\gamma^* g \to c\bar{c}$ as the integral over the $c\bar{c}$ transverse momentum is regulated by m_c. However, this in turn means that $F_2 \sim [\alpha_S \ln(Q^2/m_c^2)]^n$ at n^{th} order. In terms of the FFNS all these contributions should be absorbed in the coefficient function. However, these are just the large logarithms that are resummed[2] by DGLAP evolution, which at large Q^2 implies we should treat charm as a parton.

3.3 Variable flavour number schemes: ACOT and MRRS

The aim is to obtain a universal charm distribution $c(x, Q^2)$ all the way down to the resolution threshold $Q^2 \sim m_c^2$ in order to make predictions for other processes. To do this we must include a massive charm quark in parton evolution in a consistent way.

The ACOT [15] philosophy is to use massless splitting functions in the evolution with the number of flavours[3] n_f going from $3 \to 4$ at $Q^2 = m_c^2$ and $c = 0$ for $Q^2 < m_c^2$. All the $m_c \neq 0$ effects are assigned to the coefficient functions. In this case all the large log terms $[\alpha_S \ln(Q^2/m_c^2)]^n$ are correctly reproduced by the evolution. Here m_c plays the role of the starting scale Q_0 for the evolution with $n_f = 4$. The partons are matched to those of the FFNS at $Q = Q_0$. Nevertheless in the threshold region $Q^2 \gtrsim m_c^2$ the ACOT evolution is not quite precise enough. The contribution of the first $c\bar{c}$ pair (produced by the evolution just above threshold) looks like $\alpha_S(m_c^2)[\ln(Q^2/m_c^2) + O(1)]$ which differs from the massless ACOT approximation by a term of $O(\alpha_S(m_c^2))$. Of course this term may be calculated and for the inclusive F_2 may be absorbed into a new coefficient function.

Buza et al. [14] have evaluated the coefficient functions for $Q^2 \gg m_c^2$ in the VFNS and the FFNS to NLO. They find that for $Q^2 \gtrsim 20\text{GeV}^2$ the two schemes agree very closely. In [16] the same authors work out the matrix relation in the limit $Q^2 \gg m_c^2$ which allows a matching of all the parton four flavour densities above the scale $\mu^2 = m_c^2$, to the three flavour densities below that point. This relation may be used to calculate other inclusive cross sections (e.g. large E_T jets etc.) Again this is only for $Q^2 \gg m_c^2$, which does not solve the problem of what to do at $Q^2 \sim m_c^2$. So there has to be another matching in the threshold region. In leading order it coincides with the ACOT procedure. It remains to be investigated how to do this retaining terms $O(m_c^2/Q^2)$ at NLO where the $\alpha_S(m_c^2)$ effect becomes important.

A different approach to include $m_c \neq 0$ effects has been proposed by MRRS [17]. Their aim is to formulate the evolution procedure with $m_c \neq 0$ so as to generate universal parton distributions applicable to any inclusive or exclusive process with the (known) coefficient functions in the conventional $\overline{\text{MS}}$ renormalisation scheme. In order to do this they analyse the leading log contributions which come from the relevant

[2]The NLO corrections to the PGF structure function are known [13] and a leading twist analysis [14] has been used to perform a resummation of the $[\alpha_S \ln(Q^2/m_c^2)]^n$ terms for $Q^2 \gg m_c^2$. Unfortunately such an approach is not applicable for F_2 in the threshold region $Q^2 \sim m_c^2$.

[3]We call this a variable flavour number scheme (VFNS).

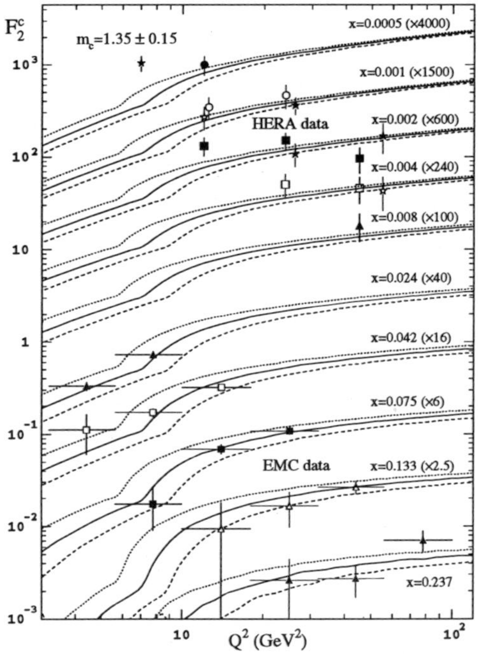

Fig. 3: The predictions [17] for F_2^c compared with the EMC [18] and HERA [11] measurements. The dotted, continuous and dashed lines correspond to $m_c = 1.2$, 1.35 and 1.5 GeV respectively. The starred data points in the HERA domain are obtained by interpolating ZEUS measurements, whereas the other HERA data correspond to the H1 measurements.

(ladder type) Feynman diagrams. It turns out that a remarkable simplification occurs. Recall that the leading logs come from the configuration where the p_T of the partons are strongly ordered along the ladder. At NLO accuracy it was found to be sufficient to take into account the $m_c \neq 0$ effect simply for the charm parton with $p_T \sim m_c$. The $m_c \neq 0$ effects in all the other parts of the ladder enter at NNLO or higher orders. In fact all the $m_c \neq 0$ effects at NLO occur only[4] in the splitting function P_{cg}. It is straightforward to generalise the MRRS to higher orders [17].

In a VFNS we have to take care to avoid double counting. The same Feynman graphs are generated by the evolution with zero-order coefficient function (corresponding to $(g \to c\bar{c}) \otimes (\gamma^* c \to c)$ or vice versa with $c \leftrightarrow \bar{c}$) and the first-order

[4]Apart from the corresponding adjustment in the $\delta(1-z)$ term in P_{gg} [17].

coefficient function which describes photon-gluon fusion $\gamma^* g \to c\bar{c}$. Thus we must subtract the contribution generated by evolution from the PGF contribution. This is done by both ACOT and MRRS.

The charm component F_2^c of F_2 is totally determined by the gluon and light quark densities. The only parameter is the value of m_c. The MRRS predictions are shown in Fig. 3. They are in good agreement with both HERA and fixed-target EMC data, which together cover a wide range of x and Q^2. Unfortunately in order to match on to conventional $\overline{\rm MS}$ coefficient functions the evolution scale is required to be $Q^2 = m_c^2 + p_T^2$ leading to a rather low charm threshold at $Q^2 = m_c^2$. The value of α_S is not sufficiently small in this region to neglect NNLO corrections. One of the main effects is to move the threshold to $Q^2 \gtrsim 4m_c^2$ which, due to the Uncertainty Principle, is the virtuality of the photon required to resolve a charm quark within the $g \leftrightarrow c\bar{c}$ fluctuations. In the NLO predictions in Fig. 3 the NNLO modification has been imposed by hand in a simplified way. This is the origin of the artificial structure at $Q^2 = 4m_c^2$. Since the NNLO corrections are available [13] they may be used to smooth out this behaviour.

Acknowledgements

I thank Bernd Kniehl for arranging such a successful meeting, and Jan Kwiecinski, Dick Roberts, Misha Ryskin, Jack Smith, Anna Stasto and James Stirling for valuable discussions and enjoyable collaborations on the work discussed here.

References

[1] S. Catani, M. Ciafaloni and F. Hautmann, *Phys. Lett.* **B242** (1990) 97; *Nucl. Phys.* **366** (1991) 657; J.C. Collins and R.K. Ellis, *Nucl. Phys.* **B360** (1991) 3; S. Catani and F. Hautmann, *Nucl. Phys.* **B427** (1994) 475.

[2] R.K. Ellis, Z. Kunszt and E.M. Levin, *Nucl. Phys.***B420** (1994) 517; Erratum-ibid. **B433** (1995) 498;
R.K. Ellis, F. Hautmann and B.R. Webber, *Phys. Lett.* **B348** (1995) 582;
R.D. Ball and S. Forte, *Phys. Lett.* **B351** (1995) 313;
J.R. Forshaw, R.G. Roberts and R.S. Thorne, *Phys. Lett.* **B356** (1995) 79.

[3] S. Catani, Proc. of Workshop on Deep Inelastic Scattering and Related Phenomena (DIS 96), Rome, 1996, eds. G. D'Agostini and A. Nigro (World Scientific) p.165.

[4] R.S. Thorne, Rutherford preprint, RAL-96-065, January 1997.

[5] A.J. Askew, J. Kwiecinski, A.D. Martin and P.J. Sutton, *Phys. Rev.* **D47** (1993) 3775; *Mod. Phys. Lett.* **A8** (1993) 3813; *Phys. Rev.* **D49** (1994) 4402; see also, I. Bojak and M. Ernst, Dortmund preprint DO-TH 96/25, hep-ph/9702282.

[6] J. Kwiecinski, A.D. Martin and A.M. Stasto, Durham preprint DTP/97/18, *Phys. Rev.* (in press).

[7] M. Ciafaloni, *Nucl. Phys.* **B296** (1988) 49; S. Catani, F. Fiorani and G. Marchesini, *Phys. Lett.* **B234** (1990) 339; *Nucl. Phys.* **B336** (1990) 18.

[8] J. Kwiecinski, A.D. Martin and P.J. Sutton, *Z. Phys.* **C71** (1996) 585; B. Anderson, G. Gustafson, H. Kharraziha and J. Samuelson, *Z. Phys.* **C71** (1996) 613.

[9] M. Ciafaloni, these proceedings

[10] J. Blumlein, *J. Phys.* **G19** (1993) 1623.

[11] K. Daum, these proceedings

[12] A.D. Martin, R.G. Roberts and W.J. Stirling, *Phys. Rev.* **D50** (1994) 6734.

[13] E. Laenen, S. Riemersma, J. Smith and W.L. van Neerven *Nucl. Phys.* **B392** (1993) 162.

[14] M. Buza, Y. Matiounine, J. Smith, R. Migneron and W.L. van Neerven, *Nucl. Phys.* **B472** (1996) 611.

[15] M.A.G. Aivazis, J.C. Collins, F.I. Olness and W.-K. Tung, *Phys. Rev.* **D50** (1994) 3102.

[16] M. Buza, Y. Matiounine, J. Smith and W.L. van Neerven, DESY report 96-258, *Z. Phys.* (in press); J. Smith, these proceedings.

[17] A.D. Martin, R.G. Roberts, M.G. Ryskin and W.J. Stirling, Durham preprint DTP/96/102, *Z. Phys.* (in press).

[18] EMC collaboration: J.J. Aubert et al., *Nucl. Phys.* **B213** (1983) 31.

SOME ASPECTS OF POLARIZED *ep* SCATTERING AT HERA

M. Stratmann
Institut für Physik, Universität Dortmund, D-44221 Dortmund, Germany

and

W. Vogelsang
Theory Division, CERN, CH-1211 Geneva, Switzerland

ABSTRACT

We study photoproduction reactions in a polarized *ep* collider mode of HERA at $\sqrt{s} = 298$ GeV. We examine the sensitivity of the cross sections and their asymmetries to the proton's polarized gluon distribution and to the completely unknown parton distributions of longitudinally polarized photons.

1. Introduction

Among the various conceivable options for future HERA upgrades is the idea to longitudinally polarize its proton beam [1] which, when combined with the already operative longitudinally polarized electron (positron) beam, results in a polarized version of the usual HERA collider with $\sqrt{S} \approx 300$ GeV. A typical conservative value for the integrated luminosity in this case should be 100 pb^{-1}.

HERA has already been very successful in pinning down the proton's unpolarized gluon distribution $g(x, Q^2)$. Several processes have been studied which have contributions from $g(x, Q^2)$ already in the lowest order, such as (di)jet, inclusive hadron, and heavy flavor production. Since events at HERA are concentrated in the region $Q^2 \to 0$, the processes have first and most accurately been studied in photoproduction[2-9]. As is well-known, in this case the (quasi-real) photon will not only interact in a direct ('point-like') way, but can also be resolved into its hadronic structure. HERA photoproduction experiments like[2-9] have not merely established evidence for the existence of such a resolved contribution, but have also been precise enough to improve our knowledge about the parton distributions, f^γ, of the photon.

Given the success of such unpolarized photoproduction experiments at HERA, it seems most promising [10] to closely examine the same processes for the situation with longitudinally polarized beams with regard to their sensitivity to the proton's polarized gluon distribution Δg, which is still one of the most interesting, but least known, quantities in 'spin-physics'. Recent next-to-leading (NLO) studies of polarized DIS [11,12] show that the x-shape of Δg seems to be hardly constrained at all by the present DIS data, even though a tendency towards a sizeable positive *total* gluon polarization, $\int_0^1 \Delta g(x, Q^2 = 4\ \text{GeV}^2) dx \gtrsim 1$, was found [11,13,12]. Furthermore, polarized photoproduction experiments may in principle allow to not only determine the parton, in particular gluon, content of the polarized *proton*, but also that of the longitudinally polarized *photon* which is completely unknown so far. Since a measurement of, e.g., the photon's spin-dependent structure function g_1^γ in polarized e^+e^- collisions is not planned in the near future, HERA could play a unique role here, even if it should only succeed in establishing the very *existence* of a resolved contribution to polarized photon-proton reactions.

Our contribution, part of which is taken from [10], is organized as follows: In the next section we collect the necessary ingredients for our calculations. In section 3 we will discuss various conceivable photoproduction reactions, namely (di)jet, inclusive hadron, and open-charm production.

2. Polarized Parton Distributions of the Proton and the Photon

Even though NLO analyses of polarized DIS which take into account all or most data sets have been published recently [11,13,12], we have to stick to LO calculations throughout this work since the NLO corrections to, e.g., polarized jet production are not yet known. This implies use of LO parton distributions, which have also been provided in the studies [11,12]. Both papers give various LO sets which mainly differ in the x-shape of the polarized gluon distribution. We will choose the LO 'valence' set of the 'radiative parton model analysis' [11], which corresponds to the best-fit result of that paper, along with two other sets of [11] which are based on either assuming $\Delta g(x,\mu^2) = g(x,\mu^2)$ or $\Delta g(x,\mu^2) = 0$ at the low input scale μ of [11], where $g(x,\mu^2)$ is the unpolarized LO GRV [14] input gluon distribution. These two sets will be called '$\Delta g = g$ input' and '$\Delta g = 0$ input' scenarios, respectively. The gluon of set C of [12] is qualitatively different since it has a substantial negative polarization at large x. We will therefore also use this set in our calculations. For illustration, we show in Fig. 1 the gluon distributions of the four different sets of parton distributions we will use, taking a typical scale $Q^2 = 10$ GeV^2. Keeping in mind that all four LO sets provide very good descriptions of the present polarized DIS data, it becomes obvious that the data indeed do not seem to be able to significantly constrain the x-shape of $\Delta g(x,Q^2)$.

In the case of photoproduction the electron just serves as a source of quasi-real photons which are radiated according to the Weizsäcker-Williams spectrum. The photons can then interact either directly or via their partonic structure ('resolved' contribution). In the case of longitudinally polarized electrons, the resulting photon will be longitudinally (more precisely, circularly) polarized and, in the resolved case, the *polarized* parton distributions of the photon, $\Delta f^\gamma(x,Q^2)$, enter the calculations. Thus one can define the effective polarized parton densities at the scale M in the longitudinally polarized electron via[a]

[a]We include here the additional definition $\Delta f^\gamma(x_\gamma,M^2) \equiv \delta(1-x_\gamma)$ for the direct ('unresolved') case.

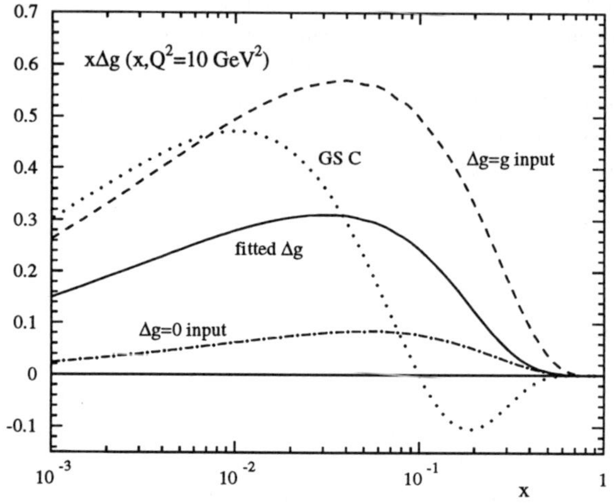

Fig. 1. *Gluon distributions at $Q^2 = 10\,GeV^2$ of the four LO sets of polarized parton distributions used in this paper. The dotted line refers to set C of* [12]*, whereas the other distributions are taken from* [11] *as described in the text.*

$$\Delta f^e(x_e, M^2) = \int_{x_e}^{1} \frac{dy}{y} \Delta P_{\gamma/e}(y) \Delta f^\gamma(x_\gamma = \frac{x_e}{y}, M^2) \tag{1}$$

($f = q, g$) where $\Delta P_{\gamma/e}$ is the polarized Weizsäcker-Williams spectrum for which we will use

$$\Delta P_{\gamma/e}(y) = \frac{\alpha_{em}}{2\pi} \left[\frac{1-(1-y)^2}{y} \right] \ln \frac{Q^2_{max}(1-y)}{m_e^2 y^2} , \tag{2}$$

with the electron mass m_e. For the time being, it seems most sensible to follow as closely as possible the analyses successfully performed in the unpolarized case, which implies to introduce the same kinematical cuts. As in [3,5,8,15] we will use an upper cut[b] $Q^2_{max} = 4$ GeV^2, and the y-cuts $0.2 \le y \le 0.85$ (for single-jet [3] and charm production) and $0.2 \le y \le 0.8$ (for dijet [5] and inclusive hadron production) will be imposed.

The polarized photon structure functions $\Delta f^\gamma(x_\gamma, M^2)$ in (1) are completely unmeasured so far, such that models for them have to be invoked. To obtain a realistic estimate for the theoretical uncertainties in the polarized photonic parton densities two very different scenarios were considered in [16] assuming 'maximal' ($\Delta f^\gamma(x, \mu^2) = f^\gamma(x, \mu^2)$) or 'minimal' ($\Delta f^\gamma(x, \mu^2) = 0$) saturation of the fundamental positivity constraints $|\Delta f^\gamma(x, \mu^2)| \le f^\gamma(x, \mu^2)$ at the input scale μ for the QCD evolution. Here μ and the unpolarized photon structure functions $f^\gamma(x, \mu^2)$ were adopted from the phenomenologically successful radiative parton model predictions in [17]. The results

---

[b]In H1 analyses of HERA photoproduction data [2,4,9] the cut $Q^2_{max} = 0.01$ GeV^2 is used along with slightly different y-cuts as compared to the corresponding ZEUS measurements [3,5,8], which leads to smaller rates.

of these two extreme approaches are presented in Fig. 2 in terms of the photonic parton asymmetries $A_f^\gamma \equiv \Delta f^\gamma / f^\gamma$, evolved to $Q^2 = 30$ GeV^2 in LO. An ideal aim of measurements in a polarized collider mode of HERA would of course be to determine the Δf^γ and to see which ansatz is more realistic. The sets presented in Fig. 2, which we will use in what follows, should in any case be sufficient to study the sensitivity of the various cross sections to the Δf^γ, but also to see in how far they influence a determination of Δg.

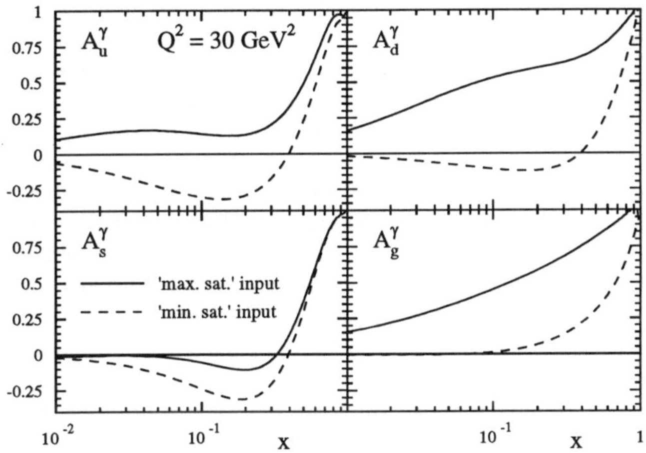

Fig. 2. *Photonic LO parton asymmetries $A_f^\gamma \equiv \Delta f^\gamma / f^\gamma$ at $Q^2 = 30\, GeV^2$ for the two scenarios considered in [16] (see text). The unpolarized LO photonic parton distributions were taken from [17].*

We finally note that in what follows a polarized cross section will always be defined as

$$\Delta\sigma \equiv \frac{1}{2}\left(\sigma(++) - \sigma(+-)\right) , \tag{3}$$

the signs denoting the helicities of the scattering particles. The corresponding unpolarized cross section is given by taking the sum instead, and the cross section asymmetry is $A \equiv \Delta\sigma/\sigma$. Whenever calculating an asymmetry A, we will use the LO GRV parton distributions for the proton [14] and the photon [17] to calculate the unpolarized cross section. For consistency, we will employ the LO expression for the strong coupling α_s with [11,12,16] $\Lambda_{QCD}^{(f=4)} = 200$ MeV for four active flavors.

3. Photoproduction Reactions at Polarized HERA

The generic LO cross section formula for the photoproduction of a single jet with transverse momentum p_T and cms-rapidity η in polarized ep collisions reads:

$$\frac{d^2\Delta\sigma}{dp_T d\eta} = \sum_{f^e, f^p, c} \Delta f^e(x_e, M^2) \otimes \Delta f^p(x_p, M^2) \otimes \frac{d^2\Delta\hat{\sigma}^{f_e f_p \to cd}}{dp_T d\eta}, \tag{4}$$

where $\otimes$ denotes a convolution and the sum is running over all properly symmetrized $2 \to 2$ subprocesses for the direct ($\gamma b \to cd$, $\Delta f^e(x_e, M^2) \equiv \Delta P_{\gamma/e}(x_e)$) and resolved ($ab \to cd$) cases. When only light flavors are involved, the corresponding differential helicity-dependent LO subprocess cross sections can be found in [18]. In all following predictions we will deal with the charm contribution to the cross section by including charm only as a *final* state particle produced via the subprocesses $\gamma g \to c\bar{c}$ (for the direct part) and $gg \to c\bar{c}$, $q\bar{q} \to c\bar{c}$ (for the resolved part). For the values of p_T considered it turns out that the finite charm mass can be safely neglected in these subprocess cross sections. In (4), $\hat{s} \equiv x_e x_p s$ and M is the factorization/renormalization scale for which we will use[c] $M = p_T$. The Δf^p stand for the polarized parton distributions of the proton. Needless to say that we obtain the corresponding unpolarized LO jet cross section $d^2\sigma/dp_T d\eta$ by using LO unpolarized parton distributions and subprocess cross sections in (4).

It appears very promising [10] to study the η_{LAB}-distribution of the cross section and the asymmetry, where η_{LAB} is the laboratory frame rapidity, related to η via $\eta \equiv \eta_{cms} = \eta_{LAB} - \frac{1}{2}\ln(E_p/E_e)$. As usual, η_{LAB} is defined to be positive in the proton forward direction. The crucial point is that for negative η_{LAB} the main contributions are expected to come from the region $x_\gamma \to 1$ and thus mostly from the direct piece at $x_\gamma = 1$. To investigate this, Fig. 3 shows our results for the single-inclusive jet cross section and its asymmetry vs. η_{LAB} and integrated over $p_T > 8$ GeV for the four sets of the polarized proton's parton distributions. For Figs. 3a,b we have used the 'maximally' saturated set of polarized photonic parton densities, whereas Figs. 3c,d correspond to the 'minimally' saturated one. Comparison of Figs. 3a,c or 3b,d shows that indeed the direct contribution clearly dominates for $\eta_{LAB} \leq -0.5$, where also differences between the polarized gluon distributions of the proton show up clearly. Furthermore, the cross sections are generally large in this region with asymmetries of a few percents. At positive η_{LAB}, we find that the cross section is dominated by the resolved contribution and is therefore sensitive to the parton content of both the polarized proton *and* the photon. This means that one can only learn something about the polarized photon structure functions if the polarized parton distributions

---

[c]The scale dependence of the theoretical LO predictions for the spin asymmetries—which are the quantities relevant in experiments—turns out to be very weak, for a discussion see [10].

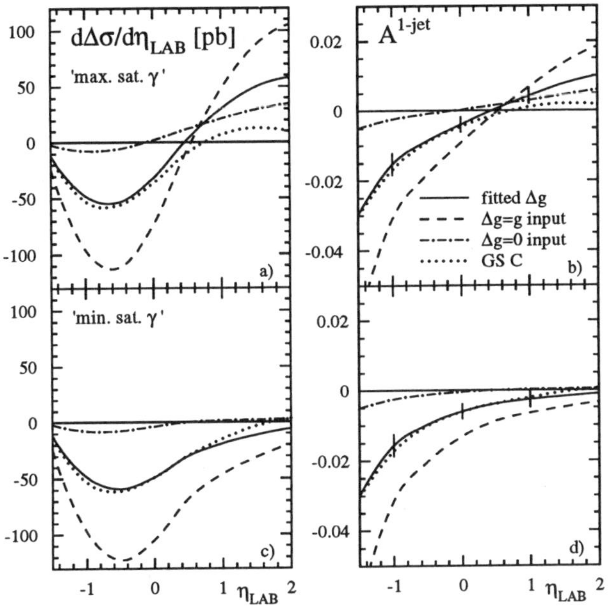

Fig. 3. **a:** *η_{LAB}-dependence of the polarized single-jet inclusive photoproduction cross section in ep-collisions at HERA, integrated over $p_T > 8$ GeV. The renormalization/factorization scale was chosen to be $M = p_T$. The resolved contribution to the cross section has been calculated with the 'maximally' saturated set of polarized photonic parton distributions.* **b:** *Asymmetry corresponding to* **a**. *The expected statistical errors have been calculated according to (5) and as described in the text.* **c,d:** *Same as* **a,b**, *but for the 'minimally' saturated set of polarized photonic parton distributions.*

of the proton are already known to some accuracy or if an experimental distinction between resolved and direct contributions can be achieved. We note that the dominant contributions to the resolved part at large η_{LAB} are driven by the polarized photonic *gluon* distribution Δg^{γ}. We have included in the asymmetry plots in Figs. 3b,d the expected statistical errors δA at HERA which can be estimated from

$$\delta A = \frac{1}{P_e P_p \sqrt{\mathcal{L}\sigma\epsilon}} , \tag{5}$$

where P_e, P_p are the beam polarizations, $\mathcal{L}$ is the integrated luminosity and ϵ the jet detection efficiency, for which we assume $P_e * P_p = 0.5$, $\mathcal{L} = 100/\text{pb}$ and $\epsilon = 1$. From the results it appears that a measurement of the proton's Δg should be possible from single-jet events at negative rapidities where the contamination from the resolved contribution is minimal.

In the unpolarized case, an experimental criterion for a distinction between direct and resolved contributions has been introduced [19] and used [5] in the case of dijet photoproduction at HERA. We will now adopt this criterion for the polarized case to see whether it would enable a further access to Δg and/or the polarized photon structure functions. The generic expression for the polarized cross section $d^3\Delta\sigma/dp_T d\eta_1 d\eta_2$ for the photoproduction of two jets with laboratory system rapidities η_1, η_2 has a form analogous to (4). Here one has

$$x_e \equiv \frac{p_T}{2E_e}\left(e^{-\eta_1}+e^{-\eta_2}\right) \ , \ x_p \equiv \frac{p_T}{2E_p}\left(e^{\eta_1}+e^{\eta_2}\right) \ , \tag{6}$$

where p_T is the transverse momentum of one of the two jets (which balance each other in LO). Following [5], we will integrate over the cross section to obtain $d\Delta\sigma/d\bar{\eta}$, where $\bar{\eta} = (\eta_1+\eta_2)/2$. Furthermore, we will apply the cuts [5] $|\Delta\eta| \equiv |\eta_1-\eta_2| \leq 0.5$, $p_T > 6$ GeV. The important point is that measurement of the jet rapidities allows for fully reconstructing the kinematics of the underlying hard subprocess and thus for determining the variable [5]

$$x_\gamma^{OBS} = \frac{\sum_{jets} p_T^{jet} e^{-\eta^{jet}}}{2yE_e} \ , \tag{7}$$

which in LO equals $x_\gamma = x_e/y$ with y as before being the fraction of the electron's energy taken by the photon. Thus it becomes possible to experimentally select events at large x_γ, $x_\gamma > 0.75$ [19,5], hereby extracting the *direct* contribution to the cross section with just a rather small contamination from resolved processes. Conversely, the events with $x_\gamma \leq 0.75$ will represent the resolved part of the cross section. This procedure should therefore be ideal to extract Δg on the one hand, and examine the polarized photon structure functions on the other.

Fig. 4 shows the results [10] for the direct part of the cross section according to the above selection criteria. The contributions from the resolved subprocesses have been included, using the 'maximally' saturated set of polarized photonic parton densities. They turn out to be non-negligible but, as expected, subdominant. More importantly, due to the constraint $x_\gamma > 0.75$ they are determined by the polarized quark, in particular the u-quark, distributions in the photon, which at large x_γ are equal to their unpolarized counterparts as a result of the Q^2-evolution (see Fig. 2), rather *independently* of the hadronic input chosen. Thus the uncertainty coming from the polarized photon structure is minimal here and under control. As becomes obvious from Fig. 4, the cross sections are fairly large over the whole range of $\bar{\eta}$ displayed and very sensitive to the shape *and* the size of Δg with, unfortunately, not too sizeable asymmetries as compared to the statistical errors for $\mathcal{L} = 100$/pb. A measurement of Δg thus appears to be possible under the imposed conditions only if luminosities clearly exceeding 100/pb can be reached. Fig. 5 displays the same results, but now for the resolved contribution with $x_\gamma \leq 0.75$ for the 'maximally' saturated set (Figs. 5a,b) and the 'minimally' saturated one (Figs. 5c,d). As expected, the results depend on the parton content of both the polarized photon and the proton, which implies that again the latter has to be known to some accuracy to allow for the extraction of some

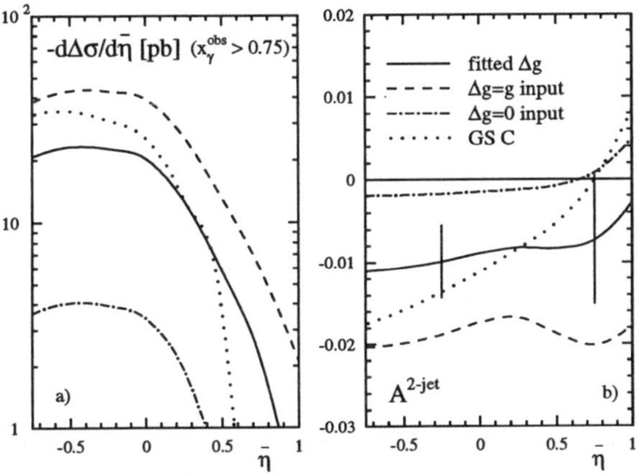

Fig. 4. **a:** *$\bar{\eta}$-dependence of the 'direct' part ($x_\gamma^{OBS} > 0.75$) of the polarized two-jet photoproduction cross section in ep-collisions at HERA for the four different sets of polarized parton distributions of the proton.* **b:** *Asymmetry corresponding to* **a**. *The expected statistical errors indicated by the bars have been calculated according to (5) and as explained in the text.*

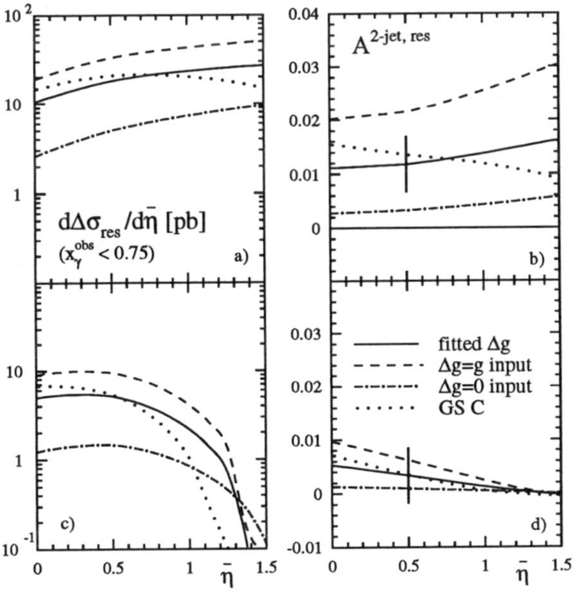

Fig. 5. *Same as Fig. 4, but for the resolved part of the cross section, defined by $x_\gamma^{OBS} \leq 0.75$ (see text). For* **a,b** *the 'maximally' saturated set of polarized photonic parton distributions has been used and for* **c,d** *the 'minimally' saturated one.*

information on the polarized photon structure. We emphasize that the experimental finding of a non-vanishing asymmetry here would establish at least the definite existence of a resolved contribution to the polarized cross section.

Let us finally discuss other photoproduction processes. From our results for one-jet production in Fig. 3 it seems worthwhile to also consider the single-inclusive production of charged hadrons. At a first glance, this process appears less interesting than jet production, as the cross section for producing a definite hadron at a given p_T will always be smaller than the one for a jet. On the other hand, in case of inclusive hadrons one can obviously go experimentally to p_T much smaller than the $p_T^{min} = 8$ GeV employed in our jet studies. Moreover, in the unpolarized case single-inclusive hadron production was successfully studied experimentally at HERA prior to jets [6,7]. The expression for the cross section for single-inclusive hadron production is similar to the one in (4), but comprises an additional convolution with the function D_c^h describing the fragmentation of particle c into the hadron h. For the D_c^h we will use the LO fragmentation functions of [20] which yield a good description of the unpolarized HERA inclusive hadron data [6,7]. Figs. 6a,b show our results for the sum of charged pions and kaons after integration over $p_T > 3$ GeV, where all other parameters were chosen exactly as for Figs. 3a,b (since the sensitivity of the results to the polarized photon structure is qualitatively similar to the one-jet case we only consider the 'maximally' saturated photon scenario here). One can see that the cross sections and their asymmetries behave similarly in shape as the corresponding results in Figs. 3a,b, but are somewhat smaller in magnitude. Nevertheless, the expected statistical errors, calculated for the rather conservative choices $P_e * P_p = 0.5$, $\mathcal{L} = 100/\text{pb}$ and $\epsilon = 0.8$ in Eq. (5) and displayed in Fig. 6b, demonstrate that inclusive hadron photoproduction remains a promising candidate.

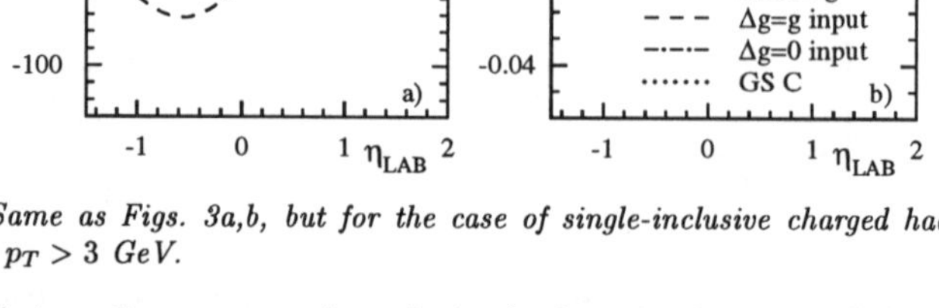

Fig. 6. **a,b:** *Same as Figs. 3a,b, but for the case of single-inclusive charged hadron production, integrated over* $p_T > 3$ *GeV.*

Another interesting process is polarized photoproduction of open charm [21,10], which in the direct case should be mainly driven by the photon-gluon fusion subprocess $\gamma g \to c\bar{c}$ and in the resolved situation by $gg \to c\bar{c}$ and $q\bar{q} \to c\bar{c}$. However, a detailed study of this process for HERA conditions reveals [10] that despite fairly large spin asymmetries for charm production the expected statistical errors for realistic luminosities and charm detection efficiencies will be too large for meaningful measurements.

To conclude, we have analyzed various photoproduction experiments in the context of a polarized *ep*-collider mode of HERA. We have found very encouraging results for jet and inclusive-hadron production which look promising tools for a determination of the polarized gluon distribution of the proton and, possibly, might even allow access to the completely unknown parton content of a polarized photon. The proposed measurements will not be easy to do—but they seem a very interesting challenge for the future at HERA.

4. References

1. For an overview see J. Feltesse and A. Schäfer, in: Proc. of the 1995/96 workshop on 'Future Physics at HERA', eds. G. Ingelman et al., p.760.
2. I. Abt et al., H1 collab.,*Phys. Lett.* **B314** (1993) 436.
3. M. Derrick et al., ZEUS collab., *Phys. Lett.* **B342** (1995) 417.
4. T. Ahmed et al., H1 collab., *Nucl. Phys.* **B445** (1995) 195.
5. M. Derrick et al., ZEUS collab., *Phys. Lett.* **B348** (1995) 665.
6. I. Abt et al., H1 collab., *Phys. Lett.* **B328** (1994) 176.
7. M. Derrick et al., ZEUS collab., *Z. Phys.* **C67** (1995) 227.
8. M. Derrick et al., ZEUS collab, *Phys. Lett.* **B349** (1995) 225.
9. S. Aid et al., H1 collab., *Nucl. Phys.* **B472** (1996) 32.
10. M. Stratmann and W. Vogelsang, *Z. Phys.* **C74** (1997) 641; Proc. of the 1995/96 workshop on 'Future Physics at HERA', eds. G. Ingelman et al., p.815.
11. M. Glück, E. Reya, M. Stratmann and W. Vogelsang, *Phys. Rev.* **D53** (1996) 4775.
12. T. Gehrmann and W.J. Stirling, *Phys. Rev.* **D53**, (1996) 6100.
13. R.D. Ball, S. Forte, and G. Ridolfi, *Phys. Lett.* **B378** (1996) 255; G. Altarelli, R.D. Ball, S. Forte, and G. Ridolfi, *Nucl. Phys.* **B496** (1997) 337.
14. M. Glück, E. Reya, and A. Vogt, *Z. Phys.* **C67** (1995) 433.
15. M. Klasen, G. Kramer and S.G. Salesch, *Z. Phys.* **C68** (1995) 113; M. Klasen and G. Kramer, *Phys. Lett.* **B366** (1996) 385; *Z. Phys.* **C72** (1996) 107.
16. M. Glück and W. Vogelsang, *Z. Phys.* **C55** (1992) 353; **C57** (1993) 309; M. Glück, M. Stratmann and W. Vogelsang, *Phys. Lett.* **B337** (1994) 373.
17. M. Glück, E. Reya, and A. Vogt, *Phys. Rev.* **D46** (1992) 1973.
18. J. Babcock, E. Monsay and D. Sivers, *Phys. Rev.* **D19** (1979) 1483.
19. J.R. Forshaw and R.G. Roberts, *Phys. Lett.* **B319**, (1993) 539.
20. J. Binnewies, B.A. Kniehl and G. Kramer, *Phys. Rev.* **D52** (1995) 4947.
21. S. Frixione and G. Ridolfi, *Phys. Lett.* **B383** (1996) 227.

Chapter 2

Low-x Phenomena

OBSERVATION OF LOW x PHENOMENA IN HADRONIC FINAL STATES

GÜNTER GRINDHAMMER
Max-Planck-Institut für Physik, (Werner-Heisenberg-Institut)
80805 München, Germany
E-mail: guenterg@desy.de

(On behalf of the H1 and ZEUS collaborations)

ABSTRACT

The expectations for and the measurements of transverse energy flows, single particle p_T spectra, and the rate of forward jets in deep inelastic ep events from the H1 and ZEUS experiments at HERA are reported and discussed. It is shown that together they offer a good chance to establish deviations from the DGLAP paradigm. At the present level of limited statistics the measurements are compatible with predictions using BFKL resummation and with the color dipole model. Models based on DGLAP evolution describe the p_T spectra and forward jets less well but are not ruled out yet.

1. Expectations and Observables

In deep inelastic scattering (DIS) at low x the simple picture of the quark parton model, where a virtual photon interacts with a point-like parton in the proton and nothing else happens, has to be modified. The probability that additional partons, particularly gluons, are radiated increases with decreasing x. An example, where a low x parton, which interacts with the photon, originates from a parton shower initiated by a gluon with large x, is shown in Fig. 1. In the approximation of just one gluon being radiated in the initial or the final state, such processes have been calculated in next to leading order (NLO). Two programs, MEPJET [1] and DISENT [2], using different methods, are available, and a new program, DISASTER++ [3], has been announced at this meeting. These programs generate partons only; no full event generators exist.

In the leading log approximation any number of more or less collinear or soft gluons may be radiated. By resumming terms of the form $(\alpha_s \ln \frac{Q^2}{Q_0^2})^n$ using the DGLAP [4] evolution equations an initial state parton shower may be evolved from the scale Q_0^2, typically between 0.3 and 4.0 GeV^2, to the scale Q^2, where the interaction with the photon takes place. DGLAP evolution implies

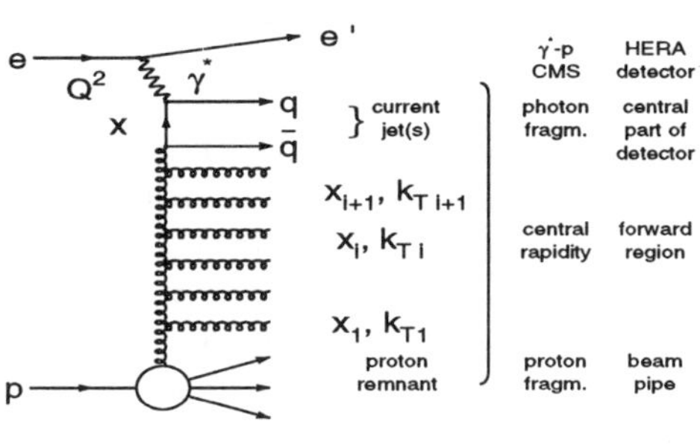

Fig. 1. Diagram of an ep collision at low x.

weakly ordered fractional gluon energies (longitudinal momenta) $x_i > x_{i+1}$ and strongly ordered gluon virtualities or transverse gluon momenta $k_{T,i} << k_{T,i+1}$ along the ladder (Fig. 1). Different implementations of "DGLAP-like" parton showers and of matching [5,6] them to leading order (LO) exact matrix elements for the QCD-Compton (QCDC) and the boson-gluon fusion (BGF) process exist in LEPTO [7], RAPGAP [8], and HERWIG [9]. For the hadronisation of the perturbative partons to the observable hadrons, these programs include phenomenological models. HERWIG uses the cluster model [10]; all other programs, including ARIADNE [11], use the LUND string model [12], as implemented in the JETSET [13] code. These fragmentation models have been well tested at LEP [14]. At HERA and at hadron colliders, the additional complication of proton remnant fragmentation arises, about which much less is known.

At low x, terms of the form $(\alpha_s \ln \frac{1}{x})^n$ become large and can be resummed using the BFKL [15] evolution equations. In a physical gauge, these terms correspond to a ladder diagram in which gluon emissions are strongly ordered in energies $x_i >> x_{i+1}$, while, in contrast to the DGLAP case, the transverse gluon momenta perform a random walk along the ladder. Analytical results at the parton level and first results including fragmentation functions exist and will be mentioned later. Unfortunately no implementations of "BFKL-like" parton showers are available in any of the event generators.

Finally we discuss ARIADNE [11], which currently gives the best overall description of the hadronic final state in DIS, but which does not fit well into any of the schemes mentioned above. It is based on the color dipole model (CDM) [16], where gluon emission originates from a color dipole stretched between the scattered quark and the proton remnant. Further gluon emission leads to a chain of independently radiating dipoles spanned between color-connected partons. The first emission is corrected by the QCDC matrix element in LO. In ARIADNE, the color charge of the proton remnant is assumed not to be point-like, leading to a phenomenological suppression of gluon radiation [11] in the remnant direction. The suppression sets in for hard gluons with wavelengths smaller than the size of the remnant. Also the color charge of the scattered quark is taken to be spread out, depending on the virtuality Q^2 of the photon [11,17]. This in turn leads to a suppression of radiation in the direction of the scattered quark. An important feature of ARIADNE is that the probability for a QCDC event and for any other gluon emission, from a dipole connected to the remnant at one end, depends on this "unorthodox suppression" [18], due to the extended color charge, while in "normal" QCD, as implemented in all other programs, it depends on the ratio of parton densities before and after the emission and on other factors which are the same for all programs. At low x this feature leads to a greatly enhanced rate of QCDC events compared to the other generators. Additionally, in contrast to the DGLAP-like programs, the gluons in ARIADNE are not ordered in k_t along the ladder. It has therefore been argued [19] that the partonic final state of ARIADNE is more closely related to the one expected from BFKL evolution.

It has been shown that all models provide a fair description of basic event properties [20,21,22].

At very low x the gluon density may possibly become so high in regions ("hot spots") of the proton that the gluons will no longer act as free particles but will interact with each other. In this saturation regime of QCD, the strong coupling is still small, but the incoherent scattering approximation is no longer valid. It is not clear whether this new regime could be observed at HERA.

What do we expect to see at HERA? Which observables will be most sensitive to the underlying parton dynamics?

As x decreases below 10^{-3} we anticipate deviations from the DGLAP predictions due to the missing $(\alpha_s \ln \frac{1}{x})^n$ terms. The most inclusive way to study the structure of the proton is to measure the probing lepton after the interaction. The resulting measurement of the proton structure function F_2 has the advantages that it is experimentally relatively easy and that it can be directly compared to analytical QCD calculations. The rise of F_2 with decreasing x as observed at HERA [23] turns out to be well describable by the DGLAP evolution equations alone, down to $Q^2 \geq 1.0$ GeV^2 and $x \geq 10^{-5}$. At this meeting, it was shown that the F_2 data at small x are also well compatible with a unified BFKL and DGLAP description [24].

Perhaps F_2 is not sufficiently sensitive to observe the change in parton dynamics at low x. The reasons might be that F_2 is a too inclusive measurement, averaging out over differences only visible in the details of the hadronic final state. There might not be enough range in x at moderate Q^2. Moreover, the theory has some flexibility due to the freedom of choice of the starting parton distributions.

A greater sensitivity to small x effects might be provided by observables based on the hadronic final state emerging from the underlying partons. It can be further enhanced by making use of the differences between the DGLAP and the BFKL cascade. However, there is a price to pay. The measurements have to be compared to Monte Carlo simulations including the QCD effects, the transition from partons to hadrons, the particle decays, and detector effects. The observables studied so far are the transverse energy flows, single particle p_T-spectra, and forward jets.

2. Transverse Energy Flows

BFKL dynamics predicts [25,26], due to the lack of ordering of k_T in the parton cascade, more average transverse energy E_T and harder tails in the E_T distribution for decreasing x than in the DGLAP scenario. These observables are measured calorimetrically in the central rapidity [a] region, between the struck quark and the proton remnant, in the hadronic center of mass (hcms), that is the rest system of the exchanged virtual photon and proton. In that frame all E_T [b] is due to either QCD

[a] Rapidity is here always pseudo-rapidity, i.e. $\eta = -\ln\tan(\theta/2)$.

[b] $E_T = \sum E_i \sin\theta_i$, i runs over calorimeter cells.

radiation or hadronisation. As indicated in Fig. 1 the central rapidity in the hcms corresponds to the forward region in the frame of the HERA detectors.

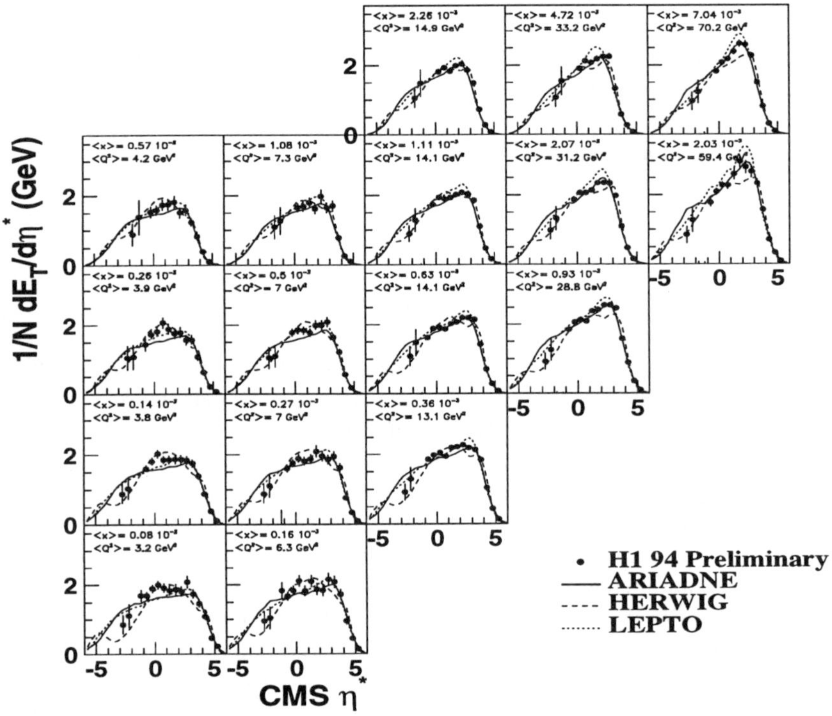

Fig. 2. Transverse energy flow as a function of rapidity η^* in the hcms. The H1 data points show statistical errors only, except for two points measured with the plug calorimeter where the statistical and systematic errors are given. The lines are the Monte Carlo predictions of three different models.

First results on the E_T-flows have been published by the H1 collaboration [27,28,29]. In ref. [29] they have been reported for DIS events with $5\ \mathrm{GeV}^2 < Q^2 < 50\ \mathrm{GeV}^2$ and with $10^{-4} < x < 10^{-2}$ in the hcms and have been compared to the then current model predictions of LEPTO 6.1 and ARIADNE 4.03. In the largest x and Q^2 bin, i.e. $x \approx 5 \times 10^{-3}$ and $Q^2 \approx 33\ \mathrm{GeV}^2$, the data and the two models were found to agree. With decreasing x for fixed Q^2 the E_T from LEPTO 6.1 was found to fall more and more below the data for rapidities away from the current jet, while ARIADNE 4.03 still managed to give a level of E_T in agreement with the data.

At this point in time one could have had the impression that models with DGLAP-like dynamics like LEPTO fail to describe the data at low x, while models with some

BFKL-like features like ARIADNE are successfull. However, soon after, two new phenomenological features have been added to LEPTO which allowed a reasonably good description of the transverse energy flows also for decreasing x. The new features are soft color interactions [30], with the intention to describe rapidity gap events without modeling a Pomeron and its structure function, and a modified sea-quark/remnant treatment, giving a smoother transition from BGF events to events where the photon interacts with a sea-quark.

The H1 collaboration in the meantime has produced new preliminary data [31,32], covering a larger range in the kinematic plane from $3 < Q^2 < 70$ GeV^2 and $8 \times 10^{-5} < x < 7 \times 10^{-3}$. The extension to lower Q^2 was achieved by analysing data from special runs, where the point of the *ep* interactions was shifted from the nominal position in the direction of the proton beam in order to have access to smaller lepton scattering angles. In addition, the E_T at two very forward rapidity bins was measured by H1 using their plug calorimeter ($0.72° < \theta_{lab} < 3.3°$), which closes the gap between the beam-pipe and the forward part of their liquid Argon calorimeter.

The data as a function of rapidity [c] in the hcms are shown in bins of x and Q^2 in Fig. 2. They are compared to predictions from ARIADNE 4.08, HERWIG 5.8, and LEPTO 6.4. The models give only a fair description of the data over the large range in x and Q^2. The average E_T in the central rapidity $-0.5 < \eta^* < 0.5$ as a function of x in bins of fixed Q^2 increases with decreasing x [31,32]. This is demonstrated in Fig. 3a for the bin $Q^2 = 14$ GeV^2, together with data from ZEUS [33], and with predictions from

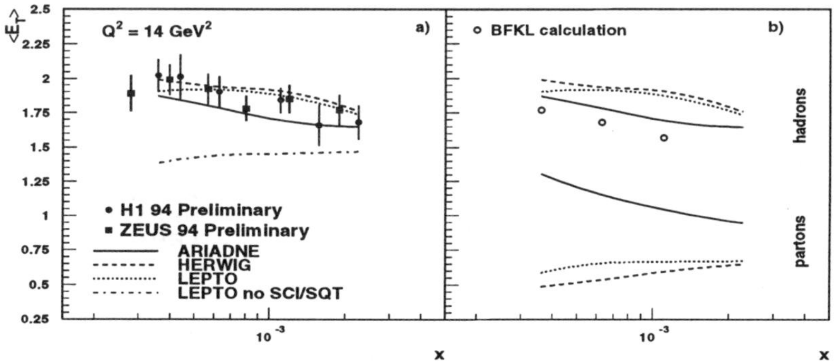

Fig. 3. Mean transverse energy in the central rapidity bin in the hcms for the bin $Q^2 = 14$ GeV^2. Besides the data from H1 and ZEUS, QCD model predictions (lines) are shown for hadrons in a) and for partons in b). An analytic BFKL calculation at the parton level is shown as open circles.

different generators at the hadron level. For LEPTO the predictions are shown with the new soft color interactions (SCI) and the new sea-quark treatment (SQT) turned on and off. The data of both experiments are in good agreement and can be described

---

[c] The direction of the proton is to the left (negative rapidity).

by all models, DGLAP and BFKL-like. In Fig. 3b the predictions of the models at the hadron level are contrasted with those at the parton level and with an analytic parton level BFKL calculation [25]. At the parton level the DGLAP-like models LEPTO and HERWIG show the opposite slope in x than ARIADNE, the BFKL calculation, the hadron level of all models, and the data. The BFKL calculation gives the highest transverse energy, 40% to 50% above the partons from the color dipole model. There are large differences in the contribution from fragmentation to the total E_T between the models. It is about 15% for the BFKL calculation (comparing it to data), 35% for ARIADNE, and about 70% for LEPTO and HERWIG.

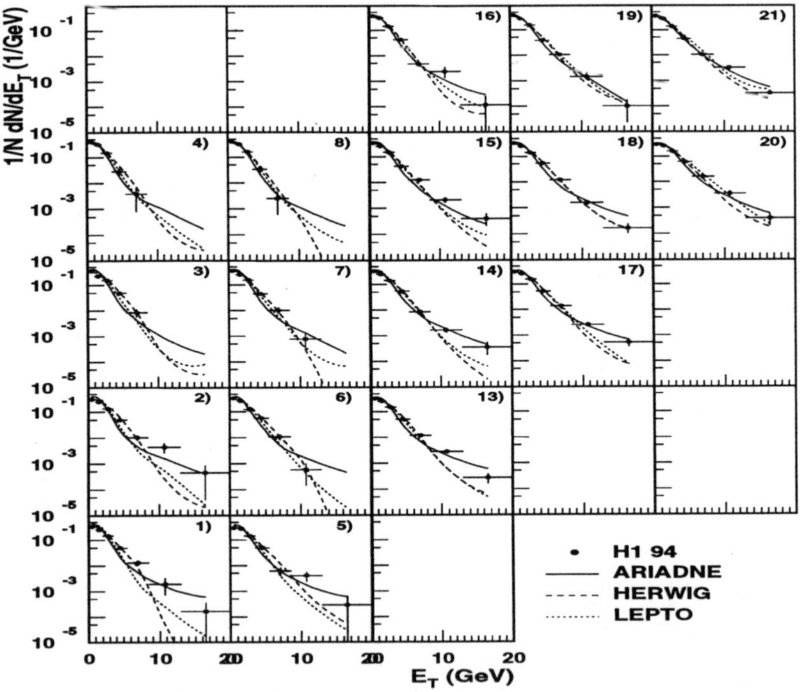

Fig. 4. Transverse energy distribution in the central rapidity bin in the hcms. Only statistical errors are shown.

Concerning the measurement of the mean E_T in the central rapidity bin, we are now in the situation that all models describe the data. Although the models differ in their underlying parton dynamics, the mean E_T can be made to agree by exploiting as yet unconstrained variations in hadronisation models.

Another observable investigated by H1 is the distribution of the E_T in the central rapidity bin per event. The E_T originating from hadronisation is expected to be limited, while high values of E_T are more likely to be produced by hard parton radiation. Preliminary data [31,32], in the same bins of x and Q^2 as in Fig. 2, are displayed in Fig. 4 together with predictions from ARIADNE 4.08, HERWIG 5.8, and LEPTO 6.4. While at the largest x and Q^2 the data and the models agree, with decreasing x and Q^2 the data and ARIADNE appear to exhibit harder tails than the DGLAP-like models. As can be seen from the figure a firm conclusion can only be drawn with more statistics extending the range in E_T to larger values. Comparing the E_T distribution for the highest and lowest x value for fixed Q^2, one finds that the data have harder tails [31,32] with decreasing x, suggesting a rise in parton activity.

3. Single Particle p_T Spectra

The E_T measured by the calorimeter can be due to many soft particles, predominantly from hadronisation, or due to a few hard particles, mainly from hard gluon emission, or due to both. Therefore it has been suggested [34] that the hard tail of the

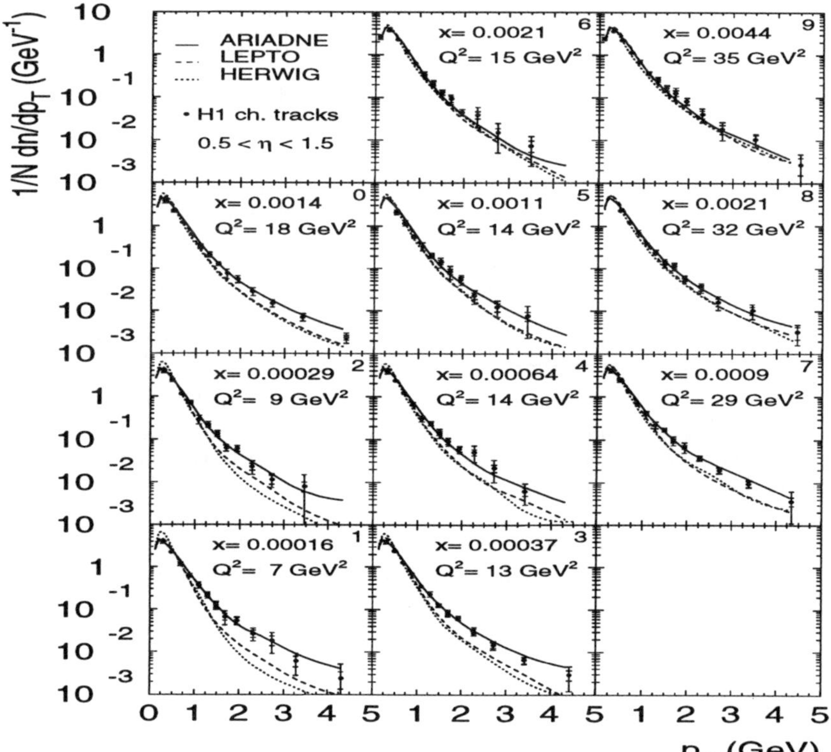

Fig. 5. The transverse momentum spectra of charged particles in the hcms. The data are shown for nine different kinematic bins and the combined sample (bin 0). Statistical and full errors are given.

p_T distribution of single charged particles in the hcms might offer better sensitivity to the basic parton dynamics since it is more directly linked to hard gluon emissions.

The charged particle p_T distributions in bins of x and Q^2 as measured by H1 [35] in the rapidity interval from 0.5 to 1.5 are shown in Fig. 5. At large x all three models presented agree with the data. With decreasing x, LEPTO and HERWIG significantly fall below the data for increasing p_T. Predictions [36] based on BFKL resummation and convolution with fragmentation functions for the transition from partons to charged hadrons [37] have been made for the three lowest bins in x, i.e. bins 1, 2, and 3. As demonstrated in Fig. 6 for bin 3, the data and the BFKL result agree quite well for $p_T \geq 1.5$ GeV. The absolute normalization was derived from a comparison of the BFKL calculation of the forward jet cross section with data from H1 to be discussed later. Also displayed in the figure is the calculation with BFKL effects turned off, which falls below the data. It is apparent from the figure

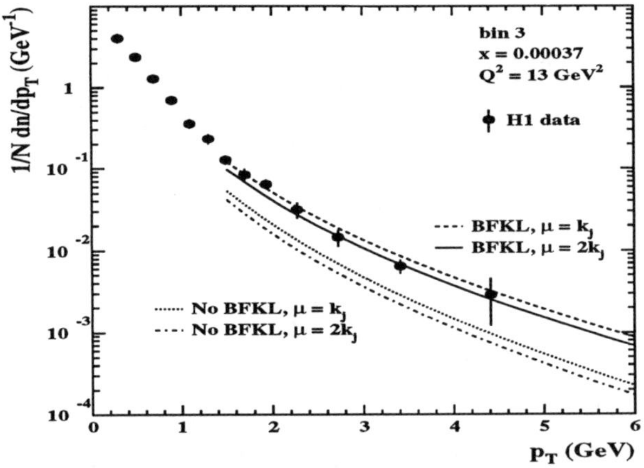

Fig. 6. The transverse momentum spectra of charged particles in the hcms for fixed low x and Q^2 (bin 3). The data and predictions from a BFKL calculation, with BFKL effects off and on, and with two different choices for the scale μ of the fragmentation function are compared.

that the significance of the agreement can be increased with more data at higher p_T. In addition, BFKL effects would become stronger, if the measurement of charged particles could be extended further in the direction of the proton remnant, i.e. to the rapidity interval $-0.5 < \eta^* < 0.5$.

4. Forward Jets

The cross section for forward jets in DIS as a function of x has been advertised [38] for some time now as an observable enhancing the effects of BFKL resummation. Diagrammatically the situation is described in Fig. 7. The forward jet is defined by the azimuthal angle θ_{jet} between the jet and the proton direction and its transverse momentum k_{Tjet}. The energy of the jet is then given by $E_{jet} = k_{Tjet}/\sin\theta_{jet}$ and $x_{jet} \approx E_{jet}/E_p$. We now require that $k^2_{Tjet} \approx Q^2$ which suppresses the phase space for forward jet production and for gluon emission between the forward and the current jet in the DGLAP case. In addition, we demand that x_{jet}/x be as large as possible which maximizes the phase space for the production of a forward jet and gluon radiation between it and the current jet in the BFKL scenario. For this particular kinematic configuration the resummation of $\alpha_s \ln x_{jet}/x$ terms should lead to a sizeable growth of the forward jet cross section with decreasing x. Another advantage of this observable is that the parton densities of the proton are probed at rather large scales, x_{jet} and $k^2_{Tjet} \approx Q^2$, where they are well known.

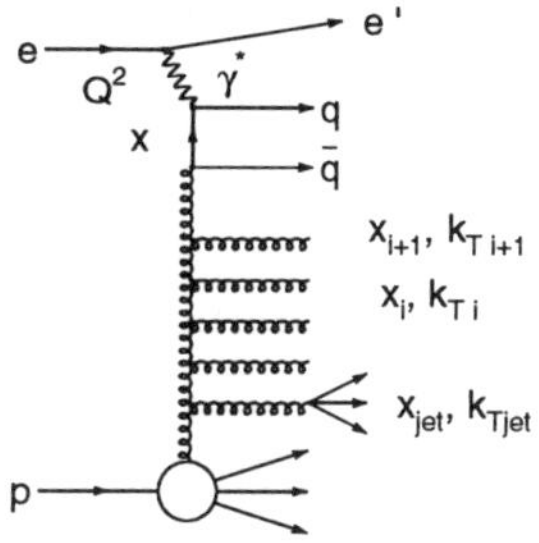

Fig. 7. Parton evolution with forward jet.

Published results on forward jets are available from H1 [29] and new preliminary results from H1 [39] and from ZEUS [40]. The cuts for the selection of DIS events and the requirements on the forward jet are summarized in Table 1. Both experiments use a cone algorithm, requiring a minimum p_{Tjet} in a cone of radius $R = \sqrt{\Delta\eta^2 + \Delta\Phi^2}$ in the HERA frame. There are differences in the current analyses between the two

Table 1. DIS event and forward jet selection

	H1	ZEUS
$E_e >$	11 GeV	10 GeV
$y_e >$	0.1	0.1
$\theta_e <$	173°	-
$\theta_e >$	160°	-
cone in lab, R	1.0	1.0
$\theta_{jet} >$	**7°**	**10°**
$p_{Tjet} >$	**3.5** GeV	**5.0** GeV
$x_{jet} >$	0.035	0.035
$\theta_{jet} <$	20°	-
$\frac{p^2_{Tjet}}{Q^2} >$	0.5	0.5
$\frac{p^2_{Tjet}}{Q^2} <$	**2.0**	**4.0**

experiments in the cuts on θ_{jet}, p_{Tjet}, and the upper limit on p^2_{Tjet}/Q^2 which will be discussed later. With the CDM the transverse energy flow around the forward jet axis can be well described for different values of p_{Tjet} [39] as well as many other distributions [41,42].

In Fig. 8 the ZEUS forward jet cross section corrected to the parton level of ARIADNE is shown and compared to several parton level calculations: an analytic BFKL calculation [43], the same calculation but without any gluon emission between the forward and the current jet system (Born BFKL), and a fixed NLO QCD calculation using MEPJET. The data show a much faster rise with decreasing x than the calculations without BFKL resummation. The BFKL calculation shows an even more dramatic rise. The authors [43], however, point out that several effects which have not been taken into account might lower their prediction. ZEUS presents their preliminary result corrected to the parton level in order to compare to parton level calculations. Using ARIADNE correction factors are found to vary between 0.6 for the lowest x bin and ≈ 1 for larger x. It will be useful to have the ZEUS cross section also at the hadron level, since it is not clear, what the relationship is between the parton level of the CDM and BFKL. It could also allow comparisons with H1 data which are corrected to the hadron level.

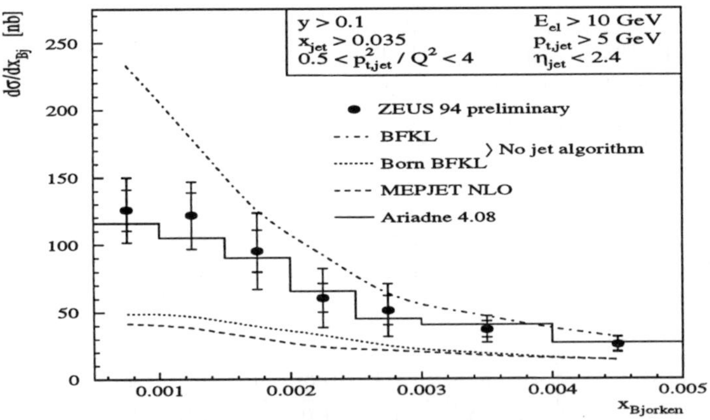

Fig. 8. The differrential forward jet cross section as a function of x. The ZEUS data have been corrected to the parton level. Statistical and systematic errors (not yet complete) are included. BFKL calculations, with BFKL effects on (dash-dotted line) and off (dotted line), calculations in NLO QCD using MEPJET, and parton level results from the color dipole model are shown.

The preliminary H1 results on the forward cross section are shown in Fig. 9a and are compared to predictions from LEPTO (MEPS) with and without soft color interactions (SCI) and to ARIADNE (CDM). Again a fast rise of the cross section is observed with the CDM falling only slightly below the data. In Fig. 9b models and calculations at the parton level are shown: ARIADNE, LEPTO, two BFKL calculation [43,36], and a NLO calculation using DISENT. ARIADNE shows a similar x dependence at the parton and hadron level with hadronisation effects amounting to less than 20%. The parton level forward jet cross sections of LEPTO and of

DISENT in NLO agree as expected and show a moderate increase. In LEPTO, with soft color interactions turned on, up to 80% of the forward jets are created in the hadronisation phase causing LEPTO to only slightly undershoot the data with decreasing x. The BFKL calculations basically can only predict the dependence on x but not the normalization. As mentioned before, the computation by Kwiecinski et al. [36] fixed the normalization to the data from H1 while the calculation by Bartels et al. [43] did not. Scaling the cross section by Bartels et al. down by a factor 0.8 brings the two BFKL calculations in agreement. The color dipole model provides the

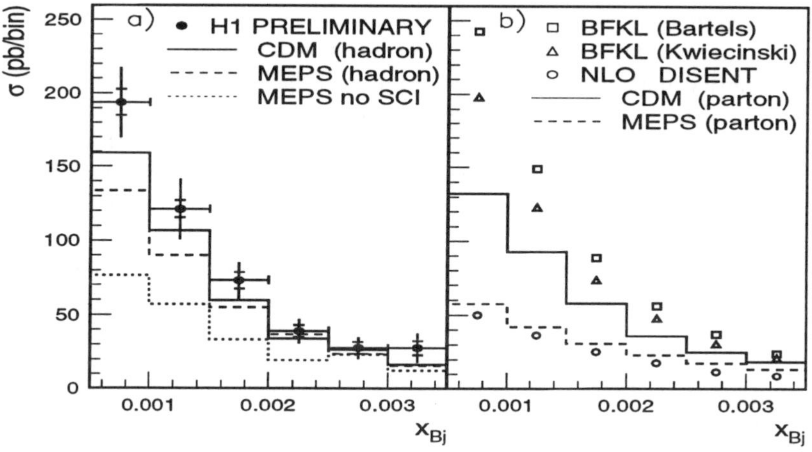

Fig. 9. The forward jet cross section as a function of x. The H1 data have been corrected to the hadron level. Statistical and systematic errors are included. QCD model predictions for hadrons are superimposed in (a) and for partons in (b). Two analytic BFKL calculation (open squares and triangles) and a NLO calculation using DISENT (open circles) are also shown in (b).

best description of the data. However, it should be pointed out that its prediction is rather sensitive to the power of soft suppression of gluon emission $(\mu/p_T)^\alpha$ due to the proton remnant. The parameter α is related to the dimensionality of the extended proton remnant and is expected to be ≈ 1. The H1 forward jets prefer $\alpha \leq 1.0$ while non-jet observables like energy flows prefer $\alpha \approx 1.5$ [20].

One may wonder about the choice for the minimum p_{Tjet}, 3.5 GeV by H1 and 5.0 GeV by ZEUS. A small value is desirable from the point of view of statistics and sensitivity to BFKL effects [44]. With increasing p_{Tjet} the slope in x for the forward jet cross section decreases. The requirement $p_{Tjet} \approx Q^2$ forces the Q^2 to increase, which in turn increases the minimum x which is probed. On the other hand, to suppress hadronisation effects, one would want p_{Tjet} to be large.

Another question concerns the choice of the upper limit on p^2_{Tjet}/Q^2 which is 2.0 for H1 and 4.0 for ZEUS. Increasing this limit from 2.0 to 4.0 increases in the lowest

x bins the contribution from $O(\alpha_s)$ matrix elements to the forward jets by roughly a factor of two. One would also expect an increased contribution from resolved photoproduction which is not included in the calculations. After this workshop the author of RAPGAP has shown [45] that for example the H1 result can also be described with direct (same as in LEPTO without SCI) and resolved contributions.

5. Conclusions

The tails of the single particle p_T spectra and of the calorimetric E_T distribution, and the forward jet cross sections offer a good chance to pin down deviations from the DGLAP paradigm. Other interesting observables in this quest, like the decorrelation of the azimuthal angle between the forward jet and the lepton [43], the forward π° cross section [46], and the production of more than one forward jet [47] should be pursued.

Along the way, we probably will and have to get a better understanding of hadronisation effects, particularly of the proton remnant. On the theory side, higher order effects [48] have to be included in the calculations and a BFKL Monte Carlo generator [49,50] is needed.

In addition to collecting more data at HERA, it would be desirable to access smaller forward angles for jets [44] and particles and to increase the HERA center of mass energy.

6. Acknowledgements

I want to thank T. Carli, M. Wobisch, and S. Wölfle for providing me with plots and J. Dainton, J. Hartmann, and D. Krücker for a careful reading of the manuscript. I am also happy to thank the organizing committee, in particular B. Kniel, for the excellent workshop and the pleasant atmosphere.

7. References

1. E. Mirkes and D. Zeppenfeld, Phys. Lett. B380 (1996) 205.
2. S. Catani and M. Seymour, Proc. Workshop on Future Physics at HERA 1995/96, Hamburg, eds. G. Ingelman, A. De Roeck, and R. Klanner, vol. 1 (1996) 519, CERN-TH/96-240 (1996) [hep-ph/9609521], and Acta Phys. Polon. B28 (1997) 863.
3. D. Graudenz, these proceedings.
4. V. N. Gribov and L. N. Lipatov, Sov. J. Nucl. Phys. 15 (1972) 438 and 675; G. Altarelli and G. Parisi, Nucl. Phys. B126 (1977) 298; Yu. L. Dokshitzer, Sov. Phys. JETP 46 (1977) 641.
5. M. Bengtsson, G. Ingelman, and T. Sjöstrand, Proc. HERA Workshop 1987, Hamburg, ed. R. D. Peccei, vol. 1 (1988) 149;

M. Bengtsson and T. Sjöstrand, Z. Phys. C37 (1988) 465.

6. M. H. Seymour, LU-TP-94-12, Nucl. Phys. B436 (1995) 443 and Comput. Phys. Commun. 90 (1995) 95.
7. G. Ingelman, Proc. Workshop on Physics at HERA 1991, Hamburg, eds. W. Buchmüller and G. Ingelman, vol. 3 (1992) 1366;
   G. Ingelman, A. Edin, and J. Rathsman, Comput. Phys. Commun. 101 (1997) 108.
8. H. Jung, Comput. Phys. Commun. 86 (1995) 147.
9. G. Marchesini et al., Comput. Phys. Commun. 67 (1992) 465.
10. B. R. Webber, Nucl. Phys. B238 (1984) 492;
    G. Marchesini and B. R. Webber, Nucl. Phys. B310 (1988) 461.
11. L. Lönnblad, Comput. Phys. Comm. 71 (1992) 15 and Z. Phys. C65 (1995) 285.
12. B. Andersson, G. Gustafson, G. Ingelman, and T. Sjöstrand, Phys. Rep. 97 (1983) 31.
13. T. Sjöstrand, Comput. Phys. Commun. 82 (1994) 74;
    T. Sjöstrand, PYTHIA 5.7 and JETSET 7.4, CERN-TH 7112/93 (1993) and LU TP 95-20 (1995).
14. I. G. Knowles et al., Proc. LEP2 Physics Workshop 1995 [hep-ph/9601212].
15. E. A. Kuraev, L. N. Lipatov, and V. S. Fadin, Sov. Phys. JETP 45 (1977) 199;
    Ya. Ya. Balitsky and L. N. Lipatov, Sov. J. Nucl. Phys. 28 (1978) 822.
16. G. Gustafson, Phys. Lett. B175 (1986) 453;
    G. Gustafson and U. Petterson, Nucl. Phys. B306 (1988);
    B. Andersson, G. Gustafson, L. Lönnblad, and U. Petterson, Z. Phys. C43 (1989) 625.
17. L. Lönnblad, M. Seymour, et al., Proc. Physics at LEP 2 Workshop, eds. G. Altarelli, T. Sjöstrand, and F. Zwirner, CERN/96-01, vol. 2, (1996) 187.
18. J. Rathsman, Phys. Lett. B393 (1997) 181.
19. A. H. Mueller, Nucl. Phys. B415 (1994) 373.
20. N. Brook et al., Proc. Workshop on Future Physics at HERA 1995/96, Hamburg, eds. G. Ingelman, A. De Roeck, and R. Klanner, vol. 1 (1996) 613.
21. T. Carli, Proc. 4th Int. Workshop on Deep Inelastic Scattering and Related Phenomena (DIS 96), Roma, eds. G. D'Agostini and A. Nigro, (1997) 415.
22. H. Jung, private communication.
23. L. Favart, these proceedings.
24. A. D. Martin, these proceedings.
25. K. Golec-Biernat, J. Kwiecinski, A. D. Martin, and P. J. Sutton, Phys. Rev. D50 (1994) 217 and Phys. Lett. B335 (1994) 220.
26. J. Bartels, H. Lotter, and M. Vogt, Phys. Lett. B373 (1996) 215.
27. H1 Collaboration, T. Ahmed et al., Phys. Lett. B298 (1993) 469.
28. H1 Collaboration, I. Abt et al., Z. Phys. C63 (1994) 377.

29. H1 Collaboration, S. Aid et al., Phys. Lett. B356 (1995) 118.
30. A. Edin, G. Ingelman, and J. Rathsman, Z. Phys. C75 (1997) 57 and Phys. Lett. B366 (1996) 371.
31. H1 Collaboration, C. Adloff et al., contrib. paper pa02-073, ICHEP '96, Warsaw, Poland, July 1996.
32. M. F. Hess, Ph.D. Thesis, University of Hamburg (1996).
33. N. A. Pavel, Proc. 4th Int. Workshop on Deep Inelastic Scattering and Related Phenomena (DIS96), Roma, eds. G. D'Agostini and A. Nigro, (1997) 502.
34. M. Kuhlen, Phys. Lett. B382 (1996) 441.
35. H1 Collaboration, C. Adloff et al., Nucl. Phys. B485 (1997) 3.
36. J. Kwiecinski, S. C. Lang, and A. D. Martin, DTP-97-56 (1997) [hep-ph/9707240].
37. J. Binnewies, B. A. Kniehl, and G. Kramer, Phys. Rev. D52 (1995) 4947.
38. A. H. Mueller, Nucl. Phys. B (Proc. Suppl.) 18C (1990) 125 and J. Phys. G17 (1991) 1443;
W. K. Tang, Phys. Lett. B278 (1992) 363;
J. Bartels, A. De Roeck, and M. Loewe, Z. Phys. C54 (1992) 635;
A. De Roeck, Nucl. Phys. B (Proc. Suppl.) 29A (1992) 61;
J. Kwiecinski, A. D. Martin, and P. J. Sutton, Phys. Lett. B287 (1992) 254, Nucl. Phys. B (Proc. Suppl.) 29A (1992) 67, and Phys. Rev. D46 (1992) 921.
39. J. G. Contreras, Proc. XXXI Rencontres de Moriond in QCD and High Energy Hadronic Interactions, ed. J. Tran Thanh Van, Edition Frontieres (1996) 411;
H1 Collaboration, C. Adloff et al., contrib. paper pa03-049, ICHEP '96, Warsaw, Poland, July 1996.
40. S. Wölfle, to be published in Proc. 5th Int. Workshop on Deep Inelastic Scattering and QCD, Chicago, April 1997, eds. D. Krakauer and J. Respond.
41. J. G. Contreras, Ph.D. Thesis, University of Dortmund (1997).
42. E. M. Lobodzinska, Ph.D. Thesis, Inst. of Nucl. Physics, Cracow (1997).
43. J. Bartels et al., Phys. Lett. B384 (1996) 300.
44. J. Bartels, A. De Roeck, and M. Wüsthoff, Proc. Workshop on Future Physics at HERA 1995/96, Hamburg, eds. G. Ingelman, A. De Roeck, and R. Klanner, vol. 1 (1996) 598.
45. H. Jung, to be published in Proc. Madrid Workshop on Low x Physics, Miraflores de la Sierra, June 1997.
46. J. Kwiecinski, S. C. Lang, and A. D. Martin, Phys. Rev. D55 (1997) 1273.
47. J. Kwiecinski, C. A. M. Lewis, and A. D. Martin, DTP-97-58 (1997) [hep-ph/9707375].
48. J. Bartels, these proceedings.
49. C. R. Schmidt, Phys. Rev. Lett. 78 (1997) 4531.
50. L. H. Orr and W. J. Stirling, DTP-97-48 (1997) [hep-ph/9706529], to be published in Phys. Rev. D.

Search for the BFKL Pomeron in the Final States at HERA

J.Bartels
II.Institut f.Theoretische Physik, Universität Hamburg
Hamburg, Luruper Chaussee 149, D-22761 Hamburg, Germany
E-mail: bartels@x4u2.desy.de

ABSTRACT

This talk discusses the present status of the experimental search for the BFKL Pomeron in deep inelastic final states at HERA.

1. Introduction

In connection with the observed rise at small x of the proton structure function, much attention has recently been given to the Balitsky-Fadin- Kuraev-Lipatov (BFKL) Pomeron [1]. Although its first derivation dates back already more than twenty years, it is still a rather "young" piece of perturbative QCD, and we are only at the beginning of searching for experimental evidence of this new object. In this talk I would like to discuss, in the context of HERA physics, the current status of the search for the BFKL Pomeron.

As it is well-known [2] the BFKL Pomeron presents, within QCD, the leading logarithmic approximation to the Regge limit (large energies and small momentum transfer) of the elastic scattering of gluons or quarks. The relevant Feynamn diagrams are t-channel ladders built from reggeizing gluons and (effective) gluon rungs. In order to justify the use of perturbative QCD, the scattering objects need to have small transverse sizes; examples are highly virtual partons (e.g. gluons with large transverse momenta) or heavy photons, γ^*, which dissociate into quark-antiquark states. The size (or the momentum scale) of all external particles should be the same; otherwise a second class of large logarithms appears, and the BFKL approximation starts to overlap with the DGLAP evolution equations. One of the main predictions of the BFKL approximation is the power-like energy dependence of the elastic scattering amplitude:

$$T_{el} \sim is^{1+\omega_{BFKL}}, \quad \omega_{BFKL} = \frac{4N_c \ln 2}{\pi} \alpha_s(\bar{Q}^2) \tag{1}$$

where the (fixed) scale $\bar{Q}^2$ of the strong coupling constant is given by the hard scale of the scattering objects. The dependence on the momentum transfer t is weak: for $|t| > \bar{Q}^2$ the t dependence of the leading singularity is obtained by simply replacing in (1) $\bar{Q}^2$ by $|t|$.

One of the important basic properties of the BFKL approximation is the diffusion in the logarithm of the transverse momenta inside the ladder diagrams. At the top and at the bottom of the exchanged gluon ladders, the internal transverse momenta are of the order of the momentum scale $\bar{Q}^2$ set by the external particles. As rapidity Y grows, in the middle of the ladders the transverse momenta diffuse more and more into the infrared and into the ultraviolet region; the mean square deviation is given by

$$\Delta^2 = \frac{28N_c\zeta(3)}{\pi}\alpha_s Y \tag{2}$$

As a result, at large energies the BFKL approximation becomes more and more sensitive to infrared physics where perturbation theory is not applicable. Therefore, even for the scattering of objects with a very small transverse extension the BFKL Pomeron remains applicable only for a finite region in Y.

A major part of the theoretical activities around the BFKL Pomeron is devoted to the computation of (perturbative) corrections and to the unitarization of the BFKL Pomeron. Using the optical theorem eq.(1) implies a strong power-like growth of the total cross section: $\sigma_{tot} \sim s^{\omega_{BFKL}}$. In hadron-hadron scattering such a strong rise cannot persist. For asympotic energies rigorous unitarity bounds [3] exist which require that

$$\sigma_{tot} \leq const \cdot \ln^2 s. \tag{3}$$

Although these bounds do not directly apply to parton parton scattering, it seems very natural to expect that also in this case such a strong power-like growth of the total cross section will not persist, i.e. higher order corrections to the BFKL Pomeron are expected to unitarize the leading logarithmic approximation. The first corrections beyond the leading logarithmic approximation are the order-α_s^2 corrections to the BFKL kernel. They have recently been completed [4,5], and they lead to the result that the BFKL power is lower than the value derived from (1). Beyond these first nonleading corrections, the BFKL Pomeron has been found to be the lowest order beginning of an effective 2+1 dimensional field theory with an underlying conformal symmetry [6]. In particular, there are generalized ladder diagrams with more than two reggeized gluons in the t-channel [7], and new vertices for the transition from two to four gluons [8]. In the large-N_c limit, substructures of this field theory have been found to be integrable [10]. Finally, an alternative formulation of the BFKL Pomeron ("color dipole picture") has been developed by A.Mueller [11] and collaborators [12], and within this picture the unitarization of the BFKL Pomeron has been investigated [13,14].

One of the main reasons why the observed rise of F_2 at HERA has stimulated so much interest in the BFKL Pomeron was the hope the BFKL calculation could be used to "compute" the small-x behavior of the input to the DGLAP evolution of the gluon structure function at the low momentum scale Q_0^2. However, from what has been said above it follows that there are fundamental difficulties in applying the BFKL calculation to deep inelastic electron proton scattering at small-x. First, the perturbative part of F_2 (and all parton distributions) has two momentum scales, the scale Q_0^2 where the perturbative DGLAP evolution starts, and the final scale Q^2. The BFKL Pomeron, on the other hand, has been derived for a single-scale problem, i.e. it describes the energy dependence of the scattering of two objects with the same size. In F_2, therefore, the BFKL approximation (leading $\ln 1/x$) competes with the DGLAP approximation (leading $\ln Q^2$). Moreover, since the lower momentum scale Q_0^2 is usually taken to be rather small, the BFKL calculation of the input gluon structure function receives a significant contribution [15] from the infrared region where perturbation theory is unreliable. Therefore, "predictions" for the gluon structure function which are based upon the BFKL approximation depend upon the way in which the infrared part of the internal phase space has been modelled (for an example see [16]). As a result, one has to conclude that deep inelastic structure functions at small-x are not well-suited to obtain a clean test of the BFKL Pomeron. On the other hand, this does not imply that deep inelastic structure functions at small-x do not contain clear "footprints" of the BFKL dynamics, e.g. effects of the diffusion. The most promising way of discovering such effects is the opening of final states contained in F_2: in the following I shall discuss two such measurements at HERA.

2. The Forward-Jet Measurement: Energy Dependence

The by far best way at HERA to search for the BFKL Pomeron is the measurement of events which contain a hard jet in the forward direction [17,18,19]. The transverse momentum squared k_t^2 of the jet should be of the same (within a factor of 2 or so) as Q^2 (the square of the photon mass), and the x_j (fraction of the proton momentum carried by the jet) as large as possible. In such a kinematic region there is no Q^2-evolution between the jet and the photon, i.e. we are in a one-scale situation to which the BFKL approximation applies. At fixed longitudinal and tansverse momenta of the forward jet, the BFKL approximation predicts a strong rise with decreasing x. The formula for the cross section [19] reads:

$$\begin{aligned} \frac{d\sigma}{dx_j dk_t^2 dy dQ^2} &= \frac{\alpha_{em}^2}{2yQ^4} \left\{ [1+(1-y)^2]\, xW_t^{(0)} + 4(1-y)\, xW_l^{(0)} \right\} \\ &\quad \cdot \frac{3\alpha_s}{4\pi^2 k_t^4} \left\{ g(x_j,k_t^2) + \frac{4}{9} \sum_f q_f(x_j,k_t^2) \right\} \end{aligned} \tag{4}$$

with

$$xW_t^{(n)} = \sum_f Q_f^2\, \alpha_s \sqrt{k_t^2 Q^2}\, \frac{\pi}{4} \int_0^\infty d\mu\, \frac{\frac{9}{4}+\mu^2}{1+\mu^2}\, \frac{\sinh(\mu\pi)}{\mu \cosh^2(\mu\pi)} \,.$$

$$\cdot \left(\frac{x_j}{x}\right)^{\chi(n,\mu)} \cos\left(\mu \ln(\frac{k_t^2}{Q^2})\right) \quad , \tag{5}$$

$$xW_l^{(n)} = \sum_f Q_f^2 \, \alpha_s \sqrt{k_t^2 Q^2} \, \frac{\pi}{4} \int_0^\infty d\mu \, \frac{\frac{1}{4}+\mu^2}{1+\mu^2} \, \frac{\sinh(\mu\pi)}{\mu \cosh^2(\mu\pi)} \cdot \left(\frac{x_j}{x}\right)^{\chi(n,\mu)} \cos\left(\mu \ln(\frac{k_t^2}{Q^2})\right) \tag{6}$$

and

$$\chi(n,\mu) = \frac{3\alpha_s}{\pi} \left\{ 2\psi(1) - \psi(\frac{n+1}{2} + i\mu) - \psi(\frac{n+1}{2} - i\mu) \right\} \quad . \tag{7}$$

where n denotes an integer starting from zero, and W_t and W_l refer to the transverse and the longitudinal part of the hadronic tensor $W^{\rho\sigma}$. The transverse momentum of the jet and its longitudinal energy fraction are given by k_t and x_j.

In the numerical calculations of [19] the production of charm was taken into account, and the first order running coupling constant was used with k_t as scale. For comparison also a fixed-order matrixelement calculation of the one-jet inclusive cross section has been performed. In all these numerical calculations, detector cuts have been taken into account. A comparison of these calculations with HERA data is shown in Figs.1 [20] and 2 [21]. In both cases, the data lie clearly above the fixed-order matrix element points; one is therefore lead to conclude that higher order processes - radiation of gluons between the forward jet and the current jet - provide a significant part of the observed cross section. In comparison with the BFKL prediction, the data points are slightly below the theoretical curves. For the H1 data the agreement is better than for the Zeus data; when comparing the two data sets one should, however, keep in mind that the experimental cuts used by both groups are different from each other. On the other hand, experience shows that the theoretical curves are rather sensitive to variations of these cuts; this may explain the difference between the two data sets.

A few general comments have to be made about these results. First of all, the theoretical calculations are done on the partonic level, i.e. no hadronization effects are included. One therefore has to rely on the assumption that hadronization will not change the x-dependence of the forward jet cross section. Secondly, as any leading-logaritmic approximation the BFKL prediction has a principal scale uncertainty: the leading-$\ln s$ approximation cannot distinguish between $\ln s$ and, for example, $\ln 2s$. Thus the most reliable prediction of the BFKL calculation is the shape of the energy dependence, and not the overall normalization. Finally, as long as we do not know the next-to-leading order corrections to the BFKL approximation, we have no precise knowledge of the region of its applicability. As mentioned in the introduction,

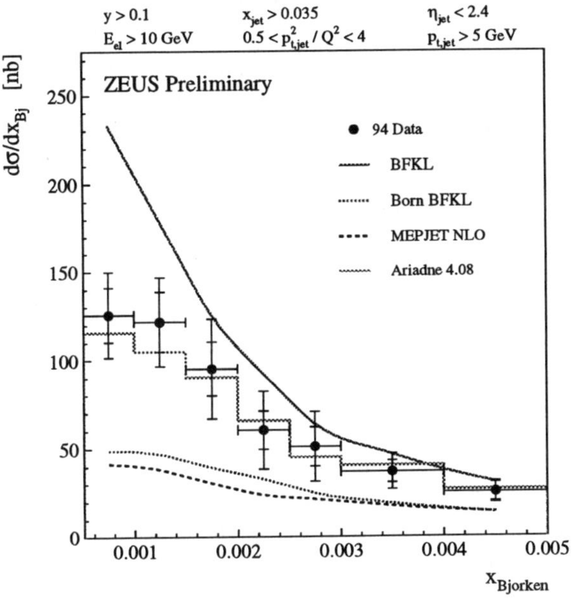

Figure 1: Comparison of ZEUS data with theoretical predictions

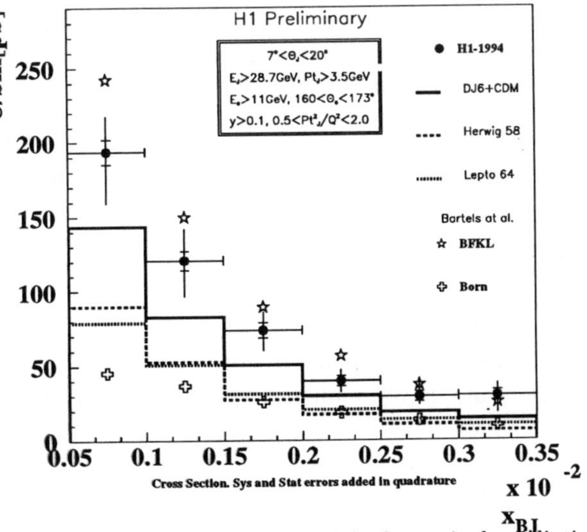

Figure 2: Comparison of H1 data with theoretical predictions

first results of the α_s^2 corrections to the BFKL kernel as well as the negative sign of the first unitarity corrections indicate that perturbative corrections to the BFKL approximation will lower the BFKL prediction. It is therefore not excluded that the observed trend of the data to lie below the theoretical BFKL curve can be explained (at least partly) by the presence of first corrections to the BFKL approximation.

Some of the difficulties of the forward jet measurement can be overcome by triggering not a jet but on a particular hadron, in particular on a π^0 [22]. The cross section for this inclusive process is obtained by folding the parton production cross section with the corresponding parton fragmentation function. Since the π^0 can be detected very close to the forward direction, the rapidity interval can be large comfortably large. However, the cross section for the single particle inclusive cross section is substantially smaller than the inclusive jet cross section. First estimates gave a suppression factor of 1/40; recent studies, however, show that this estimate might be too pessimistic.

Another signature of the BFKL Pomeron that has been discussed [23] is the angular decorrelation of the forward jet. In the γ^*-proton CM system, the scattering plane is defined through the beam axis and the momentum of the outgoing electron. The azimuthal angle ϕ is defined to be angle between the forward jet and the scattering plane. For a small rapidity interval (i.e. $x_j \approx x$), the forward jet comes mainly from the lowest order parton configuration $e + q \to e + q$ with the usual correlation $\phi = \pi$ between the electron and the current jet. As the rapidity $\eta = \ln(x_j/x)$ grows, the jet production is increasingly dominated by two- and later on three-final state partons and by gluon exchange in the crossed channel. For the three parton state ($q\bar{q}g$) the cross section is found to have a very peculiar ϕ-dependence: it peaks at $\phi = \pi/2$, i.e. the forward jet prefers to be perpendicular with respect to the scattering plane. With further increasing rapidity difference, in the BFKL calculation the ϕ-dependence completely disappears, i.e. the jet "forgets" about the azimuthal direction of the outgoing electron. In the fixed-order matrix element calculation, on the other hand, the peaking at $\phi = \pi/2$ persists. A numerical estimate of this effect is presented in [23]. The observation of this peculiar ϕ-dependence would provide a nice piece of evidence of the BFKL Pomeron.

Before ending this part of the talk, let me mention another possibility of observing the BFKL Pomeron [24,25]. It applies to electron positron colliders (LEP200 or a future Linar Collider) and, in my opinion, presents the "ideal" BFKL test: the total cross section of $\gamma^*\gamma^*$ scattering. The cross section formula is [24]:

$$\frac{d\sigma^{e^+e^-}}{dQ_1^2 dQ_2^2 dy_1 dy_2} = \frac{\alpha_{em}^2}{16\pi y_1 Q_1^4} \frac{\alpha_{em}^2}{16\pi y_2 Q_2^4} \int_{-\infty}^{\infty} \frac{d\mu}{2\pi^2} \left(s \frac{y_1 y_2}{\sqrt{Q_1^2 Q_2^2}} \right)^{\chi(0,\mu)} \qquad (8)$$

$$\cdot\left[\frac{1+(1-y_1)^2}{2}W_{t,1}^{(0)}(\mu)+(1-y_1)W_{l,1}^{(0)}(\mu)\right]\left[\frac{1+(1-y_2)^2}{2}W_{t,2}^{(0)}(\mu)+(1-y_2)W_{l,2}^{(0)}(\mu)\right]$$

where we have introduced the invariant functions:

$$W_{l,i}^{(0)}(\mu) = \sum_f Q_f^2\alpha_s\,\pi^2\sqrt{2}\,8\frac{\mu^2+\frac{1}{4}}{\mu^2+1}\frac{\sinh\pi\mu}{\mu\cosh^2\pi\mu}(Q_i^2)^{\frac{1}{2}+i\mu} \tag{9}$$

$$W_{t,i}^{(0)}(\mu) = \sum_f Q_f^2\alpha_s\,\pi^2\sqrt{2}\,4\frac{\mu^2+\frac{9}{4}}{\mu^2+1}\frac{\sinh\pi\mu}{\nu\cosh^2\pi\mu}(Q_i^2)^{\frac{1}{2}+i\mu} \tag{10}$$

Here Q_1^2 and Q_2^2 denote the masses of the two virtual photons. In the calculations of [24] the sum over the flavours goes up to three. For the (fixed) scale of the first order strong coupling constant one uses $\sqrt{Q_1^2Q_2^2}$ ($\alpha_s(M_Z^2)=0.12$). I also quote the result of the saddle point approximation near $\mu=0$:

$$\frac{d\sigma^{e^+e^-}}{dQ_1^2dQ_2^2dy_1dy_2} = \frac{\alpha_{em}^2}{16\pi y_1 Q_1^4}\frac{\alpha_{em}^2}{16\pi y_2 Q_2^4}\left(\frac{s}{s_0}\right)^{\omega_{BFKL}}\frac{e^{-\frac{\log Q_1^2/Q_2^2}{N_c\alpha_s/\pi 56\zeta(3)\log s/s_0}}}{\sqrt{N_c\alpha_s/\pi 14\zeta(3)\log s/s_0}}$$

$$\cdot\frac{1}{2\sqrt{\pi}^3}\sqrt{Q_1^2Q_2^2}\left(\sum_f q_f^2\alpha_s\pi^2\sqrt{2}\pi\right)^2 \tag{11}$$

$$\cdot\left[(1-y_1)W_{L,0}+\frac{1+(1-y_1)^2}{2}W_{T,0}\right]\left[(1-y_2)W_{L,0}+\frac{1+(1-y_2)^2}{2}W_{T,0}\right]$$

where the constants $W_{L,0}$, $W_{T,0}$ and s_0 are:

$$s_0 = \frac{\sqrt{Q_1^2Q_2^2}}{y_1y_2} \tag{12}$$

$$W_{L,0}=2,\quad W_{T,0}=9. \tag{13}$$

This formula represents an absolute prediction. By varying the energy of the outgoing leptons, the available energy of the $\gamma^*\gamma^*$ subprocess varies from a few GeV up to almost the full collider energy. An estimate of the BFKL diffusion shows that, at highest energies of the Linear Collider, corrections of the BFKL Pomeron should clearly be observable, i.e. the BFKL rise of the $\gamma^*\gamma^*$ total cross section should begin to flatten out. On the experimental side, the measurement requires the tagging of both the electron and the positron. The numerical estimates presented in [24] show that a sizable number of events can be obtained both at LEP200 and the Linear Collider, provided the outgoing leptons can be discovered sufficiently close to the

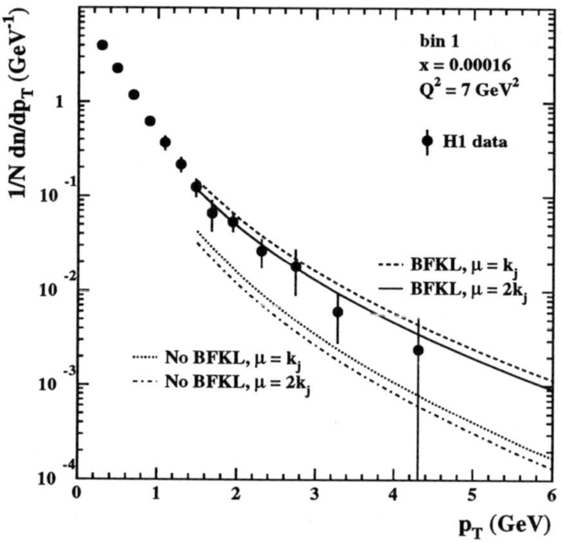

Figure 3: Comparison of H1 data with the theoretical predictions

foward direction. As for the forward jets at HERA, for these numerical estimates of event rates it is crucial to take into account kinematic limits coming from detector cuts.

3. The Forward-Jet Measurement: p_T-Dependence

The forward jet measurement described in the previous section looks for the characteristic energy dependence of the BFKL Pomeron. Another BFKL signal is the transverse momentum dependence of the single jet inclusive cross section. Starting point is the observation that, whereas in the standard DGLAP evolution of the structure function transverse momenta are strongly ordered, inside the BFKL approximation this ordering is absent. If elements of the BFKL approximation are present in deep inelastic structure functions at small x, one therefore expects to see more events with large transverse energy than in the DGLAP picture. This idea has first been proposed and worked out by the Durham group [26]: it was suggested to measure the mean transverse energy in the central rapidity region. The difference between DGLAP and BFKL does not only masnifest itself in the magnitude of $< E_T >$ but also in its characteristic dependence upon the rapidity. More recently [27] k_T-spectra of charged particles have been calculated and compared to data. For $k_T^2 \sim Q^2$ we are in the kinematic region of the forward jets, i.e. there is no DGLAP evolution between the parent parton and the photon. Therefore the DGLAP prediction for this kinematic region simply reduces to the Born digrams (i.e. no gluon production between the parent parton of the charged particle and the current jet). Using the same reasoning as discussed above, on expects the BFKL cross section to lie above the corresponding DGLAP (or Born) cross section. A Comparison between the theoretical prediction and H1 data is shown in Fig.3 (taken from [27]): in fact, the data points lie clearly above the Born curves. For two other kinematic regions of the single parton inclusive cross section the BFKL predictions does not apply: if k_T^2 is much larger than Q^2, we have (inverse) DGLAP evolution from the photon to the parton in the central rapidity region, i.e. we are probing the parton content of the pointlike photon. If, on the other hand, k_T^2 is much smaller than Q^2, we have to be careful in defining an infrared safe one-jet inclusive cross section; for a detailed discussion of BFKL-based final states I refer to G.Salams talk [28] at this meeting.

In conclusion to this part I would like to stress that, in my view, it would be extremely useful to have a Monte Carlo program that generates exclusive final states both within the BFKL and DGLAP dynamics: this would allow to compare data with the different types of QCD approximations: leading log-Q^2 (DGLAP) and leading log-s (BFKL). Very promising work in this direction is being done by the Milano Group, and I hope it will be reported by G.Salam in his talk.

4. Conclusions

In this talk I have tried to summarize the present status of the experimental verification of the BFKL Pomeron at HERA. The observation of events with a forward jet has already provided a first very promising test: the data are clearly above what one would expect from fixed order matrix element calculations. Deviations from the BFKL predictions could very well be due to BFKL corrections, and further work in this direction is certainly necessary. Also a few new suggestions have been made to refine the forward jet measurement at HERA. In my talk I have not discussed the status of the Mueller-Navelets [29,30] at the Tevatron. This is clearly an equally important place to look for the BFKL Pomeron, and. I am sure that this measurement will be discussed also at the LHC.

The cleanest BFKL test, in my view, will be the measurement of the $\gamma^*\gamma^*$ cross section at an electron-positron collider (LEP200 or the Linear Collider). Hopefully experimentalists at CERN will make all efforts to tag the outgoing leptons as close to the forward direction as possible.

After these first attempts of comparing data with the BFKl approximation we now should worry about corrections to this leading order calculation. Analytic results for the first corrections to the BFKL kernel and for unitarity corrections are available, and we have to insert them into the above cross section formulae and start to preform numerical estimates. I would expect that unitarity effects will turn out to be numerically most important: this are contributions with four gluons in the t-channel. A numerical study of their magnitude is difficult but would be very useful. It is therefore clear that both analytic and computer work will be needed to continue and improve our search for the BFKL Pomeron.

1. E. A. Kuraev, L. N. Lipatov and V. S. Fadin, Sov. Phys. JETP **45** (1977) 199; Ya. Ya. Balitskij and L. N. Lipatov, Sov. J. Nucl. Phys. **28** (1978) 822.
2. J.R.Forshaw, D.A.Ross, "Quantum Chromodynamics and the Pomeron", Cambridge Lecture Notes in Physics 1997.
3. M.Froissart, *Phys.Rev***123** (1961) 1053;
   A.Martin *Phys.Rev***129** (1963) 1432.
4. V.S.Fadin and L.N.Lipatov, *Nucl.Phys.***B 477** (1996) 767 and references therein; V.S.Fadin, L.N..Lipatov and M.I.Kotsky, hep-ph/9704267.
5. G.Camici and M.Ciafaloni, *Phys.Lett.***B 386** (1986) 341; hep-ph/9707390; M.Ciafaloni, these proceedings.
6. L.N.Lipatov, *Sov.Phys.JETP* **63** (1986) 904.
7. J.Bartels *Nucl.Phys.***B175** (1980) 365;
   J.Kwiecinski and M.Praszalowicz, *Phys.Lett.***B 94** (1980) 413.
8. J.Bartels, *Phys.Lett.***B 298** (1993) 204 and *Z.Phys.***C 60** (1993) 471;
   J.Bartels and M.Wuesthoff,*Z.Phys.***C 66** (1995) 157.
9. M.A.Braun, SPBU-97-9 and hep-ph/9706373
10. L.N.Lipatov, *JETP Lett.* **59** (1994) 596;
    L.D.Faddeev, G.P.Korchemsky, *Phys.Lett.* **B 342** (1995) 311.
11. A.H.Mueller and Bimal Patel, *Nucl.Phys.***B 425** (1994) 471;
    Zhang Chen and A.H.Mueller, *Nucl.Phys.***B 451** (1995) 579.
12. H.Navelet, R.Peschanski, C.Royon, and S.Wallon, *Phys.Lett.***B 385** (1996) 357 and hep-ph/9605389;
    A.Bialas, R.Peschanski, *Phys.Lett.***B 387** (1996) 405 and hep-ph/9605298.
13. A.H.Mueller, *Nucl.Phys.***B 437** (1995) 107;
    A.H.Mueller and G.P.Salam, *Nucl.Phys.***B 4475** (1996) 293.
14. G.P.Salam, *Nucl.Phys.***B 461** (1996) 512; Cavendish HEP 96-12 and International Workshop on Deep Inelastic Scattering and Related Phenomena, Rome, Italy, 15 -19 April 1996 and hep-ph/9608250.
15. J.Bartels, H.Lotter, and M.Vogt, *Phys.Lett.***B373** (1996) 215.
16. A.D.Martin, these proceedings.,
17. A.H.Mueller, *Nucl.Phys.B* (Proc.Suppl.) **18C** (1991) 125.

18. J. Bartels, A.DeRoeck, M.Loewe, *Z.Phys.* **C54** (1992) 635;
J.Kwiecinski, A.D.Martin, P.J.Sutton, *Phys.Lett.* **B 287** (1992) 254; *Phys.Rev.* **D 46** (1992) 921;
W.-K.Tang, *Phys.Lett.* **B 278** (1991) 363;
J.Bartels, M.Besancon, A.De Roeck, J.Kurzhoefer, in *Proceedings of the HERA Workshop 1992* (eds.W.Buchmüller and G.Ingelman), p.203.
19. J.Bartels, V.Del Duca, A.DeRoeck, D.Graudenz, M.Wuesthoff, *Phys.Lett.***B 384** (1996) 300 and hep-ph/9604272.
20. S.Woelfle, talk presented at *Deep Inelastic Scattering and Related Phenomena (DIS97)*, April 14 - 18, 1997, Chicago..
21. A.DeRoeck, in *Deep Inelastic Scattering and Related Phenomena (DIS96)*, World Scientific, p 486.
22. J.Kwiecinski, S.C.Lang, and A.D.Martin, *Phys.Rev.***D 55** (1997) 1273.
23. J.Bartels, V.DelDuca, M.Wuesthoff, DESY-96-220 and hep-ph/9610450.
24. J.Bartels, A.DeRoeck, and H.Lotter, *Phys.Lett.***B 389** (1996) 742; hep-ph/9608401.
25. S.J.Brodsky, F.Hautmann, D.E.Soper, SLAC-PUB-7480 and hep-ph/9706427; *Phys.Rev.Lett.***78** (1997) 803.
26. J.Kwiecinski, A.D.Martin, P.J.Sutton, K.Golec-Biernat, *Phys.Rev.***D 50** (1994) 217.
27. J.Kwiecinski, S.C.Lang, A.D.Martin, DTP/97/56 and hep-ph/9707240.
28. G.P.Salam, these proceedings.
29. A.H.Nueller, H.Navelet, *Nucl.Phys.* **B 282** (1987) 727.
30. V.Del Duca, C.R.Schmidt, *Phys.Rev.* **D 51** (1995) 2150.

k-FACTORIZATION AND NEXT-TO-LEADING ANOMALOUS DIMENSIONS

MARCELLO CIAFALONI
Physics Department, Università di Firenze, and INFN, Sezione di Firenze, Largo Enrico Fermi 2 50125 Firenze, Italy
E-mail: marcello.ciafaloni@cern.ch

ABSTRACT

I review recent results by Fadin, Lipatov and collaborators and by our group, leading to the almost complete calculation of the next-to-leading BFKL kernel, of its eigenvalues, and of the resummed gluon anomalous dimension. Qualitative implications for small-x structure functions include consistent running coupling effects and a sizeable decrease of the Pomeron intercept, so as to slow down the small-x rise at low values of Q^2.

1. Introduction

The small-x rise of structure functions at HERA [1] has been interpreted so far by various authors [2-4] starting from two seemingly far-apart standpoints.

The more traditional view (favoured by available fits [5,6] in the HERA range) is that such a rise, at some value of Q^2, is due to large scaling violations, i.e., to a singular anomalous dimension, starting from a flat input distribution at some lower value ${Q^2}_0$.

The other view is that, for small enough values of $x = Q^2/s$, a qualitatively new phenomenon occurs, namely a power increase $F_2 \sim x^{-\omega_{\mathcal{P}}}$ of the structure functions, due to the "hard Pomeron" singularity.[7]

On the other hand, the systematic use of high-energy, k-dependent factorization [8,9] makes it clear [10] that the two views are rather two regimes, coming from different rearrangements of the QCD perturbative series. The link joining the two regimes is given by resummation formulas which provide the α_s/ω dependence of the anomalous dimensions to all orders, where $\omega \equiv N - 1$ is the moment index, conjugate to the $\log x$ variable.

This analysis has been available for a long time [7,11] at leading $\log x$ level, and indeed shows a substantial difference between the two regimes just mentioned. On one hand, the gluon anomalous dimension departs from the DGLAP value quite slowly,

$$\gamma_{gg} \simeq \gamma_L\left(\frac{\bar{\alpha}_s}{\omega}\right) = \frac{\bar{\alpha}_s}{\omega} + 2\zeta(3)\left(\frac{\bar{\alpha}_s}{\omega}\right)^4 + \cdots, \bar{\alpha}_s \equiv \frac{N_c\alpha_s}{\pi} \tag{1}$$

and is roughly consistent with the so-called "double-scaling" picture [4] of HERA data.

On the other hand, the Pomeron singularity, at leading level, has the intercept

$$J_{\mathcal{P}} - 1 = \omega_{\mathcal{P}} = 4\bar{\alpha}_s \log 2 \tag{2}$$

which predicts a strong small-x rise ($\omega_{\mathcal{P}} = 0.4$ for $\alpha_s = 0.15$), not yet shown by the data[4], especially at low values of Q^2. Furthermore, if the scale of α_s is related to the hard scale Q^2, Eq. (2) yields a sharper increase for lower Q^2 values, a trend not confirmed by the data[12].

Therefore, the theoretical situation, at leading $\log x$ level, is somewhat embarrassing. Not only does it indicate that the Pomeron regime is still further away, but it also calls for strong unitarity corrections[13] in order to damp the small-x rise at low values of Q^2.

Here I want to show how this picture is modified at next-to-leading level [14–19], particularly by recent results [20,21] in the gluonic sector. It turns out that the NL terms are sizeable and tend to slow down the small-x rise, both at (resummed) anomalous dimension level and at Pomeron level, especially at low values of Q^2. As a consequence, the above regimes appear now to be closer, and less unitarity corrections are required to achieve a smooth matching with soft physics.

In the following, I will first use k-factorization to define the parton densities and the related BFKL equation [7,14] (Section 2), and I will then explain the features of the NL BFKL kernel (Section 3), which is the basis for the results mentioned before.

2. Parton densities and BFKL equation

The unintegrated gluon density $\mathcal{F}_\omega^g(k)$ is defined by factorization at the exchanged (Regge) gluon of transverse momentum $\mathbf{k}$, and satisfies the BFKL equation of Fig. 1.

The quark-sea density, in the DIS scheme, is defined by the F_2 structure function itself and is coupled to the unintegrated gluon density by $\mathbf{k}$-factorization of the quark loop, as in Fig. 2. The above prescriptions define the Q_0-scheme [22] for quarks and gluons.

The simple BFKL picture just mentioned is argued [16,18] to be valid up to NL $\log x$ level, while at lower levels s-channel iteration will be important also, leading to the effective action approach [23].

The leading kernel $K^{(L)}$ occurring in Fig. 1 is

$$\frac{\bar{\alpha}_s}{\omega} K_0(\mathbf{k}_1, \mathbf{k}_2) = \frac{\bar{\alpha}_s}{\omega} \left.\frac{1}{\mathbf{q}^2}\right|_R, \quad \mathbf{q} = \mathbf{k}_1 - \mathbf{k}_2, \tag{3}$$

where we have introduced the distribution

$$\left.\frac{1}{\mathbf{q}^2}\right|_R = \frac{1}{\mathbf{q}^2}\,\Theta(\mathbf{q}^2 - \lambda^2) - \delta^2(\mathbf{q}) \int_{\lambda^2}^{k_1^2} \frac{d^2q}{\mathbf{q}^2} \tag{4}$$

Due to scale invariance, the eigenfunctions of $K^{(L)}$ are simple powers $\sim (k^2)^{\gamma-1}$, and the corresponding characteristic function of the anomalous dimension has the well-known form

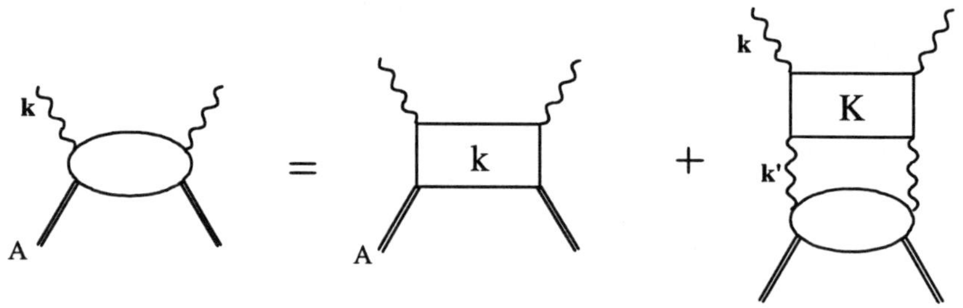

Figure 1: BFKL equation for the gluon density up to next-to-leading level. Wavy lines denote (Regge) gluon exchanges, whose initial couplings to partons a, b are understood.

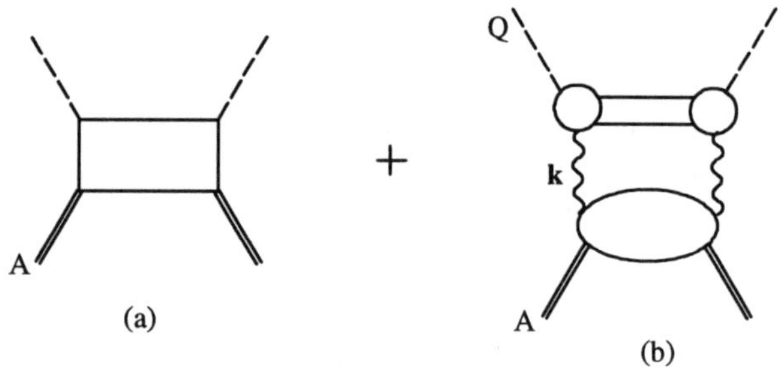

Figure 2: (a) Low energy and (b) high-energy part of the quark density in the DIS-Q_0 scheme. Dotted lines denote the electroweak probe and an F_2 projection is understood.

$$\chi_0(\gamma) = 2\psi(1) - \psi(\gamma) - \psi(1-\gamma). \tag{5}$$

The leading anomalous dimension $\gamma_{gg} = \gamma_L\left(\frac{\bar{\alpha}_s}{\omega}\right)$ is then defined by the implicit equation

$$1 = \frac{\bar{\alpha}_s}{\omega}\,\chi_0(\gamma_L), \tag{6}$$

which provides the expansion in Eq. (1). Furthermore, the ω-singularity of γ_L occurs around the minimum of χ_0 at $\gamma = \frac{1}{2}$, and yields by Eq. (6) the leading Pomeron intercept in Eq. (2).

At NL level, the kernel $K = K^{(L)} + K^{(NL)}$ is explicitly dependent on the factorization scale μ, i.e., is no longer scale invariant. I anticipate, from the results in Section 3, that the μ-dependence is logarithmic, and is dictated by the one-loop beta function, as follows:

$$\begin{aligned} K &= \frac{\bar{\alpha}_s(\mu^2)}{\omega}\left[\left(1 - b\,\alpha_s(\mu^2)\log\frac{k_1^2}{\mu^2}\right)K_0(\mathbf{k}_1,\mathbf{k}_2) + \alpha_s(\mu^2)K_1(\mathbf{k}_1,\mathbf{k}_2)\right] \\ &\simeq \frac{\bar{\alpha}_s(k_1^2)}{\omega}\left[K_0(\mathbf{k}_1,\mathbf{k}_2) + \alpha_s K_1(\mathbf{k}_1,\mathbf{k}_2)\right], \end{aligned} \tag{7}$$

where K_0 is the leading kernel, $12\pi b = 11N_c - 2N_f$, and K_1 is defined to be the scale-invariant part of the NL kernel.

The definition of K_1 in Eq. (7) depends on the choice of factorizing α_s at scale k_1. If another choice is made (e.g., the symmetrical one $\alpha_s(k_>^2)$), the form of K_1 changes accordingly (e.g., $K_1^> = K_1 + b\,\log(k_>^2/k_1^2)$).

The form and the properties of K_1 are described in Section 3. Here I just emphasize that the NL BFKL equation, due to the lack of scale-invariance embodied in the running coupling, needs a careful discussion [24,18] in order to relate anomalous dimensions and the Pomeron to the characteristic function of K_1.

We have described [18] two small-x regimes. Anomalous dimensions can be defined if ω is not too small ($\omega \to 0$ with $b\omega\,\log(\mathbf{k}_1^2/\Lambda^2 > \chi(\frac{1}{2}))$. In this regime, the larger eigenvalue $\gamma_+ \sim \gamma_{gg} + \frac{C_F}{C_A}\gamma_{qg}$ of the singlet anomalous dimension matrix is given in terms of the kernel characteristic function by

$$1 = \frac{\bar{\alpha}_s}{\omega}\,(\chi_0(\gamma_+) + \alpha_s\,\chi_1(\gamma_+)) \tag{8}$$

thus generalizing Eq. (6) to NL level.

By expanding Eq. (8) in α_s, we obtain the NL resummation formula

$$\gamma_{gg} + \frac{C_F}{C_A}\gamma_{qg} = \gamma_L\left(\frac{\bar{\alpha}_s}{\omega}\right) - \alpha_s\,\frac{\chi_1\left(\gamma_L(\frac{\bar{\alpha}_s}{\omega})\right)}{\chi_0'\left(\gamma_L(\frac{\bar{\alpha}_s}{\omega})\right)}. \tag{9}$$

The quark entry γ_{qg} is provided [(25,21)] independently by $\mathbf{k}$-factorization of the quark loop in Fig. 2, namely

$$\begin{aligned}\gamma_{qg} &= \frac{\alpha_s N_f}{2\pi}\,\gamma_L^2\,h_{ab}(\gamma_L),\\ h_{ab}(\gamma) &= \left(\frac{\pi}{\sin\pi\gamma}\right)^2\frac{\cos\pi\gamma}{1-2\gamma}\,\frac{1+\frac{3}{2}\gamma(1-\gamma)}{(1+2\gamma)(3-2\gamma)}.\end{aligned} \tag{10}$$

Therefore, the result in Eq. (9) allows the extraction of the gluon anomalous dimension with its NL corrections, to all orders.

On the other hand, the ω-singularity of Eq. (8) around $\gamma_+ = \frac{1}{2}$ yields an "α_s-dependent Pomeron" which should be interpreted as the singularity of the anomalous dimension expansion, separating in ω-space the renormalization group regime from the Pomeron regime of very small x.

Since the (true) Pomeron's intercept cannot really be predicted (being dependent [19] on the behaviour of α_s around $\mathbf{k}^2 = \Lambda^2$), the ω-singularity of Eq. (8) yields a rough estimate of the effective power behaviour in x of F_2, as a function of $\alpha_s(Q^2)$. On this basis the small-x features of the structure function can be interpreted in terms of the NL characteristic function $\chi_1(\gamma)$.

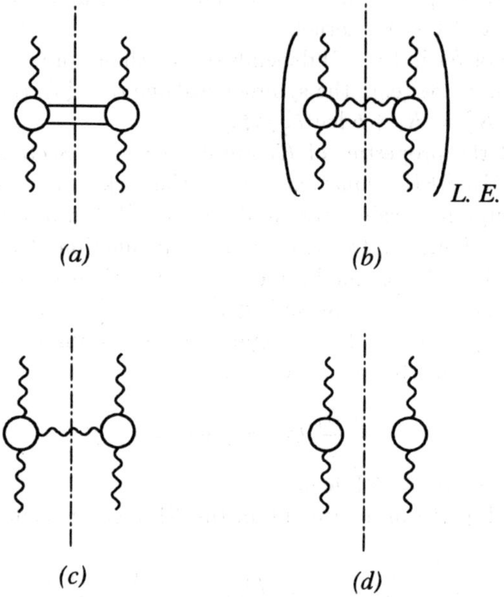

Figure 3: (a) $q\bar{q}$ and (b) low energy gg contributions to the NL kernel, together with (c) one-loop corrections to $1g$ state and (d) two-loop virtual corrections.

3. The Next-to-Leading Kernel

The method for extracting the NL kernel was set up by Fadin, Lipatov [14] and collaborators, and involves calculating the QCD partonic cross-section up to NL $\log s$ level, and arranging it in a cluster expansion, which separates leading from NL contributions (Fig. 3) which are thereby calculated [14–17,8,20].

One problem which is met in extracting the NL clusters is that one has to define an off-shell scale for the energy in order to subtract the leading terms. Changing such scale(s) yields an ambiguity in the definition of the NL terms. By using a particular prescription, Camici and I [21] have extracted the irreducible part of the NL kernel, and we have investigated its features.

Referring to the original papers [20,21] for a detailed analysis, let me just quote the results for the characteristic function $\chi_1(\gamma)$. The gluonic part is

$$\alpha_s \chi_1^{(g)}(\gamma) = \frac{\bar{\alpha}_s}{4}\left[-\frac{11}{6}\left(\chi_0^2(\gamma) + \chi_0'(\gamma)\right) + \left(\frac{67}{9} - \frac{\pi^2}{3}\right)\chi_0(\gamma) + \right.$$

$$+\left(6\zeta(3)+\tfrac{\pi^2}{3\gamma(1-\gamma)}+\tilde{h}(\gamma)\right)-\left(\tfrac{\pi}{\sin\pi\gamma}\right)^2\tfrac{\cos\pi\gamma}{3(1-2\gamma)}\left(11+\tfrac{\gamma(1-\gamma)}{(1+2\gamma)(3-2\gamma)}\right)\Big], \tag{11}$$

where

$$\tilde{h}(\gamma)\simeq\sum_{n=1}^{3}a_n\left[(\gamma+n)^{-1}+(1-\gamma+n)^{-1}\right];(a_1=0.72,a_2=0.28,a_3=0.16), \tag{12}$$

to be compared with the quark-sea part [18]

$$\alpha_s\chi_1^{(q)}(\gamma)=\frac{N_f\alpha_s}{6\pi}\left[\frac{1}{2}(\chi_0^2(\gamma)+\chi_0'(\gamma))-\frac{5}{3}\chi_0(\gamma)-\frac{1}{N_c^2}h_{ab}(\gamma)\right], \tag{13}$$

where $h_{ab}(\gamma)$ is given in Eq. (10).

Consequences for the structure functions are obtained through Eqs. (8) and (9). According to Eq. (9), the small-x behaviour of $\chi_1(\gamma)\simeq A_1/\gamma^2+A_2/\gamma+A_3+O(\gamma)$ is directly related to the low-order expansion of the anomalous dimension

$$\gamma_+=\frac{\bar{\alpha}_s}{\omega}+\alpha_s(A_1+A_2\frac{\bar{\alpha}_s}{\omega}+A_3(\frac{\bar{\alpha}_s}{\omega})^2)+\cdots. \tag{14}$$

Therefore, one can check that Eqs. (11) and (13) are consistent with the two-loop results [26] in the DIS scheme

$$A_1=-\frac{11N_c}{12\pi}-\frac{N_f}{6\pi N_c^2},A_2=-\frac{N_f}{6\pi}\left(\frac{5}{3}+\frac{13}{6N_c^2}\right). \tag{15}$$

The three-loop entry A_3 is not yet available and is difficult to obtain in the present approach also, because the scale-of-energy-dependent terms (not treated yet) are expected to contribute. It has been argued that coherence effects [27] in the CCFM equation [27,28] are important for explaining A_3 [29].

For larger γ values, $\chi_1(\gamma)$ in Eq. (11) contains non-linear effects, in particular large negative contributions, proportional to $\cos\pi\gamma(\pi/\sin\pi\gamma)^2$, which show a $\gamma=1$ double pole and are driven by the NL one-loop anomalous dimension (Fig. 4a). When the effective value of $\gamma\simeq\bar{\alpha}_s/\omega$ increases towards the saturation value $\gamma=\frac{1}{2}$, such terms decrease rapidly and tend to cancel in part [30] the large scaling violations due to the leading part of γ_{gg} and to γ_{qg}.

This picture is confirmed by the estimate of the α_s-dependent Pomeron intercept which, by Eq. (8), is given by

$$\omega_{\mathcal{P}}(\alpha_s)=\bar{\alpha}_s(\chi_0(\frac{1}{2})+\alpha_s\chi_1(\frac{1}{2}))=\bar{\alpha}_s\chi_0(\frac{1}{2})(1-a\bar{\alpha}_s), \tag{16}$$

where $a\simeq3.4$ within our present knowledge (Fig. 4b). Notice that Eq. (16) shows a maximum $\omega_{\mathcal{P}}\simeq0.2$ at $\alpha_s=0.15$, substantially smaller than the leading value in Eq. (2).

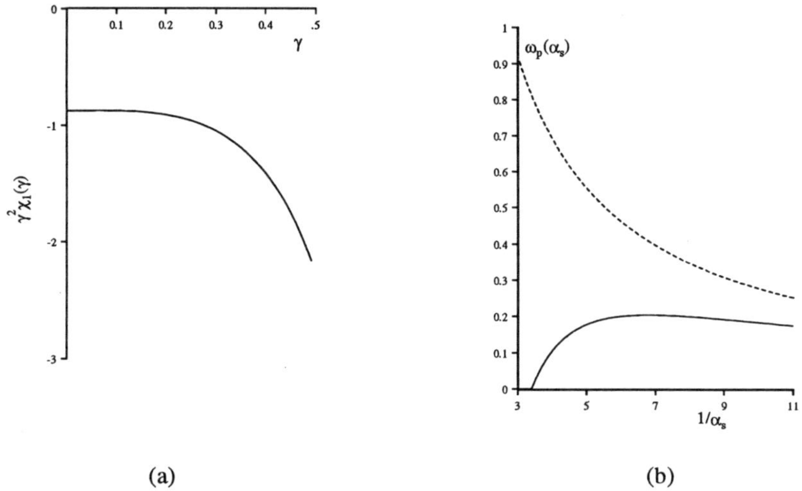

Figure 4: Plot of (a) the gluonic contribution to $\gamma^2\chi_1(\gamma)$ and (b) the α_s-dependent Pomeron with (full line) and without (dashed line) NL contributions.

Let me stress that the above indications cannot yet be taken as quantitative estimates, due to various theoretical uncertainties (scale-of-energy-dependent terms, higher-order collinear and coherence effects, and so on). However, they show a definite trend, that of slowing down the small-x rise at low values of Q^2. This is satisfactory, because it reconciles somewhat the anomalous dimension and Pomeron regimes, and helps in making a smooth transition towards soft hadronic physics.

4. Acknowledgements

I wish to thank B. Kniehl for the warm atmosphere enjoyed at the Ringberg Workshop, and J. Bartels, A. Martin, Z. Kunszt and G. Salam for a number of stimulating discussions. This work is supported in part by EC contract No. CHRX-CT96-0357 and by MURST (Italy).

5. References

1. S. Aid et al., H1 Collaboration, *Nucl. Phys.* **B470** (1996) 3;
   The Zeus Collaboration, *Z. Phys.* **C69** (1996) 607.

2. A.J. Askew, J. Kwiecinski, A.D. Martin and P.J. Sutton, *Phys. Rev.* **D47** (1993) 3775; **D49** (1994) 4402;
   C. Lopez, F. Barreiro and F.J. Yndurain, hep-ph/9605395.

3. R.K. Ellis, F. Hautmann and B.R. Webber, *Phys. Lett.* **B348** (1995) 582.

4. R. Ball and S. Forte, *Phys. Lett.* **B351** (1995) 313; **B359** (1995) 362.

5. See, e.g., M. Glück, E. Reya and A. Vogt, *Z. Phys.* **C67** (1995) 433.

6. See, e.g., A.D. Martin, W.J. Stirling and R.G. Roberts, *Phys. Lett.* **B387** (1996) 419.

7. L.N. Lipatov,*Sov. J. Nucl. Phys.* **23** (1976) 338;
E.A. Kuraev, L.N. Lipatov and V.S. Fadin, *Sov. Phys. JETP* **45** (1977) 199;
Ya. Balitskii and L.N. Lipatov, *Sov. J. Nucl. Phys.* **28** (1978) 822.

8. S. Catani, M. Ciafaloni and F. Hautmann, *Phys. Lett.* **B242** (1990) 97; *Nucl. Phys.* **B366** (1991) 135.

9. J.C. Collins and R.K. Ellis, *Nucl. Phys.* **B360** (1991) 3.

10. See, e.g., M. Ciafaloni, *Proc. VI-th Blois Workshop* (June 20-24, 1995).

11. See, e.g., A. Bassetto, M. Ciafaloni and G. Marchesini, *Phys. Rep.* **100** (1983) 201;

12. I. Bojak and M. Ernst, hep-ph/9702282.

13. See e.g. L. V. Gribov, E. M. Levin and M. G. Ryskin, *Phys. Rep.* **100** (1983) 1; For some recent approaches, see:
A.H. Mueller, *Nucl. Phys.* **B437** (1995) 107;
G. Salam, *Nucl. Phys.* **B461** (1996) 512.

14. V.S. Fadin and L.N. Lipatov, *Yad. Fiz.* **50** (1989) 1141.

15. V.S. Fadin and L.N. Lipatov, *Nucl. Phys.* **B406** (1993) 259;
V.S. Fadin, R. Fiore and A. Quartarolo, *Phys. Rev.* **D50** (1994) 2265, 5893;
V.S. Fadin, R. Fiore and M.I. Kotsky, *Phys. Lett.* **B389** (1996) 737.

16. V.S. Fadin, R. Fiore and M.I. Kotsky, *Phys. Lett.* **B359** (1995) 181, **B387** (1996) 593.

17. V.S. Fadin and L.N. Lipatov, *Nucl. Phys.* **B477** (1996) 767.

18. G. Camici and M. Ciafaloni, *Phys. Lett.* **B386** (1996) 341; *Nucl. Phys.* **B496** (1997) 305.

19. G. Camici and M. Ciafaloni, *Phys. Lett.* **B395** (1997) 118.

20. V.S. Fadin, L.N. Lipatov and M.I. Kotsky, hep-ph/9704267.

21. G. Camici and M. Ciafaloni, hep-ph/9707390.

22. M. Ciafaloni, *Phys. Lett.* **B356** (1995) 74.

23. See, e.g., L.N. Lipatov, *Proc. of DIS 96*, Rome, April 1996.

24. J.C. Collins and J. Kwiecinski, *Nucl. Phys.* **B316** (1989) 307.

25. S. Catani and F. Hautmann, *Nucl. Phys.* **B427** (1994) 475.

26. G. Curci, W. Furmanski and R. Petronzio, *Nucl. Phys.* **B175** (1980) 27; E.P. Zijlstra and W.L. van Neerven, *Nucl. Phys.* **B383** (1992) 525.

27. M. Ciafaloni, *Nucl. Phys.* **B296** (1988) 49, in particular Appendix C.

28. S. Catani, F. Fiorani and G. Marchesini, *Nucl. Phys.* **B336** (1990) 18.

29. G. Bottazzi, G. Marchesini, G.P. Salam and M. Scorletti, hep-ph/9702418.

30. Preliminary estimates have been made by J. Blümlein and A. Vogt (private communication).

ANGULAR ORDERING AND SMALL-x EVOLUTION

GAVIN P. SALAM
INFN — Sezione di Milano, Via Celoria 16
20133 Milano, Italy
E-mail: Gavin.Salam@mi.infn.it

ABSTRACT

This talks examines the effect of angular ordering on the small-x evolution of the unintegrated gluon distribution, and discusses the characteristic function for the CCFM equation, as well as some preliminary results on final-state properties.

1. Introduction

For some time now it has been known that angular ordering[1] is an essential element in any description of small-x final state properties.[2,3] As a first step of a programme to study the final state in small-x physics, one should examine the effect of angular ordering on the small-x evolution of the gluon structure function. Phenomenological studies have already been performed,[4] but this talk will examine the solutions of the CCFM equation[2,3] from a more theoretical point of view. I will also present some first results on final-state properties.

The main difference between the BFKL[5] and CCFM equations is in the collinear region: in the BFKL case, the ith emission has a transverse momentum $q_i > \mu$, with μ a cutoff put in by hand and which is taken to zero; this regulates the collinear divergence, which for the gluon structure function cancels out. But in quantities where it doesn't cancel, such as certain final properties, one obtains the wrong answer. In the CCFM equation, angular ordering of emissions leads to the following condition (see figure 1):

$$\theta_i > \theta_{i-1}\,, \qquad \Rightarrow \qquad q_i > z_{i-1}q_{i-1}\,,$$

with the corresponding gluon emission distribution being

$$d\mathcal{P}_i = \frac{d^2q_i}{\pi q_i^2}\, dz_i \frac{\bar{\alpha}_S}{z_i}\, \Delta(z_i, q_i, k_i)\, \Theta(q_i - z_{i-1}q_{i-1})\,.$$

The non-Sudakov form factor Δ, which is analogous to a probability for suppressing any further radiation, is defined by

$$\ln \Delta(z_i, q_i, k_i) = -\int_{z_i}^{1} dz'\, \frac{\bar{\alpha}_S}{z'} \int \frac{dq'^2}{q'^2}\, \Theta(k_i - q')\, \Theta(q' - z'q_i)\,.$$

The elimination of a large fraction of the small-transverse-momentum emissions means that angular ordering has a big effect on the final state. But in structure

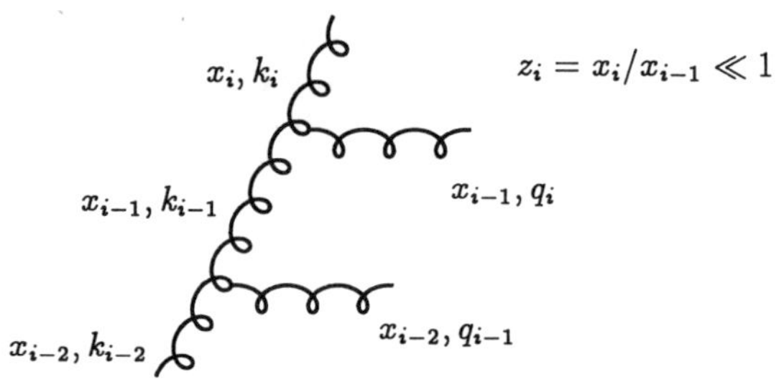

Fig. 1. Labelling of momenta.

function evolution, since collinear singularities cancel, at leading order the BFKL and CCFM structure functions are equivalent.

2. The gluon structure function

As part of a program to carry out a full investigation of the effects of angular ordering at small x, this talk examines the component of the next-to-leading order corrections to structure function evolution that arise from angular ordering. Such effects are expected to be part of the full NLL contribution.[6]

Qualitatively, since angular ordering reduces the phase space for evolution the exponent of the small-x growth ought to be reduced. The symmetry, present in the BFKL equation, between large and small scales will be broken, favouring evolution to large momentum scales. Finally diffusion will be reduced because large jumps (down) in scale are suppressed.

There are two limits in which, at small x, the effects of angular ordering should disappear: as $\alpha_S \to 0$, because the typical $z_{i-1} \sim \alpha_S$ will be very small (this justifies the assertion that for structure functions the effects of angular ordering are next-to-leading); and in the double-leading-logarithmic limit because the condition $q_i > q_{i-1}$ automatically satisfies the angular ordering condition.

The analytic treatment of the CCFM equation is more complicated than that of the BFKL equation because the gluon density contains one extra parameter, p, which defines the maximum angle for the emitted gluons. In DIS it enters through the angle of the quarks produced in the boson-gluon fusion. The equation for the CCFM density, $A(x,k,p)$, of gluons with longitudinal momentum fraction x and transverse momentum k is:

$$\mathcal{A}(x,k,p) = \mathcal{A}^{(0)}(x,k,p) + \int \frac{d^2q}{\pi q^2}\,\frac{dz}{z}\,\frac{\bar{\alpha}_S}{z}\Delta(z,q,k)\,\Theta(p-zq)\,\mathcal{A}(x/z,k',q)\,,$$

where $k' = |\,\boldsymbol{k}+\boldsymbol{q}\,|$. By analogy to the BFKL equation one can develop some understanding of it by looking for eigensolutions (strictly speaking eigensolutions of the

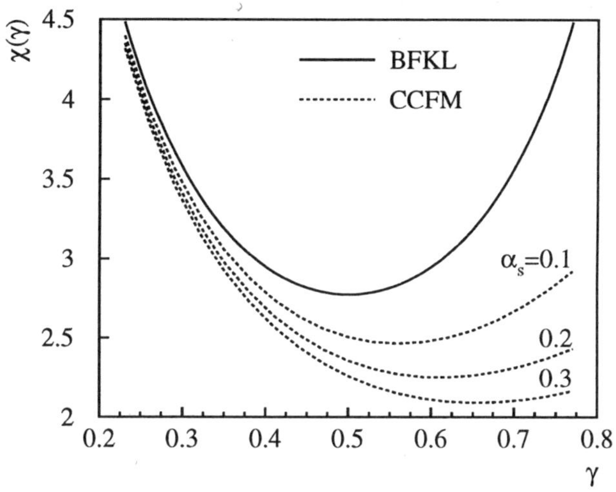

Fig. 2. The BFKL and CCFM characteristic functions as a function of γ for different values of α_S.

equation without an inhomogeneous term and with no lower limit in the z integral) of the form

$$xA(x,k,p) = x^{-\omega}\frac{1}{k^2}\left(\frac{k^2}{k_0^2}\right)^{\tilde{\gamma}} G(p/k)\,,$$

where $G(p/k)$ parameterises the unknown dependence on p. For $0 < \tilde{\gamma} < 1$, one obtains a coupled pair of equations for G and ω:

$$p\,\partial_p\, G(p/k) = \bar{\alpha}_S \int_p \frac{d^2q}{\pi q^2}\left(\frac{p}{q}\right)^{\omega} \Delta(p/q,q,k)\, G\left(\frac{q}{k'}\right)\left(\frac{k'^2}{k^2}\right)^{\tilde{\gamma}-1},$$

with the initial condition $G(\infty) = 1$ and

$$\omega = \alpha_S\tilde{\chi}(\tilde{\gamma},\alpha_S) = \alpha_S \int \frac{d^2q}{\pi q^2}\left\{\left(\frac{k'^2}{k^2}\right)^{\tilde{\gamma}-1} G\left(\frac{q}{k'}\right) - \Theta(k-q)\, G(q/k)\right\}.$$

In the second of these equations, if $G = 1$ one notes that $\tilde{\chi}$ is just the BFKL characteristic function, χ. Since $1 - G(p/k)$ is formally of order α_S, this demonstrates that angular ordering has a next-to-leading effect on structure function evolution. One can also show that in the limit of $\gamma \to 0$ the difference $\chi(\gamma) - \tilde{\chi}(\gamma,\alpha_S)$ tends to a constant, which implies corrections to the small-x anomalous dimension of the form α_S^3/ω^2.

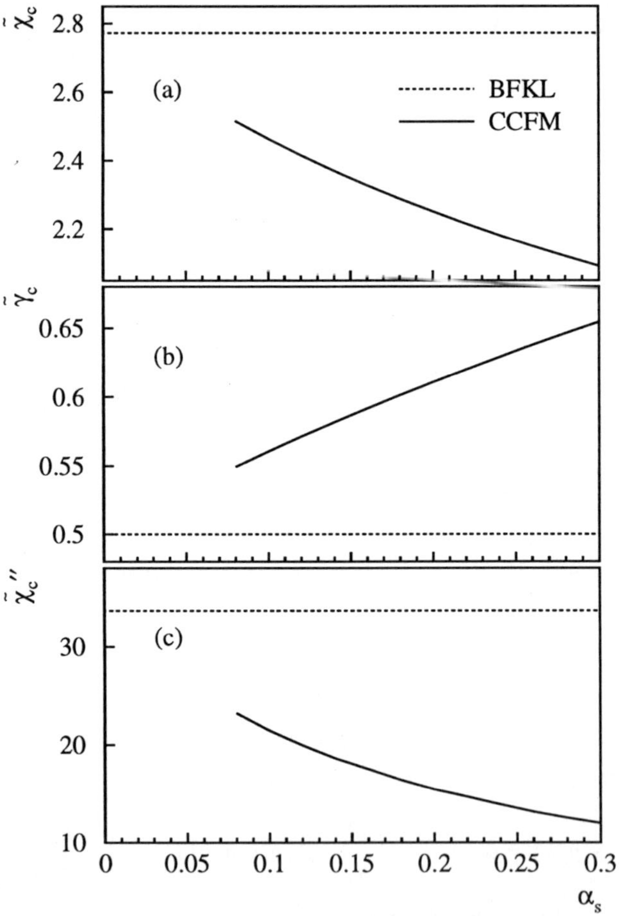

Fig. 3. (a) The value of the minimum of the characteristic function, $\tilde{\chi}_c$, as a function of α_S; (b) The position of the minimum of the characteristic function, $\tilde{\gamma}_c$, as a function of α_S; (c) The second derivative of the characteristic function, $\tilde{\chi}_c''$, at its minimum, as a function of α_S.

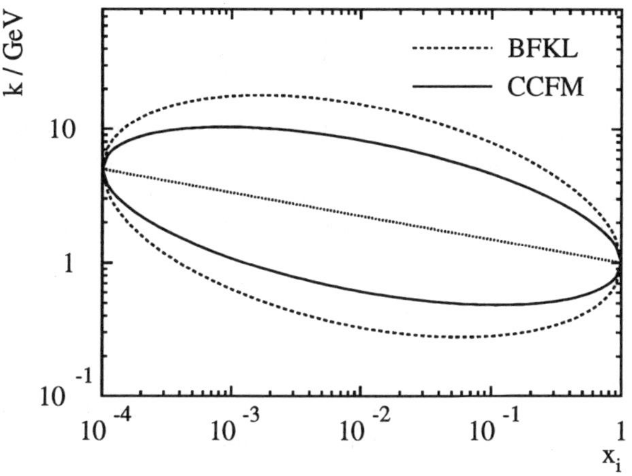

Fig. 4. "Cigars" showing the range of transverse momentum in the evolution as a function of intermediate x_i for the BFKL and CCFM equations. The evolution is to $x = 10^{-4}$, $k = 5$ GeV with $\alpha_S = 0.2$.

Though a number of asymptotic properties of $G(p/k)$ have been determined,[2,7] it has not so far been possible to obtain its full analytic form. Further understanding requires numerical analysis. This has been carried out and figure 2 shows the results for $\tilde{\chi}$ compared to the BFKL characteristic function for three different values of α_S. It illustrates that as $\alpha_S \to 0$ the two tend to coincide as happens also in the region $\gamma \to 0$ (the DLLA region).

The loss of symmetry under $\gamma \to 1 - \gamma$ relates to the loss of symmetry between small and large scales. Indeed, in contrast to the BFKL case, there is no longer even a divergence at $\gamma = 1$. Correspondingly, the minimum of the characteristic function gets shifted to the right and is lower.

Figure 3 shows the characteristics of the minimum (or critical point) of $\tilde{\chi}$ as a function of α_S: the height of the minimum, $\tilde{\chi}_c$, its position, $\tilde{\gamma}_c$ and the second derivative of $\tilde{\chi}$ at the minimum, $\tilde{\chi}''_c$. One notes that the dependence of these quantities on α_S is noticeably non-linear, indicating the presence of substantial corrections beyond next-to-leading logarithms. Indeed, for $\alpha_S = 0.2$, carrying out a simple fit to the linear component of the correction at small α_S, one sees that contributions beyond NLL are of the order of about half the total correction.

Angular ordering has a particularly large effect on the second derivative of $\tilde{\chi}$ — for $\alpha_S = 0.2$, it is reduced by a factor of two. The importance of the second derivative is that it is related to the amount of diffusion that is present. For an evolution over a range x, if one examines intermediate gluons with longitudinal momentum fraction x_i, using the saddle-point approximation one finds that the distribution of their transverse momentum has a width $\Delta \ln k$ which is given by:

$$\Delta \ln k \simeq \sqrt{\frac{\bar{\alpha}_S \chi_c''}{4} \frac{\ln x_i \ln x_i/x}{\ln x}} \tag{1}$$

Figure 4 shows the application of this formula to a particular set of evolution parameters — one sees that angular ordering leads to significant reduction of diffusion into the non-perturbative region. It has been shown by Mueller[8] that diffusion leads to a breakdown in the operator-product expansion (OPE) at a value of x defined by

$$\ln \frac{x_0}{x} \simeq \frac{1}{2\bar{\chi}_c''} \ln \frac{Q^2}{\Lambda^2},$$

where Λ is the QCD scale, x_0 some starting point for the small-x evolution, and Q^2 the hard scale of the problem. Accordingly, the effect of angular ordering can be seen as extending the x-range over which the OPE remains valid.

3. Associated quantities

The analysis of the gluon distribution is only the first step of a programme which aims to produce a Monte Carlo event generator based on the CCFM equation. An intermediate step is to examine associated final-state properties which are accessible using methods similar to that used for the structure function. Full details, including a description of the methods used, will be given in a forthcoming publication.[9] Here, preliminary results will be presented for two quantities: the probability of having n emissions with a transverse momentum above a certain minimum; and the transverse momentum flow as a function of rapidity. In neither case has the photon-gluon fusion matrix element been included, so the results are not directly comparable with experiment.

The probability of having n primary emitted gluons whose transverse momentum is larger than some minimum $q > \mu$ is an example of a quantity which differs between the BFKL and CCFM equations at leading order: in the BFKL case, for small μ, it varies in proportion to $\log \mu$, whereas in the CCFM case, for small μ it is independent of μ. Figure 5a shows the probability distribution for $\mu = 1$ GeV, while figure 5a shows it for $\mu = 0.007$ GeV. In the first case the curves are not too different (e.g. the maxima are at the same point), with the main difference being that the BFKL case has a long tail which is suppressed by phase-space constraints when angular ordering is introduced. On the other hand, for small μ the two distributions are quite different, with their maxima at very different positions: in the BFKL case there are many small-momentum emissions which are eliminated by the introduction of angular ordering.

Returning to the phenomenologically more meaningful case of $\mu = 1$ GeV, a simple analytical calculation of the DGLAP result has been also included:

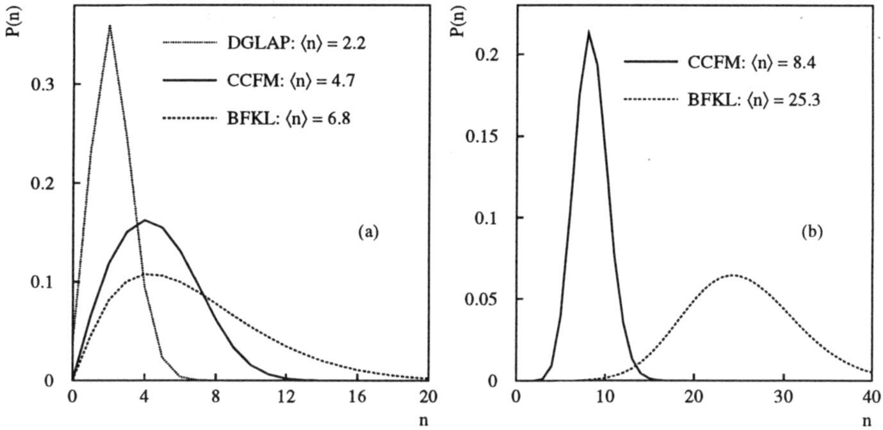

Fig. 5. The probability distribution for the number of emissions with transverse momentum $q >$ 1 GeV (a) and $q > 0.007$ GeV (b); $x = 5.10^{-5}$, $k = 5$ GeV, $\alpha_S = 0.2$

$$P(n) = \frac{(\bar{\alpha}_S \ln k^2 \ln 1/x)^n}{(n!)^2}. \tag{2}$$

As one would have expected, the CCFM result lies between the BFKL and DGLAP cases.

The other quantity for which we have a preliminary result is the transverse energy flow as a function of rapidity. This is shown in figure 6 where one sees that angular ordering significantly reduces the transverse energy flow.

4. Conclusions and outlook

In this talk, I have presented results on the effect of angular ordering on structure functions and two associated quantities. The main results are to be found in figures 2 and 3, where it is seen that angular ordering reduces the height of the minimum of the small-x characteristic function, shifts its position to larger γ and strongly reduces its second derivative (corresponding to a reduction in diffusion).

The technology developed for the study of structure functions is also being applied to the analysis of associated final-state properties and some preliminary results have been presented here, illustrating that angular ordering tends to reduce the number of emissions, particularly those with low transverse momenta, and that it also reduces the mean transverse energy flow.

To carry out phenomenology, certain extra elements are being implemented, among them the running of α_S and the inclusion of the hard matrix element, which determines the gluon-photon interaction. Also under development is a backward-evolution

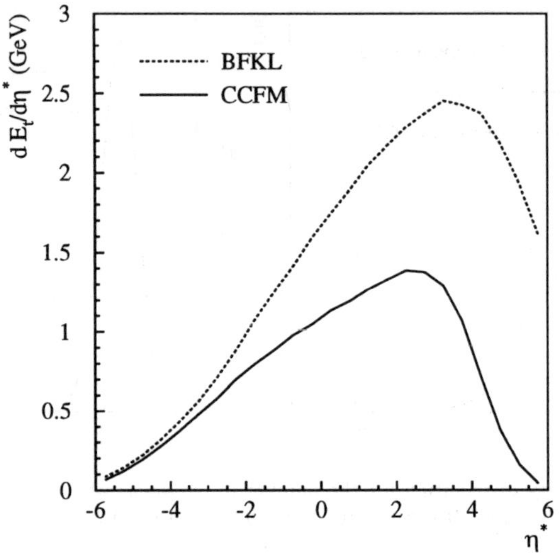

Fig. 6. The transverse energy flow in the hadronic centre of mass frame as a function of the rapidity η^*; the proton direction is to the left; $x = 5.10^{-5}$, $k = 5$ GeV, $\alpha_S = 0.2$.

CCFM Monte Carlo event generator.

5. Acknowledgements

This research was carried out in collaboration with G. Bottazzi, G. Marchesini and M. Scorletti and supported by funding from the Italian INFN. We would like to thank M. Ciafaloni, Yu.L. Dokshitzer, A.H. Mueller and B.R. Webber for helpful discussions. I am also grateful to the Fermilab theory group, where this talk was written up, for their kind hospitality.

6. References

1. Yu.L. Dokshitzer, V.A. Khoze, S.I. Troyan, and A.H. Mueller, *Basics of Perturbative QCD*, Editions Frontières, Paris, France 1991.
2. Ciafaloni M., Nucl. Phys. B296 (1987) 249.
3. S. Catani, F. Fiorani, and G. Marchesini, Phys. Lett. 234B (1990) 339; S. Catani, F. Fiorani, and G. Marchesini, Nucl. Phys. B336 (1990) 18.
4. J. Kwieciński, A.D. Martin, and P.J. Sutton, Phys. Rev. D 53 (1996) 6094; J. Kwieciński, A.D. Martin, and P.J. Sutton, Zeit. Phys. C71 (1996) 585.
5. L.N. Lipatov, Sov. J. Phys. **23** 338 (1976); E.A. Kuraev, L.N. Lipatov, V.S.

Fadin, Sov. Phys. JETP **45** 199 (1977); Ya. Balitskii, and L.N. Lipatov, Sov. J. Nucl. Phys. **28** 822 (1978).
6. M. Ciafaloni, these proceedings.
7. G. Bottazzi, G. Marchesini, G.P. Salam, and M. Scorletti, hep-ph/9702418, to appear in Nucl. Phys. B.
8. A.H. Mueller, Phys. Lett. 396B (1997) 251
9. G. Bottazzi, G. Marchesini, G.P. Salam, and M. Scorletti, in preparation.

A NON-PERTURBATIVE EFFECT IN DEEP INELASTIC SCATTERING

E.A. PASCHOS
Institut für Theoretische Physik, Universität Dortmund
D-44221 Dortmund, Germany
E-mail: paschos@hal1.physik.uni-dortmund.de

ABSTRACT

The HERA data at large Q^2 and small-x investigate *large distances on the light–cone.* At such large distances the scattered quarks can maintain their colour identity, i.e. they remain colour non–singlets and confined, by polarizing the vacuum. I calculate the probability for the creation of quark–antiquark pairs from the polarized QCD vacuum and their contribution to the structure function.

1. Introduction

The new experiments at HERA [1,2] extend the kinematic regions of deep inelastic scattering to much higher values of $Q^2 \lesssim 1000$ GeV^2 and smaller values of the Bjorken scattering variable $x \gtrsim 10^{-4}$. General arguments require that the scattering involves long distances on the light–cone[3-5].

Thus the struck quark moves with relativistic velocity to distances much larger than the radius of the proton, while at the same time remains confined. The fact that it remains confined means that it is still connected to the rest of the proton. For example, the struck quark can be one member of a virtual $q - \bar{q}$ pair. As the struck quark begins to move it transfers energy to virtual quark–antiquark pairs in the vacuum, thus polarizing the vacuum. When the deposited energy is large enough, the quark and antiquark separate and form hadrons.

In this article I describe the space–time structure of the process and assume that a chromoelectric flux tube is formed between the scattered quark and the rest of the proton. Quarks in the negative energy sea of the tube gain energy from the fields in the tube and tunnel into positive energy states leaving quark holes in the Dirac sea. The transition into positive energy states signals the breaking of the tube into two disjoint parts, which we identify with hadrons.

2. Space–time structure

Deep inelastic scattering investigates the tensor

$$W_{\mu\nu}(q \cdot p, q^2) = \frac{1}{2\pi} \int d^4y \, e^{iq \cdot y} \langle p | [J_\mu(y), \, J_\nu(0)] | p \rangle \, . \tag{1}$$

We select the momentum of the current along the 3–axis, i.e., $q_\mu = (q_0, 0, 0, q_3)$. The phase of the Fourier transform becomes stationary when

$$y_- = y_0 - y_3 \sim \pm \frac{1}{q_0 + q_3} \quad \text{and} \quad y_+ = y_0 + y_3 \sim \pm \frac{1}{q_0 - q_3} . \tag{2}$$

For the time-like distances of y these equations imply

$$y^2 = y_+ y_- - y_1^2 - y_2^2 \leq y_+ y_- \sim \frac{1}{Q^2} . \tag{3}$$

The large values of Q^2 investigated at HERA require that y^2 is very close to the light–cone. The dominant contribution of the commutator near the light–cone is singular, given by

$$W_{\mu\nu}(q \cdot p, q^2) = s_{\mu\nu\alpha\beta} \int d^4y \, e^{iq\cdot y} \frac{\partial}{\partial y_\alpha} \Delta_F(y) \langle p | \bar{q}(y) \gamma^\beta q(0) | p \rangle . \tag{4}$$

Keeping this contribution alone and following standard techniques, it can be shown that the scaling of the structure function follows. The data, however, show that scaling violations are large which means that QCD corrections are important. The new data also show that at fixed Q^2 the structure function $F_2(x, Q^2)$ increases by a factor of two as x decreases, indicating the creation of additional quanta. One method is to generate the structure function perturbatively starting with a boundary condition at $Q^2 = \mu^2$. A popular boundary condition[6] is to postulate a valence–like gluon distribution at $\mu = O(\Lambda)$ and calculate its development perturbatively. This approach reproduces the data.

When we look at the distance occurring in the commutator of eq. (1), however, we find that

$$y_3 = \pm \frac{1}{2} \frac{1}{q_0 - q_3} \approx \pm \frac{1}{2M_p x} = 2L(x) . \tag{5}$$

The distance $L(x)$ depends on the scaling variable and for small x becomes very large in comparison to the radius of the proton. Considering eq. (1) as the correlator of two currents on a proton, we are forced to accept that the distance between the currents is many times the radius of the proton. Viewing the process in the electron–proton center–of–mass system it means that the struck quark is still connected to the remaining proton through a long tube of chromoelectric flux. The process is shown schematically in figure 1. The virtual $q - \bar{q}$ pairs in the tube experience the chromoelectric field from which they gain energy and tunnel to positive energy states. This is a different picture than perturbative QCD which I shall try to develop.

3. Pair creation

In addition to the perturbative effects, hadrons can be produced non–perturbatively by the gluonic field. This means that in the process we must treat QED perturbatively (single photon exchange) and QCD non–perturbatively. The S-matrix for the problem is given by

$$S = \mathrm{T}\, exp \left\{ -ie \int A_\mu(y)\, j^\mu(y)\, d^4y - ig \int G_\nu^\alpha(y)\, j^{\nu,\alpha}(y)\, d^4y \right\} \tag{6}$$

with $A_\mu(y)$ and $G_\nu^\alpha(y)$ the electromagnetic and gluonic fields coupled to the corresponding currents $j^\mu(y) = \bar{q}(y)\gamma^\mu q(y)$ and $j^{\nu,\alpha}(y) = \bar{q}(y)\frac{\lambda^\alpha}{2}\gamma^\nu q(y)$, respectively. The capital T indicates the time-ordered product. When we treat QED perturbatively, but QCD to all orders, it can be proved that for the time–ordered product the following expansion holds

$$\langle X'|S|p\rangle = \langle X'|\,\mathrm{T}\,(-ie) \int A_\mu(y) j^\mu(y) d^4y \; exp \left\{ -ig \int G_\nu^\alpha(z) j^{\nu,\alpha}(z) d^4z \right\} |p\rangle. \tag{7}$$

We have written the matrix element of the action between a proton and a final hadronic state X', but the expansion also holds at the operator level. When we expand the QCD action in powers of g and contract the fields we obtain the perturbative QCD corrections which are frequently used in analyses of deep inelastic scattering [6]. A second contribution is obtained when we keep the QCD action intact and calculate the creation of quark–antiquark pairs. This is a non–perturbative phenomenon, which cannot be produced by a finite number of interactions of the quarks in $j^{\nu,\alpha}(y)$ with the gluonic field. The problem in the case of QED was solved by Schwinger [7,8] and we shall develop the mechanism for the creation of pairs in QCD.

For our case the scattering in the c-m system is shown in figure 1. Since the distance between the initial antiquark, $\bar{q}$, and the quark, q, is very large we will approximate the volume between them by a tube of chromoelectric flux.

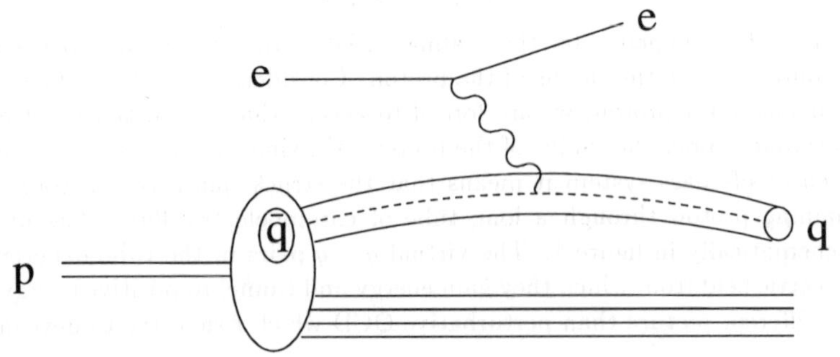

Fig. 1. The scattering of an electron on a proton with the development of the gluonic tube.

In addition to the perturbative expression for each quark distribution function, we shall show that there is now a term coming from the creation of pairs in the tube. We denote

$$q_i^{\text{tot}}(x,Q^2) = q_i^{\text{pert}}(x,Q^2) + q_i(x,Q_0^2)\, p(x,Q^2) \tag{8}$$

where $q_i^{\text{pert}}(x,Q^2)$ is the distribution function obtained from perturbative QCD and $p(x,Q^2)$ is the probability of creating pairs from the vacuum. The second quark distribution, $q_i(x,Q_0^2)$, is generated by the electroweak current without the QCD corrections. We shall see, later on, that for large and intermediate values of x, $p(x,Q^2) \approx 0$ and for small x

$$lim_{x\to 0}\, p(x,Q^2) \to 1\,. \tag{9}$$

Thus the new effect manifests itself at small values of x.

We consider a potential which produces homogeneous gluoelectric and gluomagnetic fields

$$G_\mu(y) = (0,\, y^3 gB,\, 0,\, -tgE)\,. \tag{10}$$

This is the field within the flux–tube with a chromoelectric field along the direction of the tube and a chromomagnetic field gB perpendicular to the tube. It is an Abelian field. The probability per unit time and per unit volume that a pair is created by the fields is given by [10]

$$\omega(y,m^2) = \alpha_s \frac{EB}{\pi} \sum_{n=1}^{\infty} \frac{1}{n} \coth\left(n\frac{B}{E}\pi\right) e^{-\frac{nm^2\pi}{gE}} \tag{11}$$

which in the limit $n\frac{B}{E} \ll 1$ reduces to the Schwinger solution [7,8]

$$\omega(y,m^2) = \frac{\alpha_s E^2}{\pi^2} \sum_{n=1}^{\infty} \frac{1}{n^2} e^{-\frac{nm^2\pi}{gE}}\,. \tag{12}$$

Modifications introduced by non–Abelian fields are discussed in ref. 10. The non–perturbative nature of these solutions is manifested in the exponential function, which has an essential singularity at $gE \to 0$. As we show below, these terms contribute to the function $p(x,Q^2)$ and bring in new dependence on the variables Q^2 and x, which originates from the length $L(x)$ and the variation of the gluonic fields with Q^2 and x.

The same problem was also solved for a finite size capacitor in order to study the finite size effects. In this case the authors solved [11] the one particle Dirac equation and obtained a solution which over most of the volume of the capacitor coincides with the Schwinger solution; deviations occur close to the plates.

We finally give a quantum mechanical interpretation of the result which demonstrates its generality. We consider a uniform electric field along the tube. The potential for the problem is shown in figure 2.

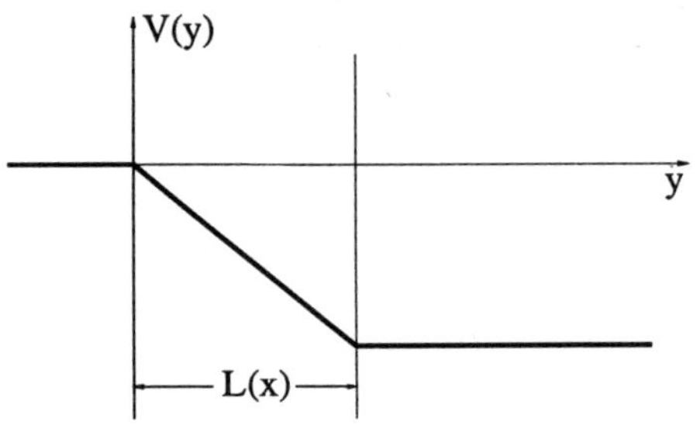

Fig. 2. The potential for a capacitor of constant electric field E and length $L(x)$.

We represent the potential

$$V(y) = \begin{cases} 0\,, & y < 0 \\ -gEy\,, & 0 < y < L \\ -gEL\,, & L < y \end{cases}$$

Consider a particle–hole pair with negative energy $-|\varepsilon|$. The probability for tunneling is given by

$$\begin{aligned} \frac{1}{\tau} &= exp\left\{-2\int_0^{y_0}\sqrt{\frac{2m}{\hbar^2}\left(|\varepsilon| - \frac{1}{2}gEy\right)}\,dy\right\} \\ &= e^{-\frac{4}{3}\frac{\sqrt{2m}}{\hbar}\frac{|\varepsilon|^{3/2}}{gE}} \end{aligned} \tag{13}$$

with $y_0 = \frac{2|\varepsilon|}{gE}$. We consider the emergence of a free quark as the creation of a pair.

4. Non–perturbative pair creation

In this section I attempt to develop a toy model for the creation of pairs in the tube. I consider a flux tube and partition it into small volume elements, as shown in figure 3. The probability of producing a pair in the time interval T at the volume element $\lambda^2 dy_i$ and no pairs elsewhere is

$$dP_1 = \lambda^2 T\omega(y_i)\,dy_i \prod_{K=1}^{\infty}\left(1 - \lambda^2 T\omega(y_K)\,dy_K\right) . \tag{14}$$

We denote the transverse dimension of the tube by λ and the time for the creation of a pair by T. These are two new parameters to be specified later on. The probability

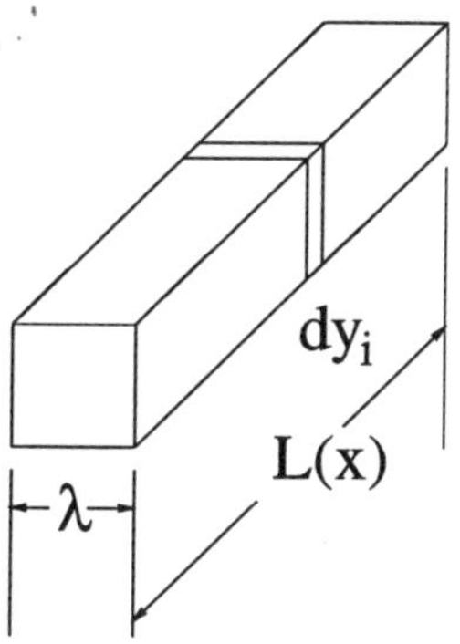

Fig. 3. A schematic drawing of the tube, where pairs are created by the gluonic field.

for producing one pair anywhere in the tube is

$$P_1 = \lambda^2 T \left[\int_0^{L(x)} \omega(y) dy \right] e^{-\lambda^2 T \int \omega(y) dy} . \tag{15}$$

We can generalize this result for n-pairs in the tube

$$P_n(x, Q^2) = \frac{1}{n!} \left[\lambda^2 T \int_0^{L(x)} \omega(y) dy \right]^n e^{-\lambda^2 T \int_0^{L(x)} \omega(y) dy} . \tag{16}$$

Finally, the sum over all possible pairs gives

$$p(x, Q^2) = \sum_{n=1}^{\infty} P_n(x, Q^2) = 1 - e^{-\lambda^2 T \int_0^{L(x)} \omega(y) dy} . \tag{17}$$

The functional form in eq. (16) is a Poisson distribution which follows from the property that the creation of pairs in each cell is independent of what happens in the other cells. One consequence is the multiplicity

$$n(x, Q^2) = \sum_{n=1}^{\infty} n P_n(x, Q^2) = \lambda^2 T \int_0^{L(x)} \omega(y)\, dy \tag{18}$$

which also determines the quark distribution function

$$q_i^{\mathrm{tot}} = q_i^{\mathrm{pert}}(x, Q^2) + q_i(x, Q_0^2) \cdot (1 - e^{n(x,Q^2)}) . \tag{19}$$

The last equation suggests a method for analysing the data. We can start with the distribution function at relatively low Q_0^2, then develop it through the Altarelli–Parisi equations to obtain $q_i^{\text{pert}}(x, Q^2)$. To this we finally add the second term with $p(x, Q^2)$ calculated as described above and $q_i(x, Q_0^2)$ is the quark distribution function at a much lower value of Q^2.

5. Numerical estimates

In the previous publication [12] I considered the gluoelectric field in the tube being independent of the Q^2 and x and determined its value from the string tension determined in particle production [13]. The field created in the tube depends on the circumstances under which the tube was created. In the c–m system the energy transferred to the struck quark is

$$E_{\text{tot}} = \frac{1}{2}\sqrt{Q^2 x}\left(\frac{1}{\sqrt{\tilde{y}}} - \sqrt{\tilde{y}}\right) + \frac{1}{2}\sqrt{Q^2\frac{\tilde{y}}{x}} \tag{20}$$

with the inelasticity $\tilde{y} = \frac{\nu}{E_e}$ and $x = \frac{Q^2}{2M\nu}$. (Note that this $\tilde{y}$ is different from the y used in the previous part of this paper). The string tension is calculated as the energy per unit length

$$k \approx \frac{E_{\text{tot}}}{L(x)} \tag{21}$$

with $L(x)$ defined in the electron–parton c–m system.

Finally we obtain the product of coupling constant times of the chromoelectric field as $gE = 4k$ from

(i) the string tension $k = \frac{1}{2}E^2 A$, and

(ii) Gauss' Law : $EA = \frac{1}{2}g$

with $A = \lambda^2$ the cross–section of the tube. We see that in general the field has a complicated dependence on Q^2, x and $\tilde{y}$.

To sum up, the increase observed in $F_2(x, Q^2)$ may originate from the creation of pairs from the vacuum. Consequently the increase of $F_2^{\text{tot}}(x, Q^2)$ from perturbative QCD can be relatively smaller. As a result a new analysis of the data is suggested in terms of two components: an increase from perturbative QCD and a faster increase from the creation of pairs. The limiting value of $F_2^{non-pert.}(x)$ at $x = 0$ is in the present theory finite. Summing the contributions from all the quarks we obtain

$$F_2^{total}(x, Q^2) = F_2^{pert.}(x, Q^2) + F_2(x, Q_0^2)(1 - e^{-n(x,Q^2)}) \tag{22}$$

with $F_2^{pert.}(x, Q^2)$ the perturbative development of the structure function and $F_2(x, Q_0^2)$ the structure function at a lower value of $Q^2 \approx Q_0^2$ where perturbative corrections are not yet dominant.

Acknowledgement

The financial support of the Bundesministerium für Bildung, Wissenschaft, Forschung und Technologie under contract No. 05 7DO93P 7 is gratefully acknowledged.

References

1. H1 Collab., T. Ahmed et al., *Nucl. Phys.* **B439** (1995) 471.
2. ZEUS Collab., M. Derrick et al., *Z. Phys.* **C65** (1995) 379.
3. B. Ioffe, *Phys. Lett.***B30** (1969) 123.
4. V. Gribov, *Sov. Phys. JETP***30** (1969) 709.
5. J.D. Bjorken, *Phys. Rev.***D7**(1973) 282 and SLAC-PUB-7096 (1996).
6. M. Glück, E. Reya and A. Vogt, *Z. Phys.* **C67** (1995) 433, and references therein.
7. J. Schwinger, *Phys. Rev.* **82** (1951) 664; **93** (1954) 615.
8. J. Schwinger, *Particles, Sources and Fields*, section 4-8, Addison-Wesley Press (1973).
9. I.A. Batalin, S.G. Matinyan and G.K. Savvidy, *Sov. J. Nucl. Phys.* **26** (1977) 214.
10. S.K. Bose, G. Cvetic and E.A. Paschos, Dortmund preprint DO–TH 97/19. The article investigates pair production in the general homogeneous non–Abelian (QCD) field.
11. R.-C. Wang and C.-Y. Wong, *Phys. Rev.* **D38** (1988) 384.
12. E.A. Paschos, *Phys. Lett.* **B389** (1996) 383.
13. A. Casher, H. Neuberger and S. Nussinov, *Phys. Rev.* **D20** (1979) 179.

Chapter 3

High-Q^2 Phenomena

RECENT RESULTS ON VERY HIGH Q^2 PHYSICS AT HERA

YVES SIROIS
LPNHE Ecole Polytechnique, IN2P3-CNRS
91128 Palaiseau, France
E-mail: sirois@polhp2.in2p3.fr

ABSTRACT

Recent observations by the H1 and ZEUS experiments at HERA of excesses of deep-inelastic ep scattering events at very high squared 4–momentum transfer Q^2 are discussed. For $Q^2 > 15000$ GeV^2, $N_{obs} = 12$ neutral current candidate events are observed by H1 for an integrated luminosity of 14.2 pb^{-1} where the expectation of the standard deep-inelastic model of lepton–nucleon scattering (DIS) is $N_{DIS} = 4.71 \pm 0.76$ events. In the same Q^2 range, $N_{obs} = 4$ charged current candidates are observed by H1 where the expectation is $N_{DIS} = 1.77 \pm 0.87$ events. The probability $\mathcal{P}(N \geq N_{obs})$ that the DIS model signal N fluctuates to $N \geq N_{obs}$ in a random set of experiments is 6×10^{-3} for neutral current and 0.14 for charged current. For $Q^2 > 15000$ GeV^2, $N_{obs} = 12$ neutral current candidate events are observed by ZEUS for an integrated luminosity of 20.1 pb^{-1} in excess of the expectation $N_{DIS} = 8.66 \pm 0.66$. For $Q^2 > 30000$ GeV^2, both experiments observe 2 neutral current candidate events. These 4 candidates are to be compared with a combined DIS expectation of $\simeq 0.6 \pm 0.06$ corresponding to $\mathcal{P}(N \geq N_{obs}) \simeq 3.5 \times 10^{-3}$. The difference in the observed and expected number of neutral current events is mostly due to events at large masses $M = \sqrt{x\,s}$ in which the positron is backscattered at large $y = Q^2/M^2$.

1. Introduction

The ep collider HERA offers the unique possibility to probe the proton at very small distances ($\simeq 10^{-16}$ cm) via t-channel exchange of highly virtual gauge bosons, and to search in the s-channel for new particles which couple to lepton-parton pairs with masses up to the kinematic limit of $\sqrt{s} \simeq 300$ GeV.

In this paper, the recent HERA results [1,2] on neutral current (NC) and charged current (CC) deep-inelastic scattering (DIS) at very high Q^2 are discussed. The NC and CC analyses from H1 [1] and the NC analysis from ZEUS [2] follow a search strategy offering best sensitivity to eventual deviations from standard DIS model expectation. The results are based on e^+p data collected at HERA from 1994 to 1996 at an e^+ beam energy of 27.5 GeV and a p beam energy of 820 GeV. The total integrated luminosities amount for H1 to 14.19 ± 0.32 pb^{-1} and for ZEUS to 20.1 ± 0.3 pb^{-1}.

2. The H1 and ZEUS Detectors

Detailed descriptions of the H1 and ZEUS detectors can be found in [3] and [4] respectively. For the high Q^2 analysis, both experiments use essentially their calorimetry to measure the final state positron or hadronic energy flow with high transverse energy.

The H1 liquid argon (LAr) sampling calorimeter [5] covers the polar angular[a] range $4^\circ \leq \theta \leq 153^\circ$. It consists of a fine granularity lead/argon electromagnetic section followed by a stainless steel/argon hadronic section which provide in total ~ 44000 read-out cells. Electromagnetic shower energies are measured with a resolution of $\sigma(E)/E \simeq 12\%/\sqrt{E/\text{ GeV}} \oplus 1\%$ and hadronic shower energies with $\sigma(E)/E \simeq 50\%/\sqrt{E/\text{ GeV}} \oplus 2\%$ after software energy weighting. These resolutions were measured [6] in test beams with electron energies up to 166 GeV and pion energies up to 205 GeV. The absolute energy scales are known to 3% and 4% for electromagnetic and hadronic energies respectively. The angular resolution on the positron induced shower in the calorimeter varies from ~ 2 mrad below 30° to $\lesssim 5$ mrad at larger angles. A lead/scintillating-fibre backward calorimeter extends the coverage at larger angles ($155^\circ \leq \theta \lesssim 178^\circ$).

The ZEUS uranium/scintillator calorimeter [7] covers the polar angular range $2.6^\circ \leq \theta \leq 176.1^\circ$ and is segmented longitudinally in ~ 6000 read-out cells. It provides a resolution of $\sigma(E)/E \simeq 18\%/\sqrt{E/\text{ GeV}}$ for electromagnetic showers and $\sigma(E)/E \simeq 35\%/\sqrt{E/\text{ GeV}}$ for pion induced hadronic showers as measured in test beams for energies up to 120 GeV. The electromagnetic and hadronic absolute energy scales are known to 3%. The angular resolution on the positron is ~ 5 mrad.

The ZEUS calorimeter provides an event time t_0 relative to the nominal time of the bunch crossing of the beams with a resolution below 1 ns. The H1 central jet chambers provides a t_0 with a resolution of $\simeq 1.6$ ns.

Both experiments use their central tracking to reconstruct the primary interaction vertex, and determine the luminosity from the rate of the Bethe-Heitler $ep \to ep\gamma$ bremsstrahlung measured in a dedicated monitor.

3. Event Selection and Kinematics

The main set of selection criteria designed by H1 and ZEUS to find NC DIS events at high Q^2 are listed in Table 1 and rely mainly on calorimetric information. Both H1 and ZEUS require an isolated positron candidate[b]. The positron to hadron (jet) separation relies on detailed shower shape analysis (isolation). A less stringent positron isolation is imposed by H1 but the finer transverse granularity of the LAr electromagnetic section allows to exploit further the transversal shower shapes. A cluster-track link matching is imposed down to $\Theta_e > 10^\circ$ by H1 and $\Theta_e > 17.2^\circ$ by ZEUS to suppress background from photoproduction of prompt photons. Below

[a]The z axis is taken to be in the direction of the incident proton, and the origin of coordinates is the nominal ep interaction point.
[b]The analyses do not explicitely distinguish between e^+ and e^-

Selection NC-like Events	
H1	ZEUS
$\mid Z_{vertex} \mid < 35$ cm	$\mid Z_{vertex} \mid < 50$ cm
Electron Identification	
shower shape analysis	shower shape analysis
isolation in $\eta - \phi$ cone ($R < 0.25$)	isolation in $\eta - \phi$ cone ($R < 0.8$)
cluster-track matching for $\Theta_e > 10°$	cluster-track matching for $\Theta_e > 17.2°$
	($E_{T,e} > 30$ GeV, $\Theta_e < 17.2°$)
Energy-momentum Conservation	
$43 < \sum E - P_z < 63$ GeV	$40 < \sum E - P_z < 70$ GeV
$(P_{T,miss}/\sqrt{E_T} < 3\sqrt{GeV})$	$(44 < \sum E - P_z$, $\Theta_e < 17.2°)$
Kinematics	
$E_{T,e} > 25$ GeV	$E_e > 20$ GeV
$0.1 < y_e < 0.9$	$0.05 < y_{2a}$
$Q^2_e > 2500$ GeV^2	$Q^2_{2a} > 5000$ GeV^2
Selection CC-like Events	
$P_{T,miss} > 50$ GeV	
$(E_T - P_{T,h})/E_T < 0.5$	

Table 1: NC DIS event selection by H1 and ZEUS and CC DIS event selection by H1.

17.2° ZEUS imposes slightly more severe $E_{T,e}$ cut (30 GeV compared to 25 GeV for H1). The positron requirements and $\sum E - P_z$ cuts eliminate the background from photoproduction of high E_T jets. From energy-momentum conservation, the $\sum E - P_z$ distribution for DIS events is peaked at $2 \times E_e^{init.} \simeq 55GeV$. Hence the $\sum E - P_z$ cuts also reject DIS events where a (undetected) very hard collinear γ is emitted by the initial state positron. This cut is less severe (above 17.2°) for ZEUS. The H1 selection criteria for CC DIS events in the high Q^2 domain are listed in Table 1 and also rely on calorimetric information. Following the above selection complemented by cuts dedicated to specific background sources and described in detail in refs. [1,2], the non-DIS background is estimated to be negligible in the high Q^2 kinematic range considered.

At lowest order two variables are needed to reconstruct the DIS Lorentz invariants[8] x, y or Q^2. In the centre of mass of the positron-parton collision $y = (1 + \cos\theta_e^*)/2$ where θ_e^* is the polar scattering angle relative to the incident proton and $M = \sqrt{xs}$ is the available energy. To reconstruct the NC kinematic, H1 and ZEUS consider both an e-method (exploiting the positron energy E_e and angle θ_e) and a $2a$-method (exploiting the positron and hadronic energy flow angles). The e-method offers high precision at large M and large y but directly depends on the absolute electron energy scale. The $2a$-method offers high precision at high Q^2 while being practically insensitive to absolute energy calibration. The two methods differ in particular when turning on higher order processes. The $2a$-method is for instance more affected at very large Q^2 in the case of the emission of a real γ from the e initial state. It always leads to a shift towards larger mass values following $M_{2a}^2 \simeq M^2 \times 1/(1-z)^2$ where $z = E_\gamma/E_e^{init.}$ is the fractional energy taken away by the radiated γ. On the contrary the e-method leads to a shift towards smaller masses for $y < 0.5$ and towards larger masses for $y > 0.5$ following $M_e^2 \simeq M^2 \times [1 + z(2y_e - 1)/y_e]$. The e-method ($2a$-method) is used as the reference method by H1 (ZEUS) while the $2a$-method (e-method) is considered for cross-checks. For the CC DIS kinematic, H1 makes use of the Jacquet-Blondel ansatz [8].

It should be noted that while it is important to control migrations in the kinematic plane (such a control is offered by the comparison of the results obtained from the e- and $2a$-methods), there is only a small probability that any given event undergoes a substantial migration (e.g. $\mathcal{O}(1\%)$ chance that the reconstructed mass differs from M by more than 10% for events at $Q^2 > 15000$ GeV^2). Note that this does **not** preclude a deviation from the true energy available to create "mass" in the hard collision ! When for instance the final state hadronic system acquires a large mass M_h through QCD radiation of gluons, the invariant $x = M^2/s$ can no longer be interpreted as the fraction of the proton's energy carried by the struck quark. Hence an hypothetical narrow lepton-quark resonance at M_{eq} would appear distorted with individual events migrating according to $M \simeq M_{eq}/(1 + M_h^2/Q^2)$ (see for instance Fig. 2d of ref. [1] for the example case of a leptoquark at $M_{LQ} \simeq 200$ GeV in a specific model).

4. Results

4.1. General properties

Fig. 1 (left) shows the two dimensional distribution of y_e against M_e for the H1 data. Fig. 1 (right) shows a similar distribution of y_{2a} against M_{2a} for the ZEUS data. H1 observes 443 NC DIS candidates satisfying their selection criteria at $Q_e^2 > 2500$ GeV^2, which is in good agreement with the expectation of 427 ± 38 from NC DIS processes. About 3.8% more events satisfy the H1 kinematic requirements when calculated with the double-angle method. ZEUS observes 191 NC DIS candidates at $Q_{2a}^2 > 5000$ GeV^2 satisfying their selection criteria; in good agreement with the expectation of 196.5 ± 9.9 from NC DIS processes. The systematic errors on the standard DIS expectations combine experimental and "theoretical" contributions. The experimental uncertainties which have been propagated come come from the integrated luminosities, the absolute energy calibrations and the angular measurements.

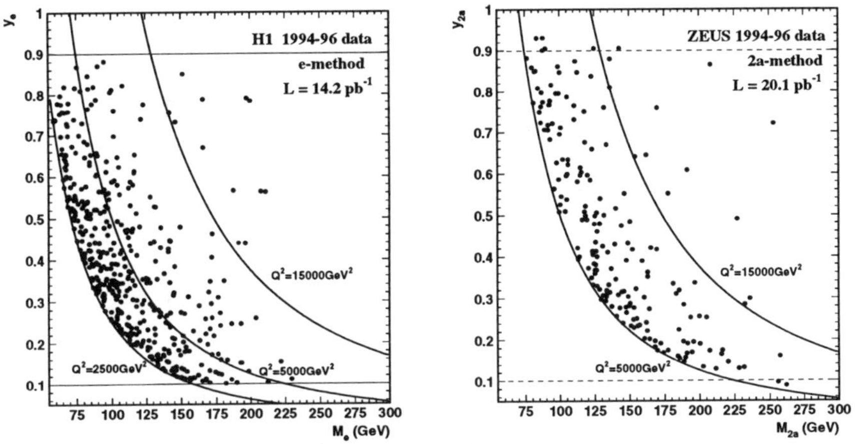

Figure 1: NC DIS candidate events selected by (left) H1 in the y_e - M_e plane and by (right) ZEUS in the y_{2a} - M_{2a} plane; also shown are contours of fixed Q^2.

The "theoretical" uncertainty on the standard DIS cross-section is common to both experiments. It is at the 7% level and reflects the remaining freedom on the parton density distributions, the modelisation of higher order QED radiations, the value of the strong coupling constant α_s and of the electroweak parameters.

Fig. 2a and 2b show the H1 projected M_e and y_e distributions of the selected events at "low" Q^2 ($2500 \text{ GeV}^2 < Q_e^2 < 15000 \text{ GeV}^2$) and Fig. 2c and 2d at "high" Q^2 ($Q_e^2 > 15000 \text{ GeV}^2$). The distributions of the data are well reproduced by standard NC DIS predictions in the low Q^2 range. At high Q^2 the data exceed the expectation, especially at $M_e \sim 200$ GeV. Fig. 2e and 2f show the y_e distributions for M_e values below and above 180 GeV. At high M_e, a difference between observation and expectation is apparent at large y_e.

4.2. Q^2 dependence

Fig. 3 shows for the H1 NC selection the measured Q_e^2 distribution in comparison with the expectation from standard NC DIS processes. Also shown is the ratio of the measured and expected distributions. The errors resulting from the convolution of the statistical error of the Monte Carlo sample and the systematic errors are correlated for different Q_e^2 bins and are indicated in Fig. 3b. They are shown as an hatched area joining the $\pm 1\sigma$ errors evaluated at the center of each bin. These errors are dominated by the uncertainty in the electromagnetic energy scale of the calorimeter

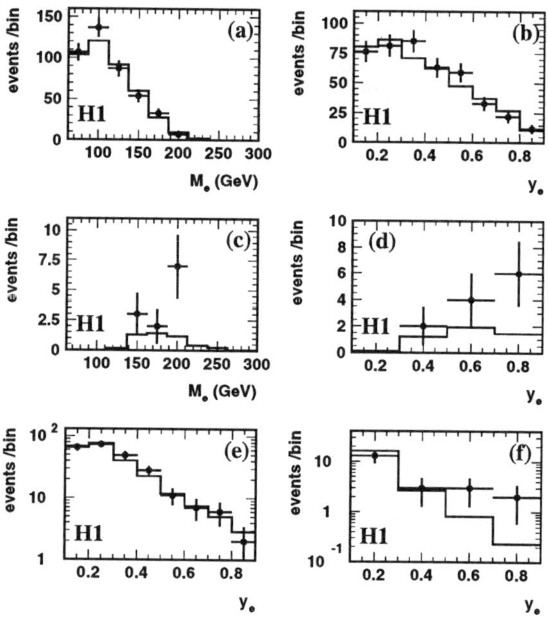

Figure 2: Distributions of M_e and y_e for selected NC DIS candidate events, (a) and (b) for 2500 GeV^2 < Q_e^2 < 15000 GeV^2, (c) and (d) for $Q_e^2 > 15000$ GeV^2; distribution of y_e (e) for $100 < M_e < 180$ GeV and (f) for $M_e > 180$ GeV; superimposed on the data points are histograms of the expectation of standard NC DIS.

and vary between 8.5% at low Q_e^2 and 30% at the highest values of Q_e^2. The NC DIS expectation agrees well with the data for $Q_e^2 \lesssim 15000$ GeV^2 while at larger Q_e^2 the number of events is in excess of the NC DIS expectation.

Fig. 4 (left) shows for the ZEUS NC selection the measured Q_e^2 distribution in comparison with the expectation from standard NC DIS processes. The data is seen to be in excess of the NC DIS expectation at the largest Q_{2a}^2.

To quantify the differences observed by both the H1 and ZEUS experiments, the numbers of observed and expected events with Q^2 above various Q_{min}^2 values are given in Table 2. Also given are the Poisson probabilities $\mathcal{P}(N \geq N_{obs})$ that in a random set of experiments the number of NC DIS events N fluctuates to values equal to or larger than the observed number of events N_{obs}. The systematic error δb on the mean

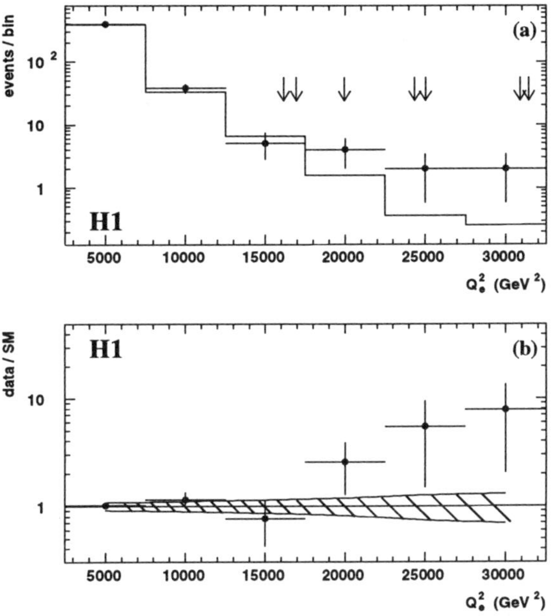

Figure 3: (a) Q^2_e distribution of the selected NC DIS candidate events for the H1 data (•) and for standard NC DIS expectation (histogram); the arrows indicate the Q^2_e values for data entries at $M_e > 180$ GeV when $Q^2_e > 15000$ GeV^2; (b) ratio of the observed and expected number of events as a function of Q^2_e; the hatched area specify the $\pm 1\sigma$ uncertainty levels combining statistical and systematic errors of the NC DIS expectation.

number of expected events b is taken into account by using the convolution :

$$\mathcal{P}(N \geq N_{obs}) = \int_0^{+\infty} dx G(x; b, \delta b) \sum_{k=N_{obs}}^{\infty} p(k; x)$$

where $p(k;x)$ is the Poisson probability to observe k events when the number of expected events is x and $G(x;b,\delta b)$ is the Gaussian probability density function for the NC DIS expectation x of mean value b and width δb. The resulting probabilities in Table 2 fall to about 1% at the largest Q^2_{min} values.

For $Q^2_e > 15000$ GeV^2, the number of observed events in H1 is $N_{obs} = 12$ for an expectation of 4.71 ± 0.76 corresponding to a probability $\mathcal{P}(N \geq N_{obs})$ of 6×10^{-3}. For $Q^2_{2a} > 15000$ GeV^2, ZEUS observes $N_{obs} = 12$ events for an expectation of 8.66 ± 0.66. In the case of ZEUS, the excess is most significant for $Q^2_{2a} > 35000$ GeV^2 where the number of observed events is $N_{obs} = 2$ for an expectation of 0.145 ± 0.013 corresponding to a probability $\mathcal{P}(N \geq N_{obs})$ of 9.6×10^{-3}.

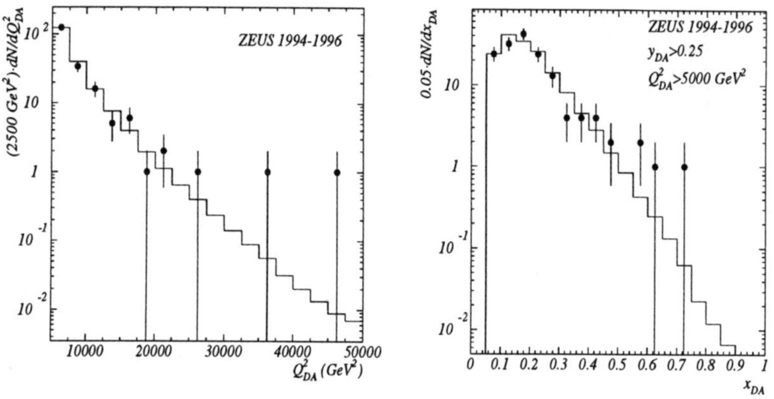

Figure 4: Q^2_{da} (left) and M_{da} (right) distributions of ZEUS for the observed events (•) compared to the standard NC DIS model expectation (histogram).

Table 3 gives the number of observed and expected CC events in H1, as well as the Poisson probability $\mathcal{P}(N \geq N_{obs})$ as defined above for increasing values of Q^2_{min}. In the kinematic region $Q^2_h > 15000$ GeV^2, there are $N_{obs} = 4$ observed events compared with an expectation of 1.77 ± 0.87 from standard CC DIS. This small excess corresponds to a Poisson probability $\mathcal{P}(N \geq N_{obs})$ of 14%, including systematic errors, that the CC DIS signal fluctuates to values equal to or larger than N_{obs} in a random set of experiments.

4.3. Mass dependence

Fig. 5a shows the measured and expected M_e distribution and Fig. 5b the ratio of the measured M_e distribution to NC DIS expectation for a minimum y_e value of $y_{min} = 0.2$. Similar distributions are shown in Fig. 5c and 5d for $y_{min} = 0.4$. An excess of events over the NC DIS expectation at the highest mass ($\sim$ 200 GeV) is seen, which becomes more visible with the larger y_{min} cut.

The range of M_e values for which the most significant excess over NC DIS expectation exists was investigated by H1 by considering "windows" of various total widths ΔM_e. The central values M_e of the mass windows were varied in steps of 1 GeV between 80 and 250 GeV and the numbers of observed and expected events are determined for different y_{min} values. For each mass window, the Poisson probability $\mathcal{P}(N \geq N_{obs})$ was determined, including the propagation of all systematic errors described above. The most significant excess is observed by H1 for a central mass value $M_e \sim 200$ GeV. Results obtained for that fixed central value with $y_{min} = 0.4$ and for various choices of ΔM_e, are given in Table 4 where the probability $\bar{\mathcal{P}}$ is calculated

H1 Results						
Q^2_{min}(GeV^2)	2500	5000	10000	15000	20000	30000
N_{obs}	443	122	20	12	5	2
N_{DIS}	426.7 $\pm$38.4	116.2 $\pm$11.6	18.3 $\pm$2.4	4.71 $\pm$0.76	1.32 $\pm$0.27	0.23 $\pm$0.05
$\mathcal{P}(N \geq N_{obs})$	0.35	0.35	0.39	6×10^{-3}	1.4×10^{-2}	2.3×10^{-2}
ZEUS Results						
Q^2_{min}(GeV^2)	5000	10000	15000	20000	30000	35000
N_{obs}	191	33	12	5	2	2
N_{DIS}	196.5 $\pm$9.87	32.2 $\pm$2.04	8.66 $\pm$0.66	2.76 $\pm$0.23	0.37 $\pm$0.04	0.145 $\pm$0.013
$\mathcal{P}(N \geq N_{obs})$	0.62	0.44	0.17	0.15	5.4×10^{-2}	9.6×10^{-3}

Table 2: Number of observed (N_{obs}) and expected NC DIS (N_{DIS}) events with 4-momentum transfer squared Q^2_e above given thresholds (Q^2_{min}); $\mathcal{P}(N \geq N_{obs})$ is the probability that the number of NC DIS events fluctuates to values equal to or larger than N_{obs} in a random set of experiments.

averaging over 3 steps. For $\Delta M_e = 25$ GeV, 7 events are observed for an expectation of 0.95 ± 0.18, corresponding to $\bar{\mathcal{P}} = 2.6 \times 10^{-4}$ ($\mathcal{P} = 10^{-4}$). The magnitude and location of this most significant excess was further verified be insensitive to the choice (e- or $2a$-) of reconstruction method.

Fig. 4 (right) shows for the ZEUS NC selection the measured M_{da} distribution at $y_{da} > 0.25$ compared with the expectation from standard NC DIS processes. An excess is found at very large masses. For $y_{da} > 0.25$, $N_{obs} = 4$ are observed at $x_{da} > 0.55$ ($M_{da} > 223$ GeV) for an expectation of 0.91 ± 0.08 corresponding to $\mathcal{P}(N \geq N_{obs})$ of 1.4×10^{-2}. Thus, the region in which the excess of NC-like events is found to be most significant appears at larger mass values for the ZEUS data than

Q^2_{min}(GeV^2)	2500	5000	10000	15000	20000
N_{obs}	31	24	10	4	3
N_{DIS}	34.2 $\pm$5.8	21.1 $\pm$4.2	5.07 $\pm$1.88	1.77 $\pm$0.87	0.74 $\pm$0.39
$\mathcal{P}(N \geq N_{obs})$	0.64	0.31	7×10^{-2}	0.14	5.4×10^{-2}

Table 3: Number of observed (N_{obs}) and expected CC DIS events (N_{DIS}) with 4-momentum transfer squared Q^2 above given Q^2_{min} values; $\mathcal{P}(N \geq N_{obs})$ is the probability that the number of CC DIS events fluctuates to values equal to or larger than N_{obs} in a random set of experiments.

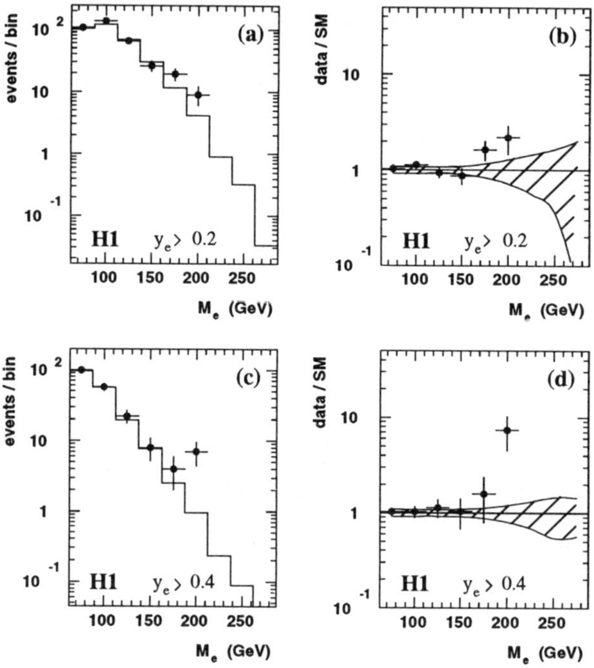

Figure 5: (a) Distribution of M_e for the observed (•) NC DIS candidates with $y_e > 0.2$; the expectation from standard NC DIS is shown as the superimposed histogram; (b) ratio of the observed and expected numbers of events as a function of M_e for $y_e > 0.2$; the hatched area specify the $\pm 1\sigma$ uncertainty levels for the standard NC DIS expectation combining statistical and systematic errors; (c) as in (a) but now with $y_e > 0.4$; (d) as in (b) but now with $y_e > 0.4$.

for the H1 data.

For $Q_e^2 > 15000$ GeV^2, the 7 events observed by H1 at $M_e > 180$ GeV appear "clusterized" in mass at an average of $< M_e >\simeq 200$ GeV with a weighted r.m.s. spread of $\sigma_{M_e} \simeq 5$ GeV. The same mean value $\simeq 200$ GeV is found using the $2a$- or the Σ-method. The Σ-method [9] is a "radiation hard" method which profits from the $\sum E - P_z$ constraint to become independent from initial state QED radiation

In the corresponding domain at $Q_{2a}^2 > 15000$ GeV^2 and $M_{2a} > 180$ GeV, the 6 events observed by ZEUS are more broadly distributed with an average of $< M_{2a} >\simeq$ 229 GeV ($< M_e >\simeq 214$ GeV) and a weighted r.m.s. spread of $\sigma_{M_{2a}} \simeq 20$ GeV ($\sigma_{M_e} \simeq 17$ GeV). Of course one cannot draw firm conclusions from the comparison of a handful of events especially given that the high Q^2 events appear to be individually indistinguishable from standard DIS events. It nevertheless is fair to say that the H1 and ZEUS observations do not mutually strengthen the statistical evidence for a

ΔM_e (GeV)	20	25	30	40
N_{obs}	5	7	7	7
N_{DIS}	0.63 $\pm$0.13	0.95 $\pm$0.18	1.10 $\pm$0.19	1.57 $\pm$0.28
$\bar{\mathcal{P}}(N \geq N_{obs})$	5.0×10^{-4}	2.6×10^{-4}	2.5×10^{-4}	1.6×10^{-3}

Table 4: Number of observed (N_{obs}) and expected (N_{DIS}) NC DIS candidates (N_{DIS}) for different mass windows of total width ΔM_e, an *a priori* choice for the central mass value of 200 GeV and for $y_e > 0.4$; also given are the probabilities $\bar{\mathcal{P}}(N \geq N_{obs})$ (see text).

mass clusterization of the excess at high Q^2.

5. Conclusions

For $Q^2 \leq 15000$ GeV^2, the Q^2 or M and y distributions are well reproduced by the expectations from standard NC DIS (H1, ZEUS) and CC DIS processes (H1).

For $Q_e^2 > 15000$ GeV^2, H1 observes 12 NC candidates for an expectation of 4.71 ± 0.76 corresponding to a probability $\mathcal{P} = 6 \times 10^{-3}$. For ZEUS, the NC DIS excess is most significant at the largest Q^2; 2 events are observed at $Q_{2a}^2 > 35000$ GeV^2 while 0.145 ± 0.013 are expected corresponding to $\mathcal{P} = 9.6 \times 10^{-3}$. Each of these H1 and ZEUS events have clean DIS-like signatures. The non-DIS background is estimated to be negligible. A small excess of CC candidate events is also observed by H1 for $Q_h^2 > 15000$ GeV^2.

The mass distribution of the events at $Q^2 > 15000$ GeV^2 for H1 data shows a most prominent excess "clusterized" around $M \sim 200$ GeV where 7 events are observed at $y_e > 0.4$ while 0.95 ± 0.18 are expected corresponding to $\mathcal{P} = 10^{-4}$. The probability to find a comparable or larger excess in a window *anywhere* in the kinematic plane was estimated to be $= \mathcal{O}(1\%)$. For $x_{da} > 0.55$ ($M > 223$ GeV) and $y_{da} > 0.25$, ZEUS observes 4 NC candidates for an expectation of 0.91 ± 0.08 corresponding to $\mathcal{P} = 1.4 \times 10^{-2}$.

The magnitude of the excess cannot be accounted for in the Standard Model given in particular the constraints from existing data and the only possible explanations are either a statistical fluctuation or new physics. Only a sizeable increase of statistics will allow a clarification of the nature of this excess.

Among the possible explanations requiring physics beyond the Standard Model (for a contemporary review see ref. [10]), considerable attention has been given so far to the hypothesis of a s-channel production of scalar leptoquarks or squarks in R-parity violating supersymmetry. Scalar resonances would lead to a flat y spectrum in sharp contrast to the $1/y^2$ dependence expected from standard DIS at fixed M. The excess at high Q^2 could also be due to a contact interaction occuring with a characteristic scale in the few TeV range or to new (yet unknown) physics interfering with the electroweak sector.

6. Acknowledgements

The H1 and ZEUS collaborations wish to thank the HERA machine group for the outstanding operation of the collider. I wish to thank the DESY directorate and my H1 and ZEUS colleagues for their support.

7. References

1. H1 Collaboration, C. Adloff et al., Z. Phys. C74 (1997) 191.
2. ZEUS Collaboration, J. Breitweg et al., Z. Phys. C74 (1997) 207.
3. H1 Collaboration, I. Abt et al., Nucl. Instr. and Meth. A386 (1997) 310; *idem* Nucl. Instr. and Meth. A386 (1997) 348.
4. ZEUS Collaboration, "The ZEUS Detector", DESY Status Report 1993.
5. H1 Calorimeter Group, B. Andrieu et al., Nucl. Instr. and Meth. A336 (1993) 460.
6. H1 Calorimeter Group, B. Andrieu et al., Nucl. Instr. and Meth. A344 (1994) 492; *idem*, Nucl. Instr. and Meth. A350 (1994) 57; *idem*, Nucl. Instr. and Meth. A336 (1993) 499.
7. M. Derrick et al., Nucl. Instr. and Meth. A309 (1992) 77; A. Andersen et al., Nucl. Instr. and Meth. A309 (1991) 101; A. Bernstein et al., Nucl. Instr. and Meth. A336 (1993) 23.
8. S. Bentvelsen, J. Engelen and P. Kooijman, Proc. of the Workshop Physics at HERA, W. Buchmüller and G. Ingelman (Editors), (October 1991, DESY-Hamburg) Vol. 1 p. 25; K.C. Hoeger, *ibid.* p. 43; and references therein.
9. U. Bassler and G. Bernardi, Nucl.Instr. and Meth. A361 (1995) 197.
10. R. Rückl, contribution to this conference.

LEPTOQUARK SEARCHES AT LEP2

SILVIA ARCELLI
for the OPAL Collaboration
University of Bologna and INFN
V.le B.Pichat 6/2, Bologna, Italy
E-mail: Silvia.Arcelli@cern.ch

ABSTRACT

Using data collected with the OPAL detector at centre of mass energies above the Z^0 resonance, limits on leptoquarks have been derived considering two production mechanisms. Assuming single leptoquark production from deep inelastic electron-photon scattering, a 95% CL lower mass limit of 131 GeV has been obtained for first generation scalar leptoquarks of charges -1/3, -5/3, in the hypothesis that the Yukawa coupling strenght λ is larger than $\sqrt{4\pi\alpha_{em}}$. Effects from the virtual exchange of a scalar leptoquark, which is expected to modify the $e^+e^- \to q\bar{q}$ cross-section relative to the Standard Model, have then been searched by analyzing both the inclusive $q\bar{q}$ sample and a $b\bar{b}$ enriched sample. Limits on the Yukawa coupling λ between typically 0.3 and 0.5 have been derived for masses below 400 GeV. Results have also been interpreted within the context of a four fermion Contact Interaction, and lower limits between 1-3 TeV on the energy scale Λ of the interaction have been derived.

1. Introduction

Several extensions of the Standard Model which connect the quark and the lepton sector foresee the existence of leptoquarks, which are coloured scalar or vector particles carrying simoultaneously leptonic and baryonic number. While the coupling to gauge bosons is fixed by leptoquark quantum numbers, the Yukawa coupling λ_{ijk} to ordinary fermions, where the indices denote the family of the incoming, exchanged and outgoing particles, is arbitrary.

At LEP, leptoquark production at centre-of-mass energies above the Z^0 resonance can be investigated through the three channels shown in figure 1. In the single

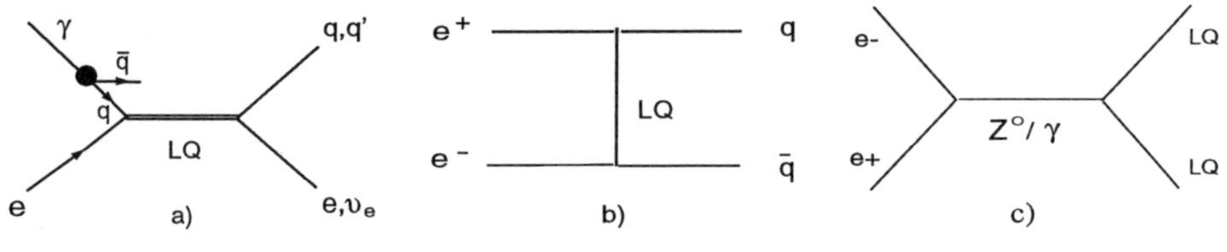

Fig. 1. Feynman diagrams for: a) single leptoquark production; b) t-channel exchange of a leptoquark and c) leptoquark pair production.

leptoquark production process (figure 1.a), a quasi-real photon radiated by one of the

LEP beams fluctuates into a $q\bar{q}$ system. First generation leptoquarks of charge [a]$-1/3$, $-5/3$ can then be produced through the interaction of the other beam electron and the up-type quarks deriving from the fluctuation of the photon. Production of charge $-2/3$ and $-4/3$ leptoquarks is suppressed by the fact that the d quark content in the photon is approximatly a factor four smaller. The cross-section for single leptoquark production depends both on the Yukawa coupling and on the mass of the leptoquark. The mass sensitivity of this channel is close to $\sqrt{s}$. Above the kinematic limit for direct production, existence of leptoquarks might be inferred through a change of the total cross-section of the process $e^+e^- \to q\bar{q}$, via the additional contribution from the t-channel virtual leptoquark exchange diagram shown in figure 1.b. Being a t-channel process, the virtual exchange channel is sensitive to leptoquark masses which can be much higher than $\sqrt{s}$. Finally, the pair production mechanism (figure 1.c) has the advantage of being independent of the leptoquark Yukawa coupling to fermions, since only the coulping to gauge bosons is involved, but its mass sensitivity, approximatly $\sqrt{s}/2$, is lower compared to the other two channels. In this report results on single leptoquark production and virtual exchange are presented, based on data collected with the OPAL Detector at energies above the Z^0 resonance.

2. Single Leptoquark Production

The total cross-section for the production of scalar leptoquarks of mass M is a function of the Yukawa coupling λ and can be expressed [1] as the convolution of the Weizsäcker-Williams effective photon distribution $f_{\gamma/e}(z)$, where z is the momentum fraction carried by the photon, and the parton distribution functions $f_{q/\gamma}(M^2/zs, \mu^2)$ of the photon evaluated at the scale $\mu = M$:

$$\sigma(e^+e^- \to LQ + X) = \frac{\lambda^2\pi}{2s}\int_{M^2/s}^{1}\frac{dz}{z}f_{\gamma/e}(z)f_{q/\gamma}(M^2/(zs), M^2). \qquad (1)$$

For the calculation of the total cross-section and selection efficiencies of the signal process, the PYTHIA5.722 [2] generator, modified to include $LQ \to \nu_e d$ decays in addition to the standard $LQ \to eu$ decays, is used. The description of the parton distribution functions is based on the GRV parametrisation [3]. At $\sqrt{s} = 172$ GeV the expected cross-section for $\lambda = \sqrt{4\pi\alpha_{em}}$ varies from 6 to 0.07 pb for leptoquark masses ranging from 60 to 160 GeV. The limits presented here refer to leptoquarks with charges of $-1/3$ or $-5/3$, for which the cross-sections in $e\gamma$ scattering are identical, and apply only to scalar states, as vector leptoquarks can currently not be simulated by PYTHIA. Both the leptoquark decays into an electron and a quark or into a neutrino and a quark are considered in the analysis.

[a]Charge conjugation is implied.

2.1. Data sample and Analysis

The analysis is based on a data sample corresponding to an integrated luminosity of 20.5 pb^{-1}, almost equally shared among centre of mass energies of $\sqrt{s} = 161$ and 172 GeV. Jets are reconstructed from tracks and calorimeter clusters using a cone finding algorithm and their transverse energy E_T is required to be greater than 15 GeV. In the LQ $\rightarrow$ eq decay channel the electron , which is usually reconstructed as a jet, is identified using a neural net algorythm [4]. Tracks and calorimeter clusters are also used to calculate invariant masses and to determine the missing transverse energy $\not{E}_T$ of the event. Contamination, which is studied using Monte Carlo generators, is expected from two-fermion events like $e^+e^- \rightarrow q\overline{q}$, $e^+e^- \rightarrow \tau^+\tau^-$, and $e^+e^- \rightarrow e^+e^-$, and from W^+W^- pair production and other 4 fermion processes. Both the signal and background Monte Carlo samples are processed through the full OPAL detector simulation.

The event topology of a leptoquark decaying into an electron and a quark is characterized by two jets, where one hadronic jet, with high transverse energy and high multiplicity, is due to the quark, while the other is essentially due to the electron. The event is required to have exactly two jets and a minimum number of five reconstructed tracks, to reject background from deep-inelastic eγ and $e^+e^- \rightarrow \tau^+\tau^-, e^+e^-$ events. The total energy measured in the electromagnetic calorimeter is also required to be less than $0.9 \cdot \sqrt{s}$ to reject $e^+e^- \rightarrow e^+e^-$ events where one of the electrons showers in the detector material. Deep-inelastic eγ scattering events are further rejected by requiring the angle of the electron to satisfy $|\cos\theta_e| < 0.85$. Candidate events are then selected according to the following conditions:

- The missing transverse energy $\not{E}_T$ is required to be less than 15 GeV to reduce contamination from $\tau^+\tau^-$ and W^+W^- pair production.
- The electron is required to be isolated by demanding $|E_e^j - E_e| < 4$ GeV, where E_e^j is the energy of the jet containing the electron and E_e is the energy of the electron. The majority of the multihadronic $e^+e^- \rightarrow q\overline{q}$ events is rejected by this cut.
- The total multiplicity n_q of the other jet in the event, which is considered to be the quark jet, is required to be $n_q \geq 7$, where n_q is the total number of tracks and calorimeter clusters associated to this jet.

The leptoquark selection efficiency varies from 38% to 55% for leptoquark masses between 60 and 140 GeV. After all cuts four events are observed in the data, with an expected background of 5.2 ± 0.4 events from all Standard Model processes. In figure 2.a the two jets invariant mass M_{jj} distribution of the four candidate events

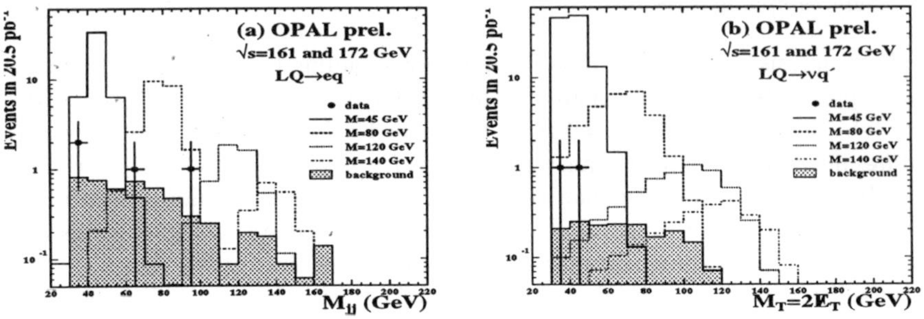

Figure 2: Number of events expected for $\lambda = \sqrt{4\pi\alpha_{\rm em}}$ and $M = 45, 80, 120$ and 140 GeV (histograms) and the candidate events (data points) in (a) LQ $\to$ eq channel and (b) LQ $\to \nu_{\rm e}$q$'$ channel. The sum of all background contributions expected from the simulation of the Standard Model processes is also shown.

is shown together with the sum of all Monte Carlo background distributions. The distribution of the signal for different leptoquark masses in case of $\lambda = \sqrt{4\pi\alpha_{\rm em}}$ is also shown.

The event topology of a leptoquark decaying into a neutrino and a quark is characterized by a single hadronic jet in the detector, whose transverse energy $E_{\rm T}^{\rm j}$ is balanced by the neutrino. To reject events where the missing transverse energy is mismeasured due to poorly reconstructed tracks, the ratio of the total electromagnetic energy to the total visible energy $E_{\rm vis}$ is required to be larger than 20 %. Exactly one jet has to be found in the event and, as in the eq channel, only events with a minimum number of five reconstructed tracks are accepted. Candidate events are then selected by the following requirements:

- The transverse energy $E_{\rm T}^{\rm j}$ of the jet is required to be greater than 15 GeV, and its direction is constrained within a pseudorapidity range $|\eta^{\rm j}| < 2$.
- The missing transverse energy $\not{E}_{\rm T}$ has to be greater than 15 GeV and should balance the energy from the hadronic jet. The two conditions $|E_{\rm T}^{\rm j} - \not{E}_{\rm T}| < 3$ GeV and $E^{\rm j}/E_{\rm vis} > 0.5$ are imposed.
- The jet is required to be an hadronic jet by asking $n_{\rm q} \geq 7$.

The efficiency to detect a leptoquark in the $\nu_{\rm e}$q$'$ channel is higher compared to that in the eq channel, as no electron identification is required, and varies from 55% to 69% for leptoquark masses between 60 and 140 GeV. Two candidates are selected, against an expected background of 1.81 ± 0.05 events. The transverse mass $M_{\rm T} = 2\not{E}_{\rm T}$, which is used to estimate the leptoquark mass, is shown in figure 2.b for the two candidate events, together with the background distribution from the Monte Carlo. The expected distribution of the leptoquark signal for different masses is also shown.

2.2. Results

The 95 % confidence level upper limit is derived taking into account the candidates, the background, the mass resolution and the systematic errors, which have however a negligible impact on the results. The dominant systematic uncertainty (4%) arises from the model dependence of the leptoquark fragmentation and is estimated by varying the cut on the average charged multiplicity by one unit in the Monte Carlo. The limit is calculated separately for $\beta = 1$ and for $\beta = 0.5$, where $\beta = \mathrm{BR}(\mathrm{LQ} \to \mathrm{eq})$ is the branching fraction of the leptoquark in charged leptons. In the case $\beta = 0.5$ both the eq and $\nu_e q'$ channels were considered in the limit calculation. The 95 % confidence level upper limit of the coupling λ as a function of the leptoquark mass M is shown in figure 3, for the two cases $\beta = 0.5$ and $\beta = 1$. Assuming $\lambda = \sqrt{4\pi\alpha_{em}}$, a lower mass limit of $M > 131$ GeV is derived both for $\beta = 0.5$ and $\beta = 1$.

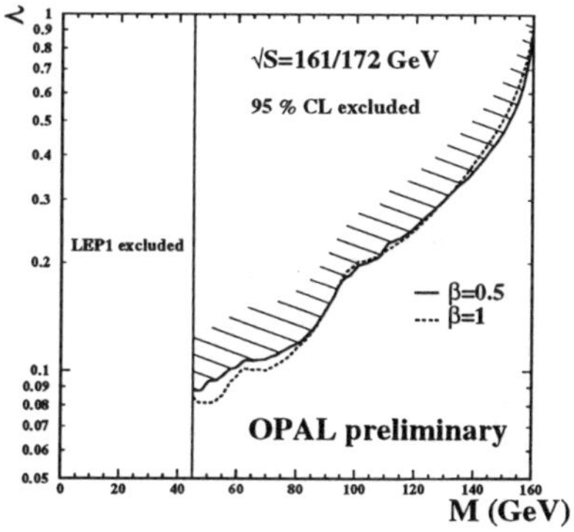

Figure 3: Upper limit at 95 % CL of the coupling λ as a function of the mass M of the leptoquark, for the two cases $\beta = \mathrm{BR}(\mathrm{LQ} \to \mathrm{eq}) = 0.5, 1$.

3. Virtual Leptoquark Exchange

The differential born-level cross-section for $e^+e^- \to q\bar{q}$ including the t-channel exchange of a scalar leptoquark with mass M and coupling λ_{1jk}, where the first index is fixed by the incoming electrons, is given by [5]:

$$\frac{d\sigma}{d\cos\theta} = \sigma_{SM} - \frac{3\alpha_{em}\lambda_{1jk}^2}{4}\left[q_e q_f + \frac{1}{4}\frac{v_e v_f + a_e a_f}{\sin^2\theta_W\cos^2\theta_W}\frac{s(s - M_{Z^0}^2)}{(s - M_{Z^0}^2)^2 + M_{Z^0}^2\Lambda_{Z^0}^2}\right]\frac{u^2}{s^2(u - M^2)} + \frac{3\lambda_{1jk}^4}{64\pi}\frac{s}{(u - M^2)^2}\left(\frac{u}{s}\right)^2 \quad (2)$$

where v_f, a_f are the vector and axial vector electroweak couplings and $u = -\frac{s}{2}(1 + \cos\theta)$, with θ being the polar angle of the outgoing fermion with respect to the e^- beam direction. The dominant contribution arises from the interference term between the Standard Model process and the new physics channel, while the pure t-channel term is usually negligible, except for very low leptoquark masses and large Yukawa couplings. For up-type (down-type) quarks the sign of this interference is negative (positive), so that the overall effect is a reduction (increase) of the cross-section.

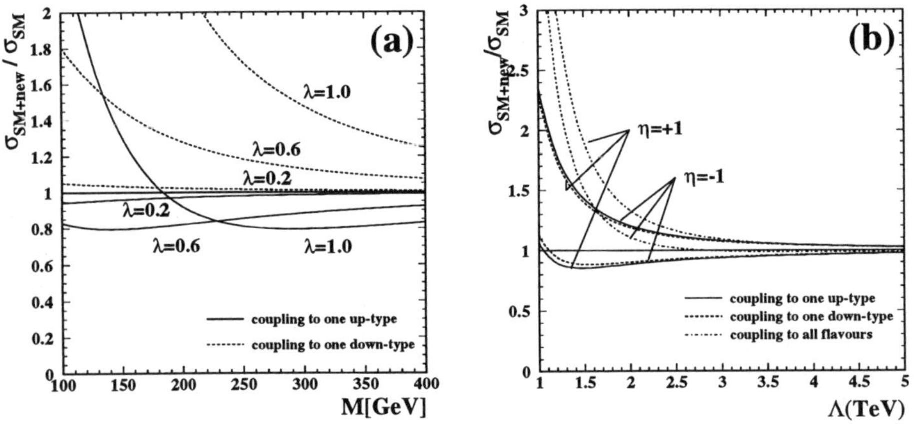

Figure 4: Relative change in the $e^+e^- \to q\bar{q}$ cross-section for the case of (a) the t-channel exchange of a leptoquark for different coupling constants λ as a function of the mass M of the leptoquark; (b) for the LL model of the Contact Interaction as a function of the energy scale Λ. Shown are the three cases of coupling to one up-type (solid line), one down-type (dashed line) and all five flavours (dash-dotted line), for the $\eta = +1$ and $\eta = -1$ case, and under the assumption that $g^2/4\pi = 1$.

In the limit of very large leptoquark masses, $M \gg \sqrt{s}$, the exchange of the leptoquark can be viewed within the very general framework of an effective 4-fermion Contact Interaction [6], characterized by a coupling constant g and an energy scale Λ. Connection to the t-channel case is given by $\lambda/M = g/\Lambda$. The differential cross-section can be expressed as [7]:

$$\frac{d\sigma}{d\cos\theta} = \sigma_{\rm SM}(s,t) + C_2^0(s,t)\frac{\eta_{ij}g^2}{(4\pi\Lambda^2)} + C_4^0(s,t)\frac{(\eta_{ij}g^2)^2}{(4\pi\Lambda^2)^2}\,. \tag{3}$$

Here $t = -\frac{s}{2}(1-\cos\theta)$. The parameters η_{ij}, where the indeces refer to the chirality of the incoming and the outgoing fermions, depend on the chiral structure of the interaction and are free; typical values are between -1 and $+1$. The C_2^0 term describes the interference between the Standard Model and the Contact Interaction, while the C_4^0 term is the pure Contact Interaction contribution. In order to compare the models with the data, the lowest order expressions given in eq. 2 and 3 are convoluted with the effect of initial state radiation [8] and electroweak corrections are taken into account by evaluating the cross-sections with the appropriate effective $\sin^2\theta_W^{\rm eff}$. The Standard model cross-sections $\sigma_{\rm SM}$ are derived from the semi-analytical program ZFITTER [9]. In figure 4 the relative change expected in the total $q\bar{q}$ cross-section from the additional contribution due to the leptoquark t-channel exchange (a) and to the Contact Interaction (b) is shown.

3.1. Data sample and Analysis

The data sample used in the analysis corresponds to an integrated luminosity of approximatly 25 $\rm pb^{-1}$. The centre of mass energies with the respective integrated luminosities are listed in table 1.

Table 1: Centre of mass energies and the respective integrated luminosities of the data considered in the analysis (top rows). The measured non-radiative $q\bar{q}$ cross-sections and $b\bar{b}$ partial widths R_b (middle rows) together with their Standard Model prediction (bottom rows), are also listed. Errors are split in the statistical and the systematic component.

$\sqrt{s}$	133 GeV	161 GeV	172 GeV
$\mathcal{L}$ ($\rm pb^{-1}$)	5.34	10.00	10.30
$\sigma_{q\bar{q}}$ (pb)	62.8±3.5±1.6	35.3±2.0±0.7	26.5±1.8±1.2
R_b	0.216 ± 0.034 ± 0.013	0.141 ± 0.028 ± 0.012	0.127 ± 0.032 ± 0.011
$R_b{}^{SM}$	0.182	0.168	0.165
$\sigma_{q\bar{q}}^{SM}$ (pb)	71.0	33.2	27.5

Events are selected as $e^+e^- \rightarrow q\bar{q}$ decays on the basis of track multiplicity and visible energy in the event, as described in [10]. In e^+e^- collisions at energies above the Z^0 the majority of the events is characterized by hard initial state radiation, due to the tendency for radiative return to the Z^0. The effective interaction energy $\sqrt{s'}$ of the e^+e^- system, which is derived from a kinematic fit to the energy of the tagged photons and to the angles of the hadronic jets, is required to be close to the actual centre of mass energy $\sqrt{s}$. Events are then defined as "non-radiative" by the condition $s'/s > 0.8$. The overall selection efficiency, which is determined from Monte Carlo, is 84%-92% depending on the centre of mass energy. The major contamination (6%-20%) comes from W^+W^- pairs, which is only relevant in the 161 GeV and 172 GeV data; small additional background (<3%) is expected from other four-fermion processes, like two-photon interactions. The dominant systematic error (2%-5%) arises from the s' reconstruction. Other systematic effects are related to uncertainties in the detector simulation (1%), and to background from two-photon (0.8%) and W^+W^- pair events (0.4%). In table 1 the measured non-radiative $q\bar{q}$ hadronic cross-sections are listed for each centre of mass energy, together with the expectations from the Standard Model.

In addition, the fraction R_b of $b\bar{b}$ to to the total non-radiative $q\bar{q}$ cross-section is used to derive limits on the coupling λ_{1j3} involving third generation quarks. The b enriched sample is prepared using the excellent vertex reconstruction capabilities of the OPAL detector. Events are devided in two hemispheres according to the event axis and secondary vertices with at least three tracks are searched within jets, according to the algorythm described in [11]. Hemispheres are tagged as b hemispheres if the maximum reconstructed decay length L divided by its error σ_L satisfies $\delta = L/\sigma_L > 3$. The effects of charm and light flavour background are reduced on a statistical basis by counting the number of hemispheres with $L/\sigma_L < -3$ and subtracting it from the number of tagged hemispheres. In figure 5 the distribution of the decay length significance is shown for the data and the Monte Carlo. The b-purity of the tagged sample is estimated from Monte Carlo to be $\approx$ 75%, for a b-tagging hemisphere efficiency of $\approx$ 40%. At 172 GeV a non-negligible fraction (13%) of the tagged sample comes from W^+W^- events and is subtracted. Systematic errors mainly arise from the

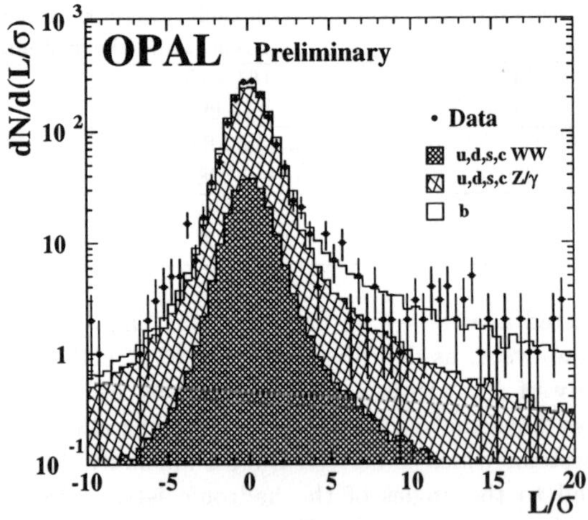

Figure 5: Distribution of the decay length significance L/σ_L in the data (points) and the Monte Carlo (line). The Monte Carlo distribution is separated into the udsc component from W^+W^- decays (densely cross hatched area), from $Z^0\gamma$ events (cross hatched) and from the b contribution (open area).

understanding of the decay properties of b,c hadrons and of the modelling of detector resolution in the Monte Carlo. The measured values of R_b and the relative Standard Model predictions are listed in table 1. The observed cross-sections and $b\bar{b}$ partial widths are in good agreement with the Standard Model expectations.

3.2. Results

The measured values for the total non-radiative $q\bar{q}$ cross-section and the $b\bar{b}$ partial width R_b are converted into a number of observed events using the efficiency and the total luminosity and taking into account the predicted background. The data are fitted under different model assumptions using a likelihood fit, and limits are derived as one-sided 95% confidence limits, equivalent to a change in the likelihood of 1.353. Systematic uncertainties, considered as fully correlated between measurements at different centre of mass energies, are taken into account in the fit.

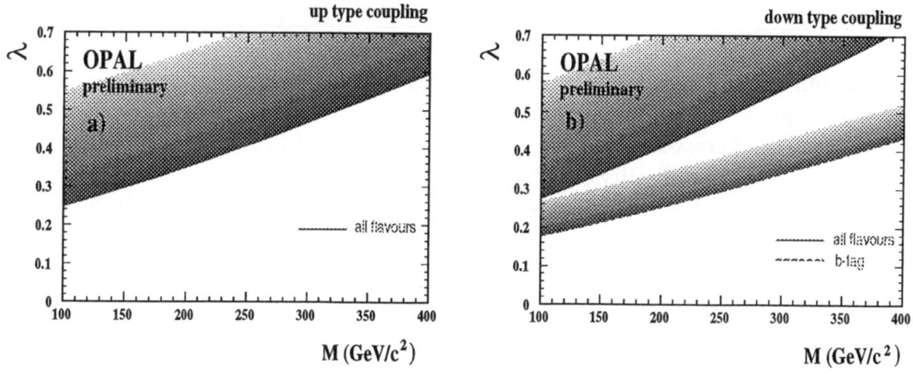

Figure 6: Plots of 95% confidence exclusion limits for λ as a function of M. The limits are shown for the two assumptions that the new particles couple exclusively to one up-type quark (a) or to one down type quark (b), considering the total non-radiative $q\bar{q}$ cross-section (solid line), and the $b\bar{b}$ partial width (dotted line).

The 95% confidence limits on the coupling constant λ are shown for the t-channel analysis in figure 6 as a function of the leptoquark mass M, separately for the two cases of couplings to one up-type (a) or one down-type quark (b). For a mass of 200 GeV and assuming that the new particle only couples to one up-type quark, a 95% confidence level limit of $\lambda_{1jk} < 0.34$ for $j,k = 1,2,3$ is obtained from the inclusive flavour sample. If instead the leptoquark couples only to one down-type quark, a limit of $\lambda_{1jk} < 0.4$ with $j,k = 1,2,3$ from the inclusive flavour sample, and $\lambda_{1j3} < 0.24$ with $k = 3$ from the b enriched sample, is derived. Despite being characterized by a larger uncertainty, the limits obtained from the $b\bar{b}$ partial width are significantly more stringent than the ones from the total cross-section analysis. The sensitivity is infact strongly enhanced by probing directly a specific flavour. Compared to previous searches this analysis improves existing limits in particular in the high mass region for $M > 250\,\mathrm{GeV}$. These limits can be interpreted also as limits for squarks in theories that violate R-parity.

Table 2: Limits on the energy scale Λ in different Contact Interaction models obtained from fits to the total non-radiative $q\bar{q}$ cross-section, under the assumption that the new interaction only couples to one up-type quark, one down-type quark, or is flavour blind.

model	coupling to one u-quark		coupling to one d-quark		coupling to all flavours	
	Λ^+(TeV) ($\eta = +1$)	Λ^-(TeV) ($\eta = -1$)	Λ^+(TeV) ($\eta = +1$)	Λ^-(TeV) ($\eta = -1$)	Λ^+(TeV) ($\eta = +1$)	Λ^-(TeV) ($\eta = -1$)
LL	1.1	2.4	2.4	1.0	2.8	2.1
RR	1.4	1.7	2.1	1.2	2.3	2.5
LR	1.5	1.6	1.7	1.4	2.4	2.4
RL	1.7	1.4	1.6	1.5	2.0	2.9

Within the framework of the Contact Interaction, which is assumed to act only upon one quark, or equally upon all flavours, limits are extracted from the fit to the total $e^+e^- \to q\bar{q}$ cross-section for four helicity structures LL, LR, RL and RR, where L and R refer to the chirality of the initial and final fermion respectively. Results are given in table 2. The 95% CL lower limits for the energy scale Λ range from 1 TeV to 3 TeV, assuming $g^2/4\pi = 1$ for the interaction coupling strenght.

4. References

1. M. A. Doncheski, S. Godfrey, Phys. Lett. **B 393** (1997) 355; hep-ph/9703285.
2. T. Sjöstrand, Comp. Phys. Commun. **82** (1994) 74.
3. M. Glück, E. Reya, A. Vogt, Phys. Rev. **D 46** (1992) 1973;
   Phys. Rev. **D 45** (1992) 3986.
4. OPAL Collaboration, G. Alexander et al., Z. Phys. **C 70** (1996) 357.
5. H.Dreiner and Morawitz, hep-ph/9703279 and private communication.
6. E. Eighten, K. Lane and M. Peskin, Phys. Rev. Lett. **50** (1983) 811.
7. OPAL Collaboration, G. Alexander et al., Phys. Lett. **B 387** (1996) 432.
8. M. Martinez and R. Miquel, Z. Phys. **C 53** (1992) 115.
9. D. Bardin et al., CERN-TH 6443/92 (May 1992);
   Phys. Lett. **B 255** (1991) 290; Nucl. Phys. **B 351** (1991) 1;
   Z. Phys. **C 44** (1989) 493.
10. OPAL Collaboration, G. Alexander et al., Phys. Lett. **B 376** (1996) 232;
    OPAL Collaboration, K. Ackerstaff et al., Phys. Lett. **B 391** (1997) 221.
11. OPAL Collaboration, R. Akers et al., Z. Phys. **C 65** (1995) 17.

THEORETICAL INTERPRETATIONS OF THE HERA HIGH-Q^2 EVENTS

R. RÜCKL
Institut für Theoretische Physik, Universität Würzburg
D-97074 Würzburg, Germany

and

H. SPIESBERGER
Fakultät für Physik, Universität Bielefeld, D-33501 Bielefeld, Germany

ABSTRACT

Theoretical interpretations of the high-Q^2 events recently observed in deep-inelastic positron-proton scattering at HERA are reviewed. We discuss leptoquarks, squarks with R-parity violating couplings, and contact interactions. Bounds from and implications on other experiments are taken into consideration.

1. The Data

Both HERA experiments, H1 [1] and ZEUS [2], have reported the observation of an excess of events in deep-inelastic positron-proton scattering at large values of Bjorken-x and momentum transfer Q^2, relative to the expectation in the standard model. Including the new data presented recently at the 1997 Lepton-Photon Symposium [3,a], H1 and ZEUS each observe 18 neutral current (NC) events at $Q^2 > 1.5 \cdot 10^4$ GeV^2, while H1 expects 8.0 ± 1.2 and ZEUS about 15 events. At H1, the excess is concentrated in the rather narrow mass range 187.5 GeV $\leq M = \sqrt{xs} \leq$ 212.5 GeV where 8 events are observed with 1.53 ± 0.29 expected. However, in the same region, ZEUS finds roughly the expected number of events. Conversely, in the region $x > 0.55$, $y = Q^2/M^2 > 0.25$ where ZEUS finds 5 events with 1.51 ± 0.13 expected, H1 observes no excess. A surplus of events is also observed in charged current (CC) scattering, although with smaller statistical significance. At $Q^2 > 10^4$ GeV^2, H1 and ZEUS together find 28 events and expect 17.7 ± 4.3.

The clustering of the H1 events at a fixed value of $M = \sqrt{xs}$ could indicate the production of a resonance with leptoquark quantum numbers and mass $M \simeq 200$ GeV. On the other hand, ZEUS has 4 events clustered at a somewhat higher mass $M \simeq 225$ GeV. Given the experimental mass resolution of 5 and 9 GeV, respectively, it appears unlikely that both signals come from a single narrow resonance [3,5]. Rather, the excess may be a continuum effect reminiscent of contact interactions. Although the anomalous number of events is not large enough to clearly exclude statistical fluctuations as the origin of the effect, and despite of the puzzling differences among the H1 and ZEUS data, it is important to investigate possible interpretations within

[a] For a detailed discussion of the 1994-96 data see, for example, Ref. [4]

and beyond the standard model.

Leaving aside very small uncertainties in electroweak parameters and radiative corrections [1,2], the main theoretical uncertainty on the high-Q^2 cross sections in the standard model come from the structure functions. The latter are obtained by extrapolation of measurements at lower Q^2 using next-to-leading order evolution equations. For presently available parametrizations the HERA collaborations have estimated an uncertainty of about 7% from this source [1,2]. Attempts [6] to add to the conventional parton densities a new valence component at very large x but low Q^2, and to feed down this enhancement to lower x by evolution to very high Q^2, fail to increase the cross sections by a sufficient amount because of the constraints put by the fixed-target data. Explanations based on a strong intrinsic charm component [7] produced by some nonperturbative effect do not seem to be more successful. In fact, up to this day no standard model mechanism is known which could explain the observed surplus of events. Moreover, no sign of a deviation from the perturbative evolution of structure functions in QCD up to $Q^2 \simeq 10^4$ GeV^2 has so far been found in the data. Whatever mechanism is responsible for the HERA anomaly, it must have quite a rapid onset.

Therefore, if the excess is not a statistical fluctuation, it is very likely produced by some new physics beyond the standard model. Then the question arises whether one is dealing with a (not necessarily single) resonance or with a continuum effect. In the following, we give a brief overview of the main hypotheses put forward.

2. Leptoquarks

The most exciting speculation is the one of a possible discovery of a new particle. Being supposedly produced as a s-channel resonance in e^+q or $e^+\bar{q}$ collisions, this new member of the particle zoo must be a boson and carry simultaneously lepton and quark quantum numbers. Such species are generically called leptoquarks. In addition to their couplings to the standard model gauge bosons, they are usually assumed to have Yukawa-type couplings to lepton-quark pairs:

$$\lambda_i \phi_{LQ} \bar{l}_i q \quad \text{or} \quad \lambda_i \phi_{LQ} \bar{l}_i q^c \ , \tag{1}$$

and similarly for vector LQs. Here, c denotes charge conjugation, and $i = L, R$ specifies the fermion handedness.

With the above couplings the resonance cross section in ep scattering is given by

$$\sigma = N \frac{\pi}{4s} \lambda^2 q_f(M^2/s, \mu^2) \ , \tag{2}$$

where $q_f(x, \mu^2)$ is the density of quarks (or antiquarks) with flavour f in the proton, and $N = 1\,(2)$ for scalars (vectors). The relevant scale μ is expected to be of order M. Obviously, leptoquarks with fermion number $F = 0\,(2)$ are dominantly produced from valence quarks in $e^+q\,(e^-q)$ fusion.

Having only couplings to standard model particles, leptoquarks decay exclusively to lepton-quark pairs. The partial width per channel is given by

$$\Gamma = \frac{a}{16\pi}\lambda^2 M = 350\,\mathrm{MeV}\,a\left(\frac{\lambda}{e}\right)^2\left(\frac{M}{200\,\mathrm{GeV}}\right)\,, \tag{3}$$

a being 1 for scalars and 2/3 for vectors. Hence, leptoquarks are very narrow for masses in the range accessible at HERA, and for couplings weaker than the electromagnetic coupling strength $e = \sqrt{4\pi\alpha}$.

Leptoquarks appear in extensions of the standard model involving unification, technicolor, compositeness, or R-parity violating supersymmetry. In the generally adopted framework described in Ref. [8], the Yukawa couplings are taken to be dimensionless and $SU(3) \times SU(2) \times U(1)$ symmetric. Moreover, they are assumed to conserve lepton and baryon number in order to avoid rapid proton decay, to be non-zero only within one family in order to exclude FCNC processes beyond the CKM mixing, and chiral in order to avoid the very strong bounds from leptonic pion decays. The allowed states can be classified according to spin, weak isospin and fermion number. They are summarized in Table 1.

The leptoquark masses and couplings are constrained by high-energy data. Direct searches for leptoquarks have been performed at the Tevatron, at HERA and at LEP. Recently, both experiments, CDF and D0, have improved their mass limits for scalar leptoquarks considerably. D0 excludes first generation leptoquarks with masses below 225 GeV assuming a branching ratio $B_{eq} = 1$ for decays into $e^{\pm}$ and a jet [9], whereas CDF quotes a limit of 213 GeV [10] (all mass limits are at 95 % CL). For branching ratios less than one, the limits are weaker, e.g., $M > 176$ GeV for $B_{eq} = 0.5$ [9]. The bounds on vector states are even stronger: 298 GeV for $B_{eq} = 1$ and 270 GeV for $B_{eq} = 0.5$ [11]. The corresponding bounds on second and third generation scalar leptoquarks are $M > 184$ GeV for $B_{\mu q} = 1$ and $M > 98$ GeV for $B_{\tau q} = 1$, respectively [12]. The above constraints follow from pair-production mainly by $q\bar{q}$ annihilation, and are therefore practically independent of the unknown Yukawa coupling λ. In contrast, the mass limits obtained at HERA depend on λ and the quantum numbers specified in Table 1. For $\lambda = e$ they range from 207 to 272 GeV [13]. These limits are lowered by about 50 GeV if $\lambda = 0.1$. Finally, the most stringent but again λ-dependent mass bound at LEP2 comes from the search for single-leptoquark production and excludes masses below 131 GeV assuming $\lambda > e$ [14]. The mass limits from leptoquark pair production [15] roughly reach half of the center-of-mass energy $\sqrt{s}/2$, and are thus much weaker than the Tevatron bounds.

Indirect bounds on Yukawa couplings and masses can be derived from t/u-channel LQ-exchange in $e^+e^- \to q\bar{q}$ [16], and from low-energy data [17]. From the very recent analysis by OPAL we infer upper limits on λ between 0.2 and 0.7 assuming $M = 200$ GeV. However, the most restrictive bounds come from atomic parity violation and lepton and quark universality, at least for first generation leptoquarks and chiral

LQ		Q	Decay Mode	BR $e^{\pm} j$	Coupling $\lambda_{L,R}$	Limits Ref. [17]	HERA estimates
S_0	$\tilde{d}_R$	$-1/3$	$e_L u$	$\frac{1}{2}$	g_L	$g_L < 0.06$	0.40
			$\nu_L d$		$-g_L$		
			$e_R u$	1	g_R	$g_R < 0.1$	0.28
$\tilde{S}_0$		$-4/3$	$e_R d$	1	g_R	$g_R < 0.1$	0.30
S_1		$+2/3$	$\nu_L u$	0	$\sqrt{2} g_L$	$g_L < 0.09$	–
		$-1/3$	$\nu_L d$	$\frac{1}{2}$	$-g_L$		0.40
			$e_L u$		$-g_L$		
		$-4/3$	$e_L d$	1	$-\sqrt{2} g_L$		0.21
$V_{1/2}$		$-1/3$	$\nu_L d$	0	g_L	$g_L < 0.09$	–
			$e_R u$	1	g_R		0.30
		$-4/3$	$e_L d$	1	g_L		0.32
			$e_R d$		g_R	$g_R < 0.05$	0.32
$\tilde{V}_{1/2}$		$+2/3$	$\nu_L u$	0	g_L	$g_L < 0.09$	–
		$-1/3$	$e_L u$	1	g_L		0.32
$S_{1/2}$		$-2/3$	$\nu_L \bar{u}$	0	g_L	$g_L < 0.1$	–
			$e_R \bar{d}$	1	$-g_R$		0.052
		$-5/3$	$e_L \bar{u}$	1	g_L		0.026
			$e_R \bar{u}$		g_R	$g_R < 0.09$	0.026
$\tilde{S}_{1/2}$	$\bar{\tilde{d}}_L$	$+1/3$	$\nu_L \bar{d}$	0	g_L	$g_L < 0.1$	–
	$\bar{\tilde{u}}_L$	$-2/3$	$e_L \bar{d}$	1	g_L		0.052
V_0		$-2/3$	$e_L \bar{d}$	$\frac{1}{2}$	g_L	$g_L < 0.05$	0.080
			$\nu_L \bar{u}$		g_L		
			$e_R \bar{d}$	1	g_R	$g_R < 0.09$	0.056
$\tilde{V}_0$		$-5/3$	$e_R \bar{u}$	1	g_R	$g_R < 0.09$	0.027
V_1		$+1/3$	$\nu_L \bar{d}$	0	$\sqrt{2} g_L$	$g_L < 0.04$	–
		$-2/3$	$e_L \bar{d}$	$\frac{1}{2}$	$-g_L$		0.080
			$\nu_L \bar{u}$		g_L		
		$-5/3$	$e_L \bar{u}$	1	$\sqrt{2} g_L$		0.019

Table 1: *Scalar (S) and vector (V) leptoquarks, and their electric charges Q, decay modes, branching ratios into charged lepton + jet channels, and Yukawa couplings. Given are also the most stringent low-energy bounds and the couplings deduced from the 1994-96 HERA data* [18]. *Inclusion of the 1997 data decrease the couplings by about 15%. Using the H1 data alone would roughly give the couplings shown above. Also shown are the possible assignments of squarks with R-parity violating couplings.*

couplings. The maximum allowed couplings for $M = 200$ GeV are given in Table 1 [18].

In order to explain the observed excess of high-Q^2 events at HERA by the production and decay of a 200 GeV leptoquark, one roughly needs $\lambda \simeq e$ for $F = 2$ states and $\lambda \simeq e/10$ for $F = 0$. The factor 10 difference in λ simply reflects the factor 100 difference in the sea and valence quark densities in the region of x and Q^2 where the signal resides. Similarly, the coupling of $F = 0$ leptoquarks to the d quark has to be two times larger than the coupling to the u quark in order to compensate the factor four difference in the corresponding quark densities. These simple rules of thumb describe the main pattern in the couplings found in detailed analyses [19], and shown in the last column of Table 1.

Whereas the coupling strength λ required for $F = 0$ leptoquarks is compatible with all existing bounds, the coupling necessary for $F = 2$ leptoquarks is already excluded by the low-energy constraints, and also at the borderline of getting in conflict with LEP2 data. Moreover, with such strong couplings, $F = 2$ leptoquarks should have shown up in e^-p scattering at HERA [20], where they can be produced off the valence quark component, despite of the low luminosity of the previous e^-p run. Since vector leptoquarks cannot be made responsible for an excess of events at $M \simeq 200$ to 225 GeV because of the high Tevatron mass bounds, only the two scalar doublets $S_{1/2}$ and $\tilde{S}_{1/2}$ remain from the whole Table 1 as a possible interpretation. However, also these solutions have difficulties. Firstly, the Tevatron mass limits require scalar leptoquarks of the first generation with $M \simeq 200$ GeV to have branching ratios into $e + jet$ final states less than about 0.7, whereas in the simple framework considered in Table 1, $S_{1/2}$ and $\tilde{S}_{1/2}$ are expected to have $B_{eq} = 1$. Secondly, as an immediate consequence, these states cannot give rise to CC events.

Thus it seems that the leptoquark interpretation of the HERA high-Q^2 events points to more complicated and maybe more realistic scenarios. Several possibilities have been suggested allowing for $B_{eq} < 1$ and, at the same time, providing CC final states: leptoquarks with interfamily couplings [21,22], leptoquark mixing [23], LQ models with additional vector-like fermions [24], and squarks with R-parity violating couplings [25].

3. Squarks

Supersymmetry with R-parity violation provides an attractive theoretical framework in which squarks can have direct couplings to lepton-quark pairs, and therefore act as leptoquarks. However, because of the usual R-parity conserving couplings one naturally expects the branching ratio for $\tilde{q} \to e + jet$ to be smaller than unity. In addition, one can get CC-like final states, e.g., through the decay chain $\tilde{q} \to q\chi \to q\nu + \cdots$, χ being either a neutralino or chargino. Thus it appears possible to avoid the two main problems encountered in the simplest leptoquark models can be avoided.

ijk	C	n,m	source
111	0.004	$2,\frac{1}{2}$	ν-less $\beta\beta$ decay
112 113	0.04	1,0	CC universality
121 131	0.07	1,0	atomic P-violation
122 133	0.08 0.003	$\frac{1}{2}$,0	ν_e mass
123	0.28	$\frac{1}{2}$,0	$D-\overline{D}$ mixing
132	0.68	1,0	R_e (LEP)

Table 2: *Low-energy constraints on R-parity violating couplings* λ'_{ijk} *relevant for* e^+p *scattering* ($M_{\tilde{q}} = 200$ *GeV*, $m_{\tilde{g}} = 1$ *TeV*) [29].

In the minimal supersymmetric extension of the standard model, one can have a renormalizable, gauge invariant operator in the superpotential that couples squarks to quarks and leptons:

$$W_{\not{R}} = \lambda'_{ijk} L^i_L Q^j_L \overline{D}^k_R. \tag{4}$$

Here, L and Q denote doublets of lepton and quark superfields, respectively, D stands for singlets of d-quark superfields, and i, j, and k are generation indices. This interaction term violates global invariance of R-parity, defined as $R = (-1)^{3B+L+2S}$ which is $+1$ for particles and -1 for superpartners. In general, there are other R-odd operators in the superpotential that couple sleptons to leptons and squarks to quarks. Together, they may induce rapid proton decay. This can be avoided by requiring conservation of R-parity, or a strong hierarchy in the various couplings. Generally, these two options lead to very different phenomenology.

Expanding the superfields in (4) in terms of matter fields, one finds interaction terms which allow for resonance production of squarks at HERA [26]:

$$e^+ d^k_R \to \tilde{u}^j_L \qquad (\tilde{u}^j = \tilde{u}, \tilde{c}, \tilde{t}), \tag{5}$$

$$e^+ \bar{u}^j_L \to \bar{\tilde{d}}^k_R \qquad (\tilde{d}^k = \tilde{d}, \tilde{s}, \tilde{b}). \tag{6}$$

The cross sections are determined by the coupling constants λ'_{1jk}. Similarly as the leptoquark Yukawa couplings $\lambda_{L,R}$, these couplings are strongly constrained by existing data. The relevant bounds are summarized in Table 2. As already pointed out, since the excess of events was observed in e^+p but not in e^-p scattering, the process of class (6) involving the $\bar{u}$ sea is unlikely. Moreover, the coupling strength $\lambda'_{11k} \simeq e$, required for production off sea quarks, is incompatible with the existing bounds. This

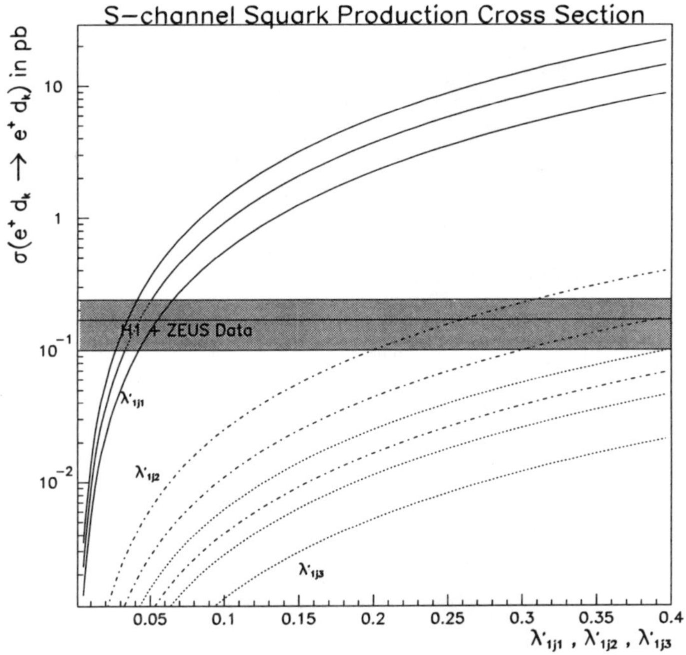

Figure 1: *Cross section for* $e^+d_k \to \tilde{u}_j \to e^+d_k$ *as a function of the coupling* λ'_{1jk} *for* d *valence quarks (full), s quarks (dash-dotted) and b quarks (dotted) assuming* $B_{eq} = 1$. *The curves from top to bottom correspond to* $M_{\tilde{u}_j} = 200$, *210, and 220 GeV. The shaded region shows the HERA excess cross section* $\sigma_{exp} = (0.17 \pm 0.07)$ *pb (from Ref.* [27]*).*

also applies to the $e^+\bar{c}$ channel with the marginal exception of the subprocess $e^+\bar{c} \to \bar{\tilde{t}}$ [27]. The top sea plays no role. Turning to the processes of class (5), one finds three possible explanations of the HERA anomaly [25,b] :

$$e^+d \to \tilde{c} \ \ (\lambda'_{121}), \tag{7}$$

$$e^+d \to \tilde{t} \ \ (\lambda'_{131}), \tag{8}$$

$$e^+s \to \tilde{t} \ \ (\lambda'_{132}). \tag{9}$$

The corresponding cross sections are plotted in Fig. 1 for $M_{\tilde{q}} = 200$ to 220 GeV, and setting $B_{eq} = 1$. As can be seen, within the limits on λ' quoted in Tab. 2 one

[b]For a discussion of the strange stop scenario see in particular Ref. [28]

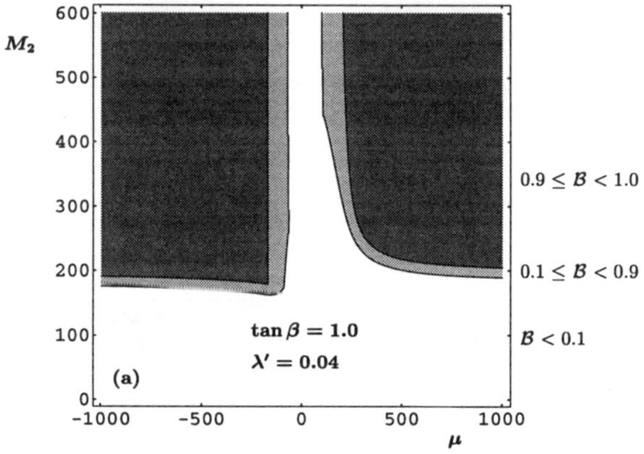

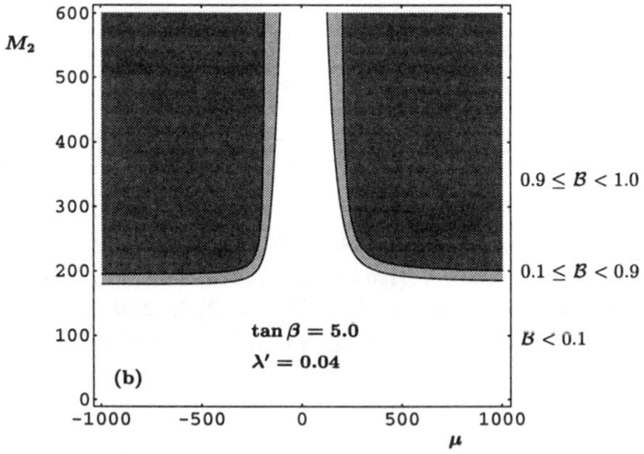

Figure 2: *Contours of $B(\tilde{t} \to e^+ d)$ in the $\mu - M_2$ plane assuming vanishing stop left-right mixing. The LEP2 bound of 85 GeV for the chargino mass is taken into account. From Ref. [31]).*

can still afford a branching ratio below 0.7, necessary in order to avoid the D0/CDF mass bounds. Studies [30,31] have shown that one can find allowed regions in the supersymmetry parameter space in which $B_{eq} < 0.7$. This is exemplified in Fig. 2 for $B(\tilde{t} \to e^+d)$. However, there is not too big a room for a consistent squark interpretation of the HERA anomaly.

Note that the NC events from $\tilde{t}, \tilde{c} \to e^+d$ have the same visible particles as the DIS-NC events. This is not expected for the CC events originating from cascade decays of squarks on the one hand, and DIS-CC events on the other.

Finally, the difficulty to interpret the excess of events as a single-resonance effect may also find a simple solution. In the MSSM each fermion has two superpartners which mix in general. In the case of stop this mixing may be sizeable and lead to two mass eigenstates with a small but pronounced mass difference. Such a case is illustrated in Fig. 3. The resulting mass distribution can apparently mimic a continuum effect.

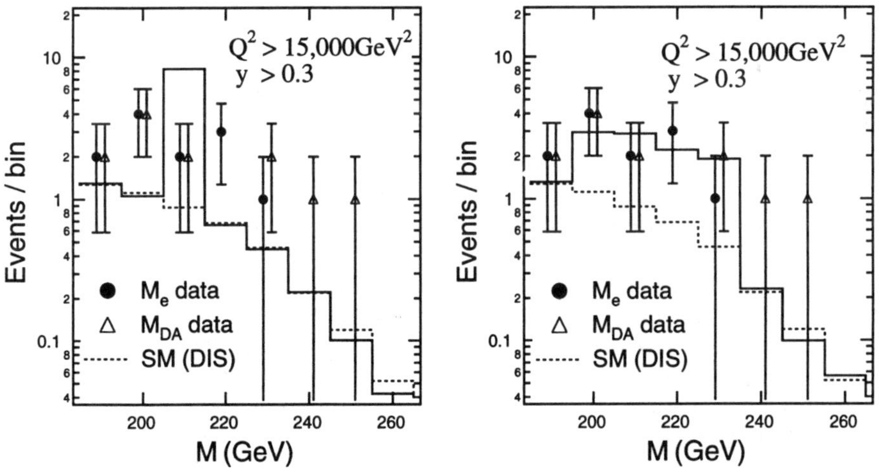

Figure 3: *Distributions in* $M = \sqrt{xs}$ *of the HERA data in comparison with the distributions in a single stop scenario (left,* $M_{\tilde{t}} = 210$ *GeV,* $\lambda'_{131} = 0.04$*) and in a mixed left-right stop scenario (right,* $M_{\tilde{t}_1} = 205$ *GeV,* $M_{\tilde{t}_2} = 225$ *GeV,* $\theta_t = 0.95$, $\lambda'_{131} = 0.045$*). From Ref.* [32].

4. Contact Interactions

With the new 1997 HERA data it has become somewhat more likely that the

Source	Limit
Tevatron (CDF)[36]	$\Lambda > 2.5 \div 6.0$ TeV
HERA (H1)[37]	$\Lambda > 1.0 \div 2.5$ TeV
LEP2 (OPAL)[16]	$\Lambda > 1.1 \div 5.2$ TeV
atomic P-violation[38]	$\Lambda > 7.4 \div 12.3$ TeV
CCFR (for $\nu_\mu \nu_\mu qq$)[39]	$\Lambda > 1.8 \div 7.9$ TeV
CKM unitarity[22]	$\Lambda > 10 \div 90$ TeV

Table 3: *Bounds on the scale of contact terms.*

observed excess of events is due to some continuum mechanism. An appropriate and very general description is provided by contact interactions. They could originate from the exchange of a new heavy particle, or from lepton and quark substructure. For NC lepton-quark scattering, one may use the following effective Lagrangean:

$$\mathcal{L}_{\text{eff}} = \sum_{\substack{i,k=L,R \\ q=u,d,\cdots}} \eta^q_{ik} \frac{4\pi}{(\Lambda^q_{ik})^2} \left(\bar{e}_i \gamma^\mu e_i\right) \left(\bar{q}_k \gamma_\mu q_k\right) \ . \tag{10}$$

At $Q^2 \ll \Lambda^2$, the interference of contact terms with the standard model amplitudes leads to an enhancement or suppression of the NC cross section depending on the sign η_{ik}. The x and Q^2 dependence of the deviations is expected to be rather smooth. In Ref. [33] it was shown that e^+p scattering is particularly sensitive to LR and RL contact terms, while e^-p scattering is better for probing LL and RR helicity structures. Furthermore, it was pointed out that destructive interference leads to a rather sharp onset of the deviations with Q^2. Explanation of the HERA data by contact terms is possible with Λ of the order of 3 TeV [34].

As is obvious from (10), contact interactions in $eq \to eq$ also modify the potential in atoms and affect the crossed channels $e^+e^- \to q\bar{q}$ and $q\bar{q} \to e^+e^-$. Therefore, the existence of contact terms is strongly constrained by atomic parity violation (APV) experiments, hadron production at LEP, and Drell-Yan production at the Tevatron. Moreover, $SU(2) \times U(1)$ symmetry implies the existence of contact interactions involving neutrinos. Typical bounds on Λ from the various sources are summarized in Table 3. The existing constraints on $eeqq$ contact terms still allow a reasonable fit to the excess of NC events as shown in Fig. 4, provided the APV bound is avoided by choosing an P-even combination of contact terms[34]. However, an analogous explanation of the possible excess of CC events by $e\nu qq'$ contact interactions is ruled out[31]. Finally, if this interpretation of the signal in NC e^+p scattering is correct, one should observe a similar effect in e^-p scattering.

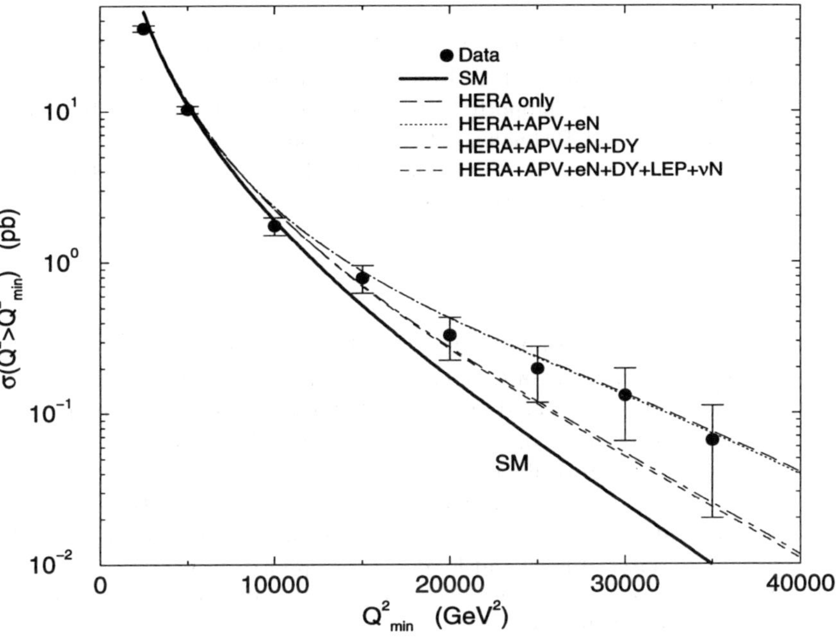

Figure 4: *Fits of contact interactions to the measured cross section* $\sigma(Q^2 > Q^2_{min})$ *for* $e^+p \to e^+X$ *including bounds from low-energy and collider experiments (for details see Ref.* [35]*).*

5. Conclusions

For the time being, it is an open question whether or not the excess of high-Q^2 events observed at HERA is a statistical fluctuation or a physical effect. If it is a real signal, then it very likely originates from new physics beyond the standard model. Making this assumption, the present data slightly favour some continuum mechanism, but do not yet allow to rule out a resonance effect. Both kinds of interpretations are tightly constrained by measurements at LEP2 and the Tevatron, as well as by low-energy data. These bounds rule out the simplest leptoquark scenarios and do also not leave much room for the squark interpretation. Particularly difficult would be the explanation of an excess of CC events. At any rate, if the excess of high-Q^2 events is confirmed by future data it is likely that related signals will soon show up in other experiments.

6. Acknowledgments

We wish to thank J. Kalinowski and P. Zerwas for the fruitful collaboration on the subject of this talk, and T. Kon for providing us with an updated version of Fig. 3. This work was supported by the Bundesministerium für Bildung, Wissenschaft, Forschung und Technologie, Bonn, Germany, Contracts 05 7BI92P (9) and 05 7WZ91P (0). R. R. wants to thank B. Kniehl and G. Kramer for their kind invitation to and hospitality at the workshop.

7. References

1. C. Adloff et al., H1 Collaboration, Z. Phys. C74 (1997) 191.
2. J. Breitweg et al., ZEUS Collaboration, Z. Phys. C74 (1997) 207.
3. B. Straub, Talk presented at the XVIII International Symposium on Lepton Photon Interactions, July 28 – August 1, 1997, Hamburg.
4. Y. Sirois, these proceedings.
5. U. Bassler and G. Bernardi, DESY 97-136 (hep-ex/9707024).
6. S. Kuhlmann, H. L. Lai and W. K. Tung, hep-ph/9704338; K. S. Babu, C. Kolda, J. March-Russell and F. Wilczek, hep-ph/9705399;
7. J. F. Gunion and R. Vogt, hep-ph/9706252; W. Melnitchouk and A. W. Thomas, hep-ph/9707387.
8. W. Buchmüller, R. Rückl and D. Wyler, Phys. Lett. B191 (1987) 442.
9. B. Abbott et al., D0 Collaboration, FERMILAB-PUB-97/252-E (hep-ex/9707033).
10. F. Abe et al., CDF Collaboration, FERMILAB-PUB-97/280-E.
11. J. A. Valls, FERMILAB-CONF-97/135-E, XXXII Rencontres de Moriond, QCD and Hadronic Interactions, Les Arcs, France, 1997.
12. F. Abe et al., CDF Collaboration, Phys. Rev. Lett. 78 (1997) 2906, Phys. Rev. Lett. 75 (1995) 1012; B. Abbott et al., D0 Collaboration, paper 109 submitted to the International Europhysics Conference on High Energy Physics, Jerusalem, 1997.
13. S. Aid et al., H1 Collaboration, Phys. Lett. B369 (1996) 173; M. Derrick et al., ZEUS Collaboration, Phys. Lett. B306 (1993) 173, Z. Phys. C73 (1997) 613.
14. OPAL Collaboration, paper LP138 submitted to the XVIII International Symposium on Lepton-Photon Interactions, Hamburg, 1997.
15. B. Adeva et al., L3 Collaboration, Phys. Lett. B261 (1991) 169; G. Alexander et al., OPAL Collaboration, Phys. Lett. B263 (1991) 123; P. Abreu et al., DELPHI Collaboration, Phys. Lett. B316 (1993) 620.

16. K. Ackerstaff et al., OPAL Collaboration, CERN-PPE/97-101.
17. M. Leurer, Phys. Rev. D49 (1994) 333; *ibid.* D50 (1994) 536; S. Davidson, D. Bailey and B. Campbell, Z. Phys. C61 (1994) 613.
18. J. Kalinowski, R. Rückl, H. Spiesberger and P. M. Zerwas, Z. Phys. C74 (1997) 595.
19. G. Altarelli et al., hep-ph/9703276; J. Blümlein, Z. Phys. C74 (1997) 605; J. Kalinowski, R. Rückl, H. Spiesberger and P. M. Zerwas, Z. Phys. C74 (1997) 595; K. S. Babu, C. Kolda, J. March-Russell and F. Wilczek, Phys. Lett. B402 (1997) 367; J. L. Hewett and T. G. Rizzo, hep-ph/9703337; Z. Kunszt and W. J. Stirling, Z. Phys. C75 (1997) 453.
20. S. Aid et al., H1 Collaboration, Phys. Lett. B369 (1996) 173.
21. Z. Kunszt and W. J. Stirling, Z. Phys. C75 (1997) 453.
22. G. Altarelli, G. F. Giudice and M. L. Mangano, hep-ph/9705287.
23. K. S. Babu, C. Kolda and J. March-Russell, hep-ph/9705414.
24. J. L. Hewett and T. G. Rizzo, hep-ph/9708419.
25. D. Choudhury and S. Raychaudhuri, Phys. Lett. B401 (1997) 54; G. Altarelli et al., hep-ph/9703276; H. Dreiner and P. Morawitz, hep-ph/9703279; J. Kalinowski, R. Rückl, H. Spiesberger and P. M. Zerwas, Z. Phys. C74 (1997) 595; T. Kon and T. Kobayashi, hep-ph/9704221; J. Ellis, S. Lola and K. Sridhar, hep-ph/9705416; T. Kon, T. Matsushita and T. Kobayashi, hep-ph/9707355; M. Carena et al., hep-ph/9707458.
26. J. L. Hewett, Proceedings, *1990 Summer Study on High Energy Physics – Research Directions for the Decade* (Snowmass 1990), ed. E. Berger (World Scientific, 1992); T. Kon, K. Nakamura and T. Kobayashi, Z. Phys. C45 (1990) 567, Acta Phys. Polon. B21 (1990) 315; J. Butterworth and H. Dreiner, Nucl. Phys. B397 (1993) 3; H. Dreiner, E. Perez and Y. Sirois, hep-ph/9703444.
27. H. Dreiner and P. Morawitz, hep-ph/9703279.
28. J. Ellis, S. Lola and K. Sridhar, hep-ph/9705416.
29. H. Dreiner, hep-ph/9707435.
30. D. Choudhury and S. Raychaudhuri, Phys. Lett. B401 (1997) 54.
31. G. Altarelli et al., hep-ph/9703276.
32. T. Kon and T. Kobayashi, hep-ph/9704221.
33. R. Rückl, Phys. Lett. B129 (1983) 363, Nucl. Phys. B234 (1984) 91; P. Haberl, F. Schrempp and H.-U. Martyn, in the Proceedings, *Physics at HERA* (Hamburg, 1991), eds. W. Buchmüller and G. Ingelman.
34. K. S. Babu, C. Kolda, J. March-Russell and F. Wilczek, Phys. Lett. B402 (1997) 367; V. Barger et al., Phys. Lett. B404 (1997) 147, hep-ph/9707412; N. Di Bartolomeo and M. Fabbrichesi, Phys. Lett. B406 (1997) 237; N. G. Deshpande, B. Dutta and X.-G. He, hep-ph/9705236; F. Cornet and J. Rico, hep-ph/9707299.
35. V. Barger et al., hep-ph/9707412.

36. X. Wu, Talk presented at the International Europhysics Conference on High Energy Physics, August 19 – 26, 1997, Jerusalem.
37. S. Aid et al., H1 Collaboration, Phys. Lett. B353 (1995) 578.
38. A. Deandrea, hep-ph/9705435.
39. E. Gallas, Talk presented at the International Europhysics Conference on High Energy Physics, August 19 – 26, 1997, Jerusalem.

Chapter 4

Final States in Deep Inelastic Scattering

Jet Production in Deep-inelastic Scattering at HERA

Tancredi Carli
Max-Planck-Institut für Physik, Werner-Heisenberg-Institut
Föhringer Ring 6, D-80805 München, Germany,
E-mail: h01rtc@rec06.desy.de

ABSTRACT

The main experimental results on jet production at HERA are reviewed. A study of jet shapes shows that the internal structure of jets is well understood. The potential to accurately determine the strong coupling constant using jet rates is discussed. First attempts to extract the gluon density in next-to-leading order are presented. Although fixed order perturbative QCD is able to describe the shape of most variables associated to the hard subprocess leading to jets, it does not reproduce the absolute jet cross sections over a large phase space region.

1. Introduction

HERA colliding 27.5 GeV positrons on 820 GeV proton offers an ideal testing ground for perturbative QCD in deep-inelastic scattering (DIS).

Hard processes can be investigated as function of a variable scale, such that the transition from one phase space region to another or the interplay between two hard scales, like e.g. the photon virtuality Q^2 or the transverse energy of a parton, can be studied. The center of mass energy of $\sqrt{s} \approx 300$ GeV leads to a large phase space for hadron production and to the possibility to observe clean 'jets' in the hadronic final state.

Jets are event properties relating the unobservable quarks and gluons to the measurable hadronic final state. Quantitative studies of jets require an exact definition of how to combine the spray of hadrons observed in the detector to jets. The definition should be suitable for the experimental analysis and for the theoretical calculation. Moreover, it should lead to infrared and collinear safe results, which do not change, if a low energetic particle is added or if a particle is split into two. Jets are usually defined either by cone type algorithms maximizing the transverse energy flowing through a cone in the (η, ϕ) plane[a] or by algorithms successively recombining closest particles. Clustering algorithms iteratively merge pairs of particles until only a few well separated objects are left. The decision to assign a given object to a jet is based on a distance measure d_{ij} and a resolution parameter $y_{cut} = d_{ij}/\text{scale}$. Table 1 summarizes choices for commonly used clustering algorithms. Also shown are various procedures for the addition of particle 4-momenta to obtain the jet four-momentum (recombination schemes).

[a] The pseudo-rapidity is $\eta = -\ln\tan(\theta/2)$, where θ is defined with respect to the incident proton. The proton defines the negative z axis. ϕ is the azimuthal angle.

Name	distance measure d_{ij}	scale	remnant treatment
JADE	$2E_iE_j(1-\cos\theta_{ij})$	W^2	$\sum_h E - P_z$
W	$(p_i+p_j)^2$		
K_T	$2\min(E_i^2, E_j^2)(1-\cos\theta_{ij})$	Q^2 or	$d_{ip} = 2\ E_i^2\ (1-\cos\theta_i)$
long.inv. K_T	$2\min(E_{Ti}^2, E_{Tj}^2)(\Delta\eta_{ij}^2 + \Delta\phi_{ij}^2)$	fixed E_T^2	$d_{ip} = 2\ E_{Ti}^2$

scheme	E or JADE	P	E_0
combination procedure	$p_k = p_i + p_j$	$E_k = \lvert\vec{p_i}+\vec{p_j}\rvert$ $\vec{p_k} = \vec{p_i}+\vec{p_j}$	$E_k = E_i + E_j$ $\vec{p_k} = (E_i+E_j)\,(\vec{p_i}+\vec{p_j})/(\lvert\vec{p_i}+\vec{p_j}\rvert)$

Fig. 1. Summary of the main characteristics of popular jet clustering algorithms. For the K_T algorithm the remnant is considered as a particle p with infinite momentum. In the JADE and W algorithm a pseudo-particle with longitudinal components calculated from the hadrons in the final state is introduced to account for the remnant. The combination procedure defines how the four-momentum $p_i = (E_i, \vec{p}_i)$ of two objects i and j are combined to an object k.

Traditionally cone algorithms have been applied in $p\bar{p}$ collisions where special care has to be taken to separate particles belonging to the beam remnant from particles associated with the hard subprocess. Cone algorithms are invariant under longitudinal boosts. In e^+e^- collisions, where the initial state is free of color charges and where every particle can be assigned to jets, clustering algorithms have been used. In DIS or $p\bar{p}$ collisions they can also be used if an additional particle to mimic the hadron remnant is put in by hand.

At HERA, events with two jets in the central part of the detector plus the jet associated with the proton remnant ($2+1$ jet events) can be produced in a quark ($q\gamma \to qg$) or a gluon ($g\gamma \to q\bar{q}$) initiated hard subprocess (see Fig. 2). The $2+1$ jet cross section is given by[1]:

$$\frac{d^2\sigma_{2+1}}{dx dQ^2} \sim \alpha_s(\mu_r) \int \frac{d\xi}{\xi} (C_{qg}(x/\xi, z_q, x, Q^2)\cdot q(\xi, \mu_f^2) + C_{q\bar{q}}(x/\xi, z_q, x, Q^2)\cdot g(\xi, \mu_f^2)) \quad (1)$$

where α_s is the strong coupling constant, μ_r the renormalisation and μ_f the factorisation scale. x and Q^2 are the usual variables to inclusively describe deep-inelastic scattering (see Fig. 2 for definition). The coefficient functions $C_{q\bar{q}}$ and C_{qg} can be calculated in perturbative QCD. In addition to x and Q^2 they depend[b] on the two variables ξ and z_q characterizing the short distance subprocess which is responsible for the dijet system. The fraction of the longitudinal proton momentum carried by the incoming parton ξ, can be calculated in leading order α_s (LO) from the invariant mass of the jet system $\sqrt{\hat{s}}$:

$$\xi = x\ (1 + \hat{s}/Q^2) \quad (2)$$

The variable z_q is given by:

$$z_q = \frac{P\cdot j_1}{P\cdot q} \approx \frac{1}{2}(1-\cos\theta^*) \approx \frac{E_j(1-\cos\theta_j)}{\sum_{j=1,2} E_j(1-\cos\theta_j)} \quad (3)$$

[b]The dependence of C_{qg} and $C_{q\bar{q}}$ on the azimuthal angle ϕ between the lepton and the parton scattering plane is not given in the formula.

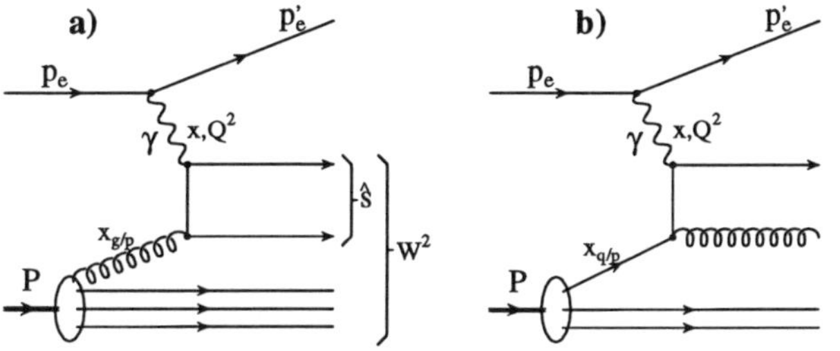

DIS variables:

$Q^2 = -q^2$

$x = Q^2 / (2P \cdot q)$

$W^2 = Q^2 (1-x)/x$

$\hat{s} = (j_1 + j_2)^2$

$\xi = x (Q^2 + \hat{s})/Q^2$

$z_q = (P \cdot j_1)/(P \cdot q)$

Fig. 2. Feynman diagrams for the production of $2+1$ jet events to first order of α_s in ep -collisions. q (P) denote the four-momentum of the photon (proton). j_1 (j_2) is the four-momentum of one of the jets associated to the hard subprocess.

where q (P) is the 4-momentum of the photon (proton) and θ^* is the polar angle[c] of one of the jets with 4-momentum j_1 in the photon-parton center of mass frame. E_j and θ_j are the energy and polar angle of the jets in the laboratory frame. The parton distribution functions in the proton $q(\xi, Q^2)$ and $g(\xi, Q^2)$ absorb the collinear and infrared singularities which occur in the calculations of C_{qq} and C_{qg}, and they have to be extracted from experiments. Once measured at some scale Q_0^2, they can be perturbatively evolved to any other scale, e.g. Q^2, using the DGLAP equations[2]. By convoluting them with the appropriate coefficient functions, they can be used to calculate the cross section in any other process. In this sense they are universal.

In leading order of α_s, the coefficient functions diverge like $C_{q\bar{q}} \sim 1/(z_q(1-z_q))$ and $C_{qg} \sim 1/((1-x/\xi)(1-z_q))$ and have to be regulated by a cut-off. In Fig. 3 is shown how the cross section $d^2\sigma/dx_p dz_q$ (where $x_p = x\ /\xi$) increases towards low z_q and x_p and towards $z_q = 1$. It is clear that in the regions where the LO cross section diverges, large contributions from higher orders can be expected.

The dijet cross section including higher order parton emissions are difficult to calculate using exact expressions for the coefficient functions. Since the phase space integrals cannot be solved analytically, numerical methods have to be applied. Recently several Monte Carlo integration programs computing the cross section to next-to-leading order (NLO) became available. Early attempts like PROJET[5] or DISJET[6] used a semi-analytical approach which only allowed jets to be defined by the W algorithm with the JADE scheme[7] and were restricted to a small set of calculable distributions. Moreover, in these programs approximations were made which turned out not to be valid over the full phase space[8]. The programs MEPJET[8], based on a phase space slicing method[9], and DISENT[10], based on a method to subtract the divergences from individual contributions[11] and on an analytical treatment of specific divergent terms[12], are flexible NLO Monte Carlo programs which allow any jet definition scheme and arbitrary experimental cuts to be analyzed. For $Q^2 > 40$ GeV^2 and for jets defined by the k_T algorithm[13] the agreement of the $2+1$ and $3+1$ jet

[c]The polar angle is defined with respect to the incoming parton direction.

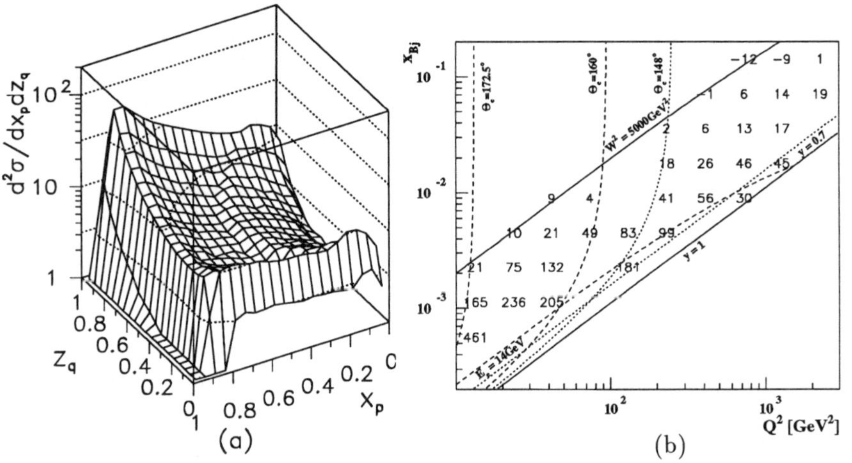

Fig. 3. a) Differential cross section $d^2\sigma/dx_p dz_q$ in first order of α_s for $x = 10^{-3}$ and $Q^2 = 10$ GeV^2. The figure is taken from ref.[3]. b) Relative change in percent of the dijet rates when adding to the leading order cross section higher orders based on a leading logarithm parton shower model in the LEPTO Monte Carlo. The lines represent typical selection cuts on the scattered electron. In addition, a cut on $z_q > 0.1$ has been applied. The jet rate is based on the JADE algorithm for $y_{cut} = 0.02$. The figure is taken from ref.[4].

cross sections between the two programs is found to be on the 3% level[14].

Higher order parton emissions and models for the transition of partons into hadrons are only included in Monte Carlo programs incorporating coefficient functions to leading order such as ARIADNE[15], LEPTO[16], HERWIG[17], PYTHIA[18] and RAPGAP[19]. They model multi-parton emission by either parton showers[20] (PS) based on the DGLAP equations or by a chain of independently radiating colour dipoles[21] (CDM). HERWIG uses the cluster hadronisation model[22]. In all other programs hadronisation is based on the LUND string model[23] as implemented in JETSET[24].

The relative change in percent of the dijet rates defined with the JADE algorithm for $y_{cut} = 0.02$ when adding to the LO cross section higher orders based on a leading logarithm parton shower model (MEPS) is shown in Fig. 3 in the x -Q^2 plane applying typical analysis cuts on the scattered electron. While for $Q^2 > 100$ GeV^2, the relative change only exceeds 20% in the low x region, parton showers change the dijet rate by a large factor for $10 < Q^2 < 100$ GeV^2 and $1 \cdot 10^{-4} < x < 2 \cdot 10^{-3}$. The influence of parton emissions beyond leading order does also depend on the jet variables z_q and ξ. Since different jet algorithms cover different phase space regions, it is always important to specify which jet definition scheme was used. Fig. 4 illustrates the rate of $2+1$ jet events as function of the invariant mass of the jet system $\sqrt{\hat{s}}$ and

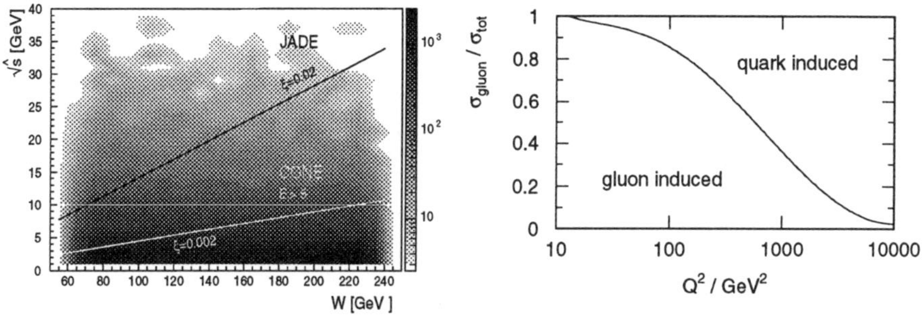

Fig. 4. a) Rate of $2+1$ jet events as function of the invariant mass of the two jet system $\sqrt{\hat{s}}$ and the invariant mass of the hadronic final state W. Indicated are the phase space restrictions introduced when using the cone or the JADE algorithm. b) Fraction of gluon and quark induced processes leading to $2+1$ jet events with $E_T > 5$ GeV and $-1 < \eta < 2$ defined by a K_t algorithm in the Breit frame as function of Q^2. The cross section has been calculated in NLO in the kinematic range $x > 1 \cdot 10^{-4}$ and $0.15 < y < 0.65$.

the invariant mass of the whole hadronic final state W. The rate rapidly increases towards low $\sqrt{\hat{s}}$. Indicated as lines are the phase space boundaries introduced by the JADE and the cone algorithm. With a cut on the transverse energy of the jets $E_T > 5$ GeV the minimum $\sqrt{\hat{s}}$ which can be reached is 10 GeV, since (in the limit of massless partons) from $E_T^2 = \hat{s}\, z_q\,(1 - z_q)$ follows $\sqrt{\hat{s}}/2 > E_T$. When using the JADE algorithm with a resolution parameter $d_{ij}/W^2 \approx \hat{s}/W^2 \approx \xi$, low values of $\sqrt{\hat{s}}$ can only be reached for low (high) values of W (x). If one takes $y_{cut} = 0.02$, $\sqrt{\hat{s}}$ is below 10 GeV for $W \lesssim 70$ GeV. The JADE algorithm with a high resolution parameter is therefore more restrictive than the cone algorithm. In addition to the region of low $\sqrt{\hat{s}}$, where collinear and infrared divergences become important, it naturally avoids the region of high W (low x) where a large phase space for multi-parton emission is available. However, the divergences at small and large z_q can only be avoided by a large y_{cut}.

Multi-jet production allows a variety of quantitative tests of our understanding of QCD dynamics. In a phase space region where the data are well described by perturbative QCD, HERA offers the opportunity to extract α_s over a wide range of Q^2 in a single experiment. This can be done at large x and Q^2 where the parton densities in the proton are well constrained. At low Q^2 and correspondingly low values of ξ, gluon initiated processes become increasingly important. Here, the measured $2+1$ cross section can be used to determine the gluon density assuming a value of α_s. The sensitivity to both α_s and $g(\xi, Q^2)$ could also be used to simultaneously extract these quantities. Such an analysis is however theoretically and experimentally difficult. In the region where the data significantly differ from what is expected by a fixed order calculation of perturbative QCD, jet observables are useful tools to gain

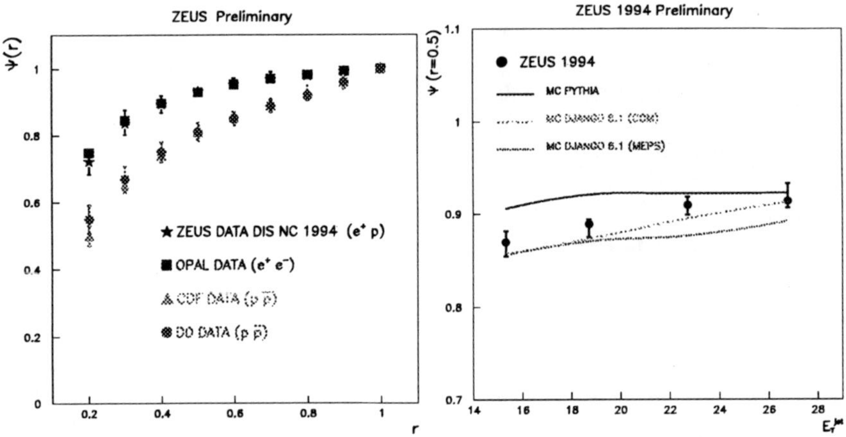

Fig. 5. a) Fractional transverse energy $\Psi(r)$ in a jet defined by a cone algorithm with radius $R = 1$ and $37 < E_t < 45$ GeV in the laboratory frame as function of $r = \sqrt{\Delta\eta^2 + \Delta\phi^2}$. For OPAL jets with $E_T > 35$ GeV were selected, for CDF (D0) the considered E_T range of the jets is $40\,(45) < E_T < 60\,(70)$ GeV. b) $\Psi(r = 0.5)$ as function of the transverse energy of the jet. QCD Monte Carlo models are superimposed. They either calculate the dijet cross section in LO and add multi-gluon emissions based on the colour dipole model (CDM) or by parton showers (MEPS) or treat QCD effects only by parton showers (PYTHIA).

insights in the complex parton dynamics occurring in electron proton collisions.

2. Jet shapes in DIS

The internal structure of jets provides useful information on the transition of a parton to the complex aggregate of hadrons which can be observed in the detector. The capability to describe detailed properties of jets is also an important ingredient when comparison of perturbative calculations with experimental data.

To study the internal structure of jets in the sample of neutral current DIS events with $Q^2 > 100$ GeV^2 jets are found in the laboratory system using the CDF-cone algorithm[25] with a cone radius $R = \sqrt{\Delta\eta^2 + \Delta\phi^2} = 1$. Only jets are considered which have in the laboratory frame a transverse energy $E_T > 14$ GeV and lie in the rapidity range $-1 < \eta < 2$. The jet shape is defined as the average fraction of the transverse energy of the jets inside an inner cone r concentric to the outer jet cone with radius R:

$$\Psi(r) = \frac{1}{N_{\rm jet}} \sum_{\rm jets} \frac{E_T(r)}{E_T(R)} \;, \quad E_T(r) = \int_0^r dr' dE_T(r')/dr' \tag{4}$$

where $E_T(r)$ is the transverse energy within an inner cone of radius r and $N_{\rm jet}$ is the total number of selected jets. By definition $0 < \Psi < 1$ and $\Psi(R) = 1$. The steepness of the rise of Ψ describes the collimation of the jet.

H1 preliminary

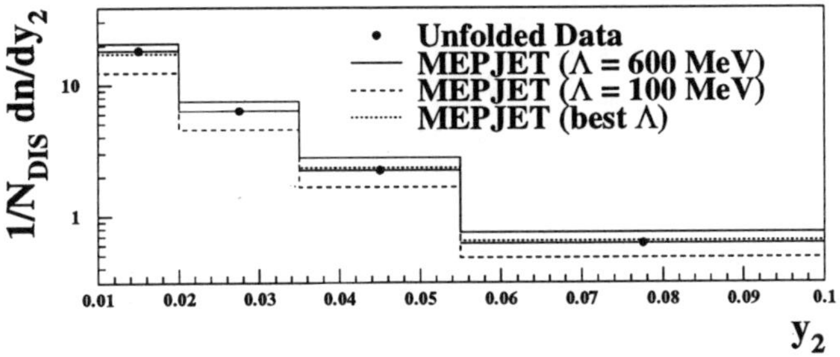

Fig. 6. Distribution of the resolution parameter y_2 in the JADE algorithm where a $1+1$ jet event switches to a $2+1$ jet event. Shown are unfolded data and the next-to-leading order prediction for some values of $\Lambda^{(4)}_{QCD}$. $N_{\rm DIS}$ is the total number of DIS events in the considered kinematic region.

The jet shape Ψ corrected to hadron level for jets with $E_T > 37$ GeV as function of r is shown in Fig. 5a. 90% of the jet transverse energy is already contained in a cone with $r = 0.4$ around the jet axis. This is very similar to jets produced in e^+e^- collisions[26] with $E_T > 35$ GeV. The jets selected from $p\bar{p}$ collisions for $40\,(45) < E_T < 60\,(70)$ GeV measured by CDF (D0)[27,28] are considerably broader. This might be explained by the fact that in e^+e^- and ep collisions the partons initiating the cascade leading to the jet are mainly quarks while in $p\bar{p}$ collisions they are gluons.

The jet shape depends on the transverse energy of the jet. Jets get narrower with increasing E_T as is shown in Fig. 5b.

Monte Carlo models based on a hard subprocess in LO plus additional multi-gluon emissions are able to describe the data. Also PYTHIA where QCD effects are only treated by parton showers reproduces the main features of the data. The best description of the E_T dependence of Ψ is obtained when partons are emitted according to the colour dipole model. The internal jet structure in DIS at HERA is fairly well reproduced by all QCD models.

3. Determination of the strong coupling constant

The $2+1$ jet rate defined with the JADE algorithm depends on the jet resolution parameter y_{cut}. For each event y_2 is defined as the value of y_{cut} where a $1+1$ jet event switches to a $2+1$ jet event.

A region of phase space is selected where NLO QCD can provide a reasonable description of the data, and where effects from multi-gluon emissions are small. In the H1 analysis[29] this region is defined by: $Q^2 > 200$ GeV^2 and $W^2 > 5000$ GeV^2. To avoid events where the invariant mass of the two jet system is small and where the cross-section diverges, y_2 is restricted to $y_2 > 0.01$. The y_2 distribution, corrected to

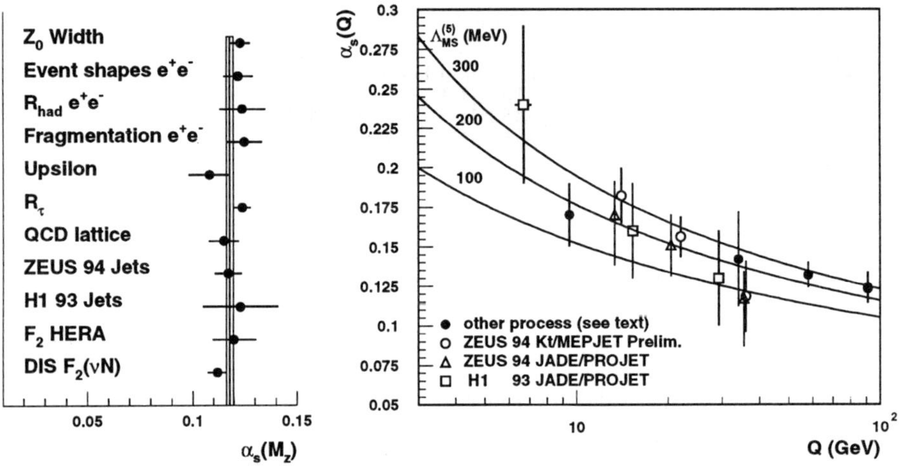

Fig. 7. Left: Values and total error of $\alpha_s(M_Z)$ from various processes. The solid line indicates the world average and the band its total error. Right: $\alpha_s(Q)$ from HERA (open symbols) and other processes with increasing Q (closed circles): Γ_Υ and $\sigma_{\text{had}}/\sigma_{\text{tot}}$, event shapes and $\Gamma_{\text{hadron}}/\Gamma_{\text{lepton}}$ in e^+e^-. The preliminary points using the k_T algorithm include the statistical errors only. The predictions of the renormalisation group equation for three values of $\Lambda_{\overline{MS}}$ are superimposed as lines.

parton level by an unfolding method[30], is shown in Fig. 6. Overlayed is the prediction of NLO QCD for three values of α_s (respectively Λ_{QCD}). In all bins a strong sensitivity to α_s is observed. For one given value of α_s NLO QCD gives an excellent description of the data. The precision which can be expected from this measurement is of the order of $\Delta\alpha_s(M_Z)/\alpha_s(M_Z) \approx 5 - 10\%$.

Keeping $y_{cut} = 0.02$ fixed, the number of $2+1$ jet events (N_{2+1}) normalized to the sum of $1+1$ and $2+1$ jet events ($N_{2+1} + N_{1+1}$) can be measured and is sensitive to α_s. Only jets in the acceptance region of $10^o < \theta_{jet} < 145^o$ and with $z_q > 0.1$ are counted. The fractional jet rate $R_2 = N_{2+1}/(N_{1+1} + N_{2+1})$ can be expressed by:

$$R_2(Q^2, y_{cut}) = A(Q^2, y_{cut})\,\alpha_s(Q^2) + B(Q^2, y_{cut})\,\alpha_s^2(Q^2) \tag{5}$$

From the coefficients A and B which can be calculated in NLO QCD and the measured R_2, $\alpha_s(Q^2)$ can be extracted in different Q^2 bins. Using the renormalisation group equation these independent measurements can be used to fit $\alpha_s(M_Z)$.

For this measurement only the kinematic range of high Q^2 and high x is considered in order to reduce the uncertainties stemming from higher order gluon emissions and from the hadronisation of a complex partonic configuration. The theoretically and experimentally difficult region towards the proton remnant is excluded. Three values of $\alpha_s(Q^2)$ extracted by the H1[31] and ZEUS[32] collaboration are shown in Fig. 7. For increasing values of Q^2, $\alpha_s(Q^2)$ decreases as predicted by the renormalisation group equation shown for $\Lambda_{\overline{MS}}$ between 100 and 300 MeV (solid lines). An extrapolation to $\alpha_s(M_z)$ yields:

$$\text{H1 93}: \quad \alpha_s(M_z) = 0.123 \pm 0.012\,(\text{stat}) \pm 0.013\,(\text{syst.})$$
$$\text{ZEUS 94}: \quad \alpha_s(M_z) = 0.117 \pm 0.005\,(\text{stat}) \pm 0.005\,(\text{exp.}) \pm 0.007\,(\text{th.})$$

which is consistent with other values obtained from a large variety of different processes as shown in Fig. 7 (for ref. see[33]). The agreement found in e^+e^-, ep and $p\bar{p}$ collision for reactions at very different scales is an important and successful test of QCD. The error on the HERA measurements are even with the present limited statistics already competitive. The dominant experimental errors stem from the uncertainty on the energy scale for hadrons, the model dependence of assigning partons to jets and the phenomenological description of the hadronisation process. Furthermore the renormalisation scale and the choice of the input parton density systematically influences the result. These results have been obtained with the NLO program PROJET or DISJET.

When particles are combined to jets, the details of how the 4−momenta are added are important, in particular, because higher order QCD calculations are only carried out for massless partons whereas the jets acquire a mass during the combination procedure. In the E-scheme 4-momenta are added to massive jets while in the E_0 and P-schemes jets remain massless by rescaling either the 3-momenta or the energy (see table 1 for details). The dependence of the jet cross section on the recombination schemes (RS) only appears in the calculation when jets are composed out of more than one parton. Corrections of $\mathcal{O}(\alpha_s^3)$ due to the different recombination schemes are potentially large. When applying different recombination schemes to the data, the measured 2+1 jet cross sections defined with the E and E_0 algorithms agree within few percent, but are lowered by up to 20% for the P-scheme[34]. In PROJET only the E scheme is implemented. It is therefore not surprising that the extracted $\alpha_s(M_Z)$ agrees within 3% when the E and E0 scheme is only applied to the data, but gives an inconsistent result when using the P scheme[34]. An additional problem is that in the NLO calculation the W algorithm is implemented, while in the experimental analysis the JADE algorithm has to be used, i.e. mass terms in d_{ij} are neglected.

The new NLO programs MEPJET and DISENT reveal a strong dependence on the recombination scheme[35] when using the JADE or W algorithm. In addition lead these algorithms to large NLO corrections of the dijet cross section.

The cone or k_T jet algorithms seem to be better suited for precision QCD tests, since the variation of the cross section when going from leading to next-to-leading order is much smaller[36]. Moreover, these jet algorithms are less dependent on the choice of the renormalisation and factorisation scale and allow a larger phase space to be covered.

The ZEUS collaboration has reanalyzed their data[37] using the k_T algorithm. The preliminary values of $\alpha_s(Q)$ obtained in three bins of Q are shown[d] in Fig. 7. They are consistent with the results obtained with the JADE algorithm. The value extrapolated to the Z^0 boson mass is: $\alpha_s(M_Z) = 0.118 \pm 0.008\,(\text{stat})$.

4. Determination of the gluon density

The parton distributions are well constrained for larger x where a lot of data are available. However, for low x, where gluons play a crucial rôle, they are not well known. In this regime, HERA has the unique possibility to investigate the distribution of the gluon density in the proton. The scaling violation of the proton structure function F_2 offers an indirect way to get a handle on the gluon density. This measurement[38,39] is possible because of the extremely successful description of F_2 by the NLO DGLAP equations. Observables based on $2+1$ jet event can - by including them in the global analysis - provide a further constraint on the gluon density. Dijet rates can be calculated in next-to-leading order and are sensitive to the gluon density, since for $10 \lesssim Q^2 \lesssim 100$ GeV^2 the $2+1$ jet cross section is dominated by gluon initiated processes. In this kinematic region less than about 20% of the $2+1$ jet events are induced by quarks (see Fig. 4).

An alternative approach is to only base the analysis on the $2+1$ jet rate. Since large invariant jets masses ($\hat{s} > 100$ GeV^2) are experimentally required to define clean jets, the very low x regime can not be reached. These measurements are nevertheless important to fill the gap between the results from small-x scaling violations in F_2 and the data on direct photon and jet production in $p\bar{p}$ collisions reaching to large-x[40].

The H1 collaboration[41] has already exploited the $2+1$ jet production to extract the gluon density in LO in the range $2 \cdot 10^{-3} \lesssim \xi \lesssim 0.18$. When processes where only one gluon is involved are considered, ξ is directly observable and the determination of $g(\xi, Q^2)$ is straight forward. This is however only of limited theoretical interest, since the LO dijet cross section is subject to large higher order corrections. There is moreover a large dependence on the arbitrary choice of the renormalisation and factorisation scale. Only, in NLO a reliable prediction of the dijet cross section is possible. The $2+1$ jet cross section, for instance, defined with a cone algorithm for jets with $E_T > 5$ GeV remains constant within 5% when varying the scale by 3 orders of magnitude while in LO variations of 40% are calculated[36].

A NLO extraction of $g(\xi, Q^2)$ requires, however, an iterative procedure where the input gluon density used in the NLO calculation is adjusted until the theoretical dijet cross sections agree with the measurement. This is most efficiently done by expanding the gluon density for a discrete set of points $(\xi_1, ..., \xi_n)$ using appropriate test functions $g(\xi, Q^2) = \sum_{n=1}^{n_{max}} a(\xi_n, Q^2)\, T_n(\xi)$ such that the integral $\sigma_g = \sum_{n=1}^{n_{max}} a(\xi_n, Q^2) \int d\xi/\xi \, C_{q\bar{q}}(x/\xi, Q^2)\, T_n(\xi)$ in equation 1 must only be calculated once. The coefficient a can then be used in the fitting procedure. Preliminary results

[d] Only the statistical errors are included in the figure.

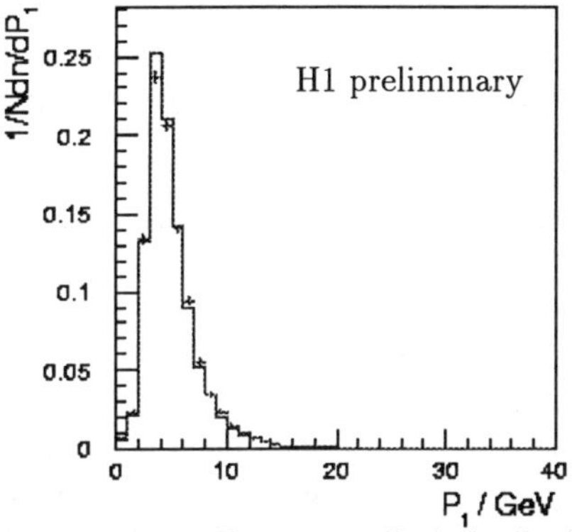

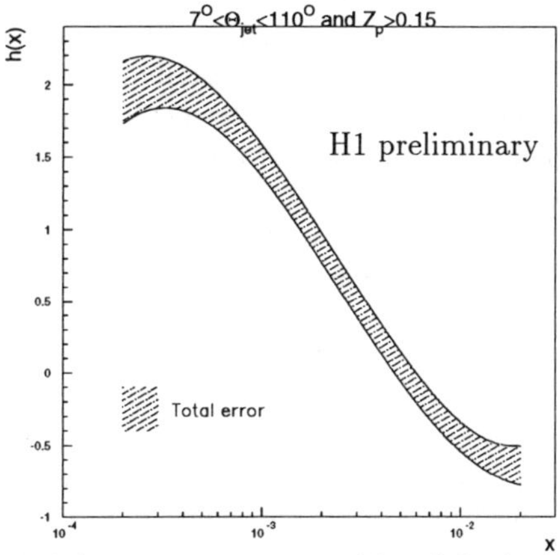

Fig. 8. Left: Shape normalised distribution of the transverse energy of jets defined with the k_t algorithm (in the Breit frame) using $Q^2/2$ as resolution parameter. The data have not been corrected for detector effects. Superimposed is a simulation based on LEPTO. Right: The function $h(x)$ describing the non-perturbative contribution of the dijet cross section as defined in the H1 analysis to extract the gluon density in NLO.

using an elegant technique based on Mellin moments[42], where the test functions take the form $T_n = (x/\xi)^n$, have been presented by the H1 collaboration[34] covering the range $0.01 < \xi < 0.1$ using the JADE algorithm.

Another approach also pursued by the H1 collaboration[43] is to use as test functions cubic splines[44] with the property $T_i(\xi_j) = \delta_{ij}$ and to introduce the $2+1$ jet rate as additional observable in a global fit analysis minimizing $\chi^2 = \chi^2(F_2) + \chi^2(R^{det}_{2+1})$. This has the advantage that no assumption on the quark densities is needed. The correlation of quarks and gluons via the DGLAP equation is naturally taken into account. R^{det}_{2+1} as measured in the detector is for each considered kinematic bin related to the NLO calculation of the dijet cross section $\sigma^{\rm pert}_{2+1}$ performed by DISENT:

$$R^{\rm det}_{2+1} = \frac{{\rm MIG}^{1+1}_{2+1} \cdot (\sigma^{\rm pert}_{2+1}(x,Q^2) + h^{\rm non-pert.}_{2+1}(x,Q^2))}{\sigma_{tot}(x,Q^2)} \tag{6}$$

The coefficients ${\rm MIG}^{1+1}_{2+1}$ correct for the migration of $1+1$ to $2+1$ jet events from hadron to detector level and can be determined using a fully simulated event sample based on LEPTO. The transition from partons to hadrons is modeled by a simple ansatz: $h^{\rm non-pert.}_{2+1}(x,Q^2) = (\alpha + \beta \log x + \gamma \log^2 x + \delta \log^3 x)/Q^2$ When comparing NLO calculations to measurements, usually the data are 'corrected to parton level' by assuming that this transition is correctly described in a LO (improved by parton showers) Monte Carlo simulation program. To introduce a simple hadronisation model with additional free parameters in the fit procedure has the advantage that no a priori assumption on the underlying parton level is needed to relate the NLO calculation to the measurement. Recently, similar models based on power behaved terms[45] have been very successful in describing event shapes in DIS[46]. In this context theoretical studies suggest that the terms $1/Q$ may also take into account multi-emissions of soft gluons associated with the behaviour of the running coupling constant at small scales[45].

In the kinematic range $25 < Q^2 < 100$ GeV^2 and $2 \cdot 10^{-4} < x < 2 \cdot 10^{-2}$ jets are selected using the k_T algorithm with $d_{ij} = 0.5$ and Q^2 as scale. The jet rate for jets with $7^o < \theta_{jet} < 110^o$ and $z_q > 0.15$ is measured. LEPTO is able to describe the shape of most distributions associated to the hard subprocess. As an example Fig. 8 shows the E_T distribution of the jets as measured in the detector. LEPTO, however, fails to describe the absolute jet rate. The measurement is always about 20% above the predicted jet rate. This is true in the whole considered (x,Q^2) plane, but most pronounced for low Q^2 ($10 < Q^2 < 14$ GeV^2) and low x ($2 \cdot 10^{-4} < x < 1 \cdot 10^{-2}$). The observed difference can not be accounted for by systematic effects. The most important systematic error of about $3 - 13\%$ on the dijet rate is mainly due to the uncertainty of the hadronic energy scale.

The fit to the F_2 data and the jet rates leads to a good $\chi^2(R_{2+1}) = 10$ for 11 degrees of freedom, but no influence of the jet data on the extracted parton density functions is found in this analysis. Also the error band on the extracted gluon density is only marginally improved. This means that the extracted gluon density is essentially constrained by the inclusive F_2 data. The fitted non-perturbative function $h(x)$ is shown in Fig. 8. Without a strong x dependence of $h(x)$ no satisfactory fit can be performed.

A possible explanation of this results is that the way in which the k_T jet algorithm is used in this analysis leads to jets with relatively low transverse energy of $3 - 5$ GeV (see Fig. 8) such that the found jet system does not always stem from the hard subprocess. Another explanation is that a fixed order calculation is not sufficient to account for the data in this phase space region. The difference between the NLO calculation and the data has therefore to be absorbed in the non-perturbative function $h(x)$ and consequently no sensitivity of the jet rates to the gluon density is found.

5. Comparision of jet rates to NLO-QCD

A similar observation about the incapability of NLO QCD to describe the dijet rates has been made by the ZEUS collaboration[47]. In this study jets are selected using a cone algorithm in the laboratory frame requiring $E_T > 4$ GeV in the hadronic center of mass and the laboratory frame and excluding the forward region by $\eta_{\text{lab}} < 2$. The data are then corrected to parton level using LEPTO 6.3. The correction factors are between 1.2 and 1.8 for $4 \cdot 10^{-3} < \xi < 8 \cdot 10^{-2}$, but get very large (up to $3 - 4$) for ξ around 10^{-3}. The shape of variables connected to the hard subprocess like

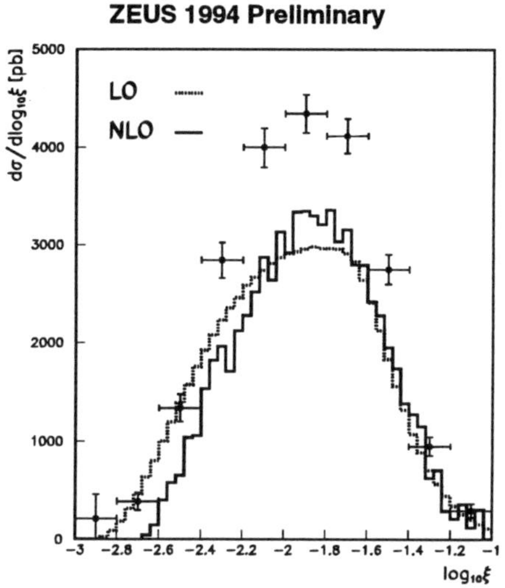

Fig. 9. 2 + 1 jet rate cross section in function of the momentum fraction of the parton entering the hard subprocess ξ. Data have been corrected to parton level using the QCD Monte Carlo program LEPTO 6.3. Superimposed is a prediction from the QCD matrix element in LO and in NLO in α_s.

E_T, η and ξ are well described by NLO programs. The absolute value of the 2 + 1 cross section is, however, found to be about 30% higher in the data. This large difference cannot be explained by experimental effects like variation of the energy scale or the model dependence in the correction procedure nor by varying theoretical choices like different parameterizations of the parton densities or different factorisation and renormalisation scales.

That NLO QCD is not able to describe the dijet rates has recently been reported by the H1 collaboration[48,49]. The dijet rate is measured with a cone algorithm using a cone radius $R = 1$ and requiring $E_T > 5$ GeV in the hadronic center of mass system. The rapidity difference between the two jets has to be $\Delta\eta < 2$. The 2 + 1 jet rate corrected for detector effects is shown in Fig. 10 as function of x and Q^2. NLO QCD (DISENT) as well as a conventional QCD model based on the exact LO coefficient functions plus additional DGLAP parton shower (RAPGAP 'DIR') clearly fail to describe the data. This is true for the whole kinematic range $5 \lesssim Q^2 \lesssim 100$ GeV^2 and $10^{-4} \lesssim x \lesssim 10^{-2}$, but is most pronounced in the low x and Q^2 region. In this region the prediction of the NLO QCD calculation is a factor of 2 higher than the QCD model, but is by the same factor below the data. Varying the input parton density functions leads to a change in the jet rate by up to 30%, but cannot account for the large discrepancy with the data.

The CDM model as implemented in ARIADNE is able to describe the data. In

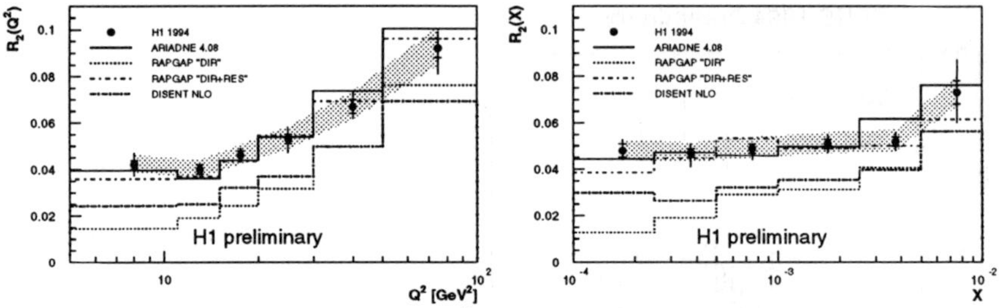

Fig. 10. 2 + 1 jet rates as function of Q^2 and x corrected to hadron level. Jets are defined by the cone algorithm with $E_T > 5$ GeV in the hadronic center of mass system. The difference in pseudorapidity between the two jets has to be smaller than 2. The band indicates the systematic error. Superimposed are the predictions from ARIADNE, a next-to-leading calculation (DISENT) and from RAPGAP (the direct and resolved contribution are shown separately).

this model gluons are emitted from a chain of colour dipoles spanned between the current quark and the proton remnant. The first emission in the cascade is corrected by the matrix element to first order[50]. It has been observed[51] that quark induced processes have a up to a factor of 8 higher cross section than expected from conventional QCD, because the parton distribution functions in equation 1 are replaced by a phenomenological expression better adapted to the dipole picture. Partons with short wavelengths having high transverse momenta (P_T) only resolving part of the extended colour charge distribution of the proton remnant are suppressed by $(\mu/P_T)^\alpha$ where μ and α are phenomenological parameters usually set to 1. A similar suppression depending on the virtuality Q^2 is introduced in the direction of the scattered quark. It has been argued that the partonic state as encountered in ARIADNE is closely related to the one expected from a BFKL scenario[52]. In ARIADNE the hardest parton emission is not bound to the photon vertex, but can happen anywhere between the photon and the proton remnant.

Such a feature can also be artificially introduced in the conventional QCD model by assuming that the photon emitted from the incoming electron can fluctuate into quarks and gluons before entering the hard subprocess. This can be modeled by assuming a structure in the virtual photon like e.g. parameterized in the SaS model[53]. Fig. 10 shows that adding this contribution (RAPGAP 'RES') to the processes where the photon is assumed to be point-like (RAPGAP 'DIR') does also reproduce the measured jet rate.

Only models where the hardest parton emission is not bound to occur at the photon vertex are able to describe the data. This indicates that in the phase space region tested by 2+1 jet production higher order processes which cannot be described by NLO QCD are important. Whether it is sufficient to simply add processes in the photon direction without being in conflict with other HERA data, is not yet clear. The phenomenological treatment of higher orders in the colour dipole model has been proven to be extremely successful in describing all the data on the hadronic final state which have been measured at HERA[54,55]. It remains however a puzzle why the shape of all distributions associated with the hard dijet system, like E_T, x_p, η etc., is well described by conventional QCD. It is hard to imagine that the importance of higher orders only affects the normalisation and not the shape of the distributions.

6. Conclusions

The understanding of jet production in DIS remains one of the main physics goals at HERA. While with the first HERA data, topics like the measurement of the strong coupling constant and the gluon density have been attacked with great enthusiasm, it turned out that these goals can only be achieved in a (very) restricted phase space region.

Existing data on jet production have always been tailored to a specific goal and present themselves today as a loose collection of scattered information. While conventional QCD is able to describe the data for $Q^2 > 100$ GeV^2 and $\theta_{jet} > 10^o$ using the JADE algorithm, a discrepancy of about 30% is found for the low Q^2 region when using the cone or the k_T algorithm and excluding the forward region. The fully acceptance corrected jet rates in this region disagree by about a factor of two.

It seems that before precisely pinning down the strong coupling constant and the gluon density, one has to systematically investigate the applicability of NLO QCD in terms of kinematic phase space and the way how the jet observables are defined (jet algorithm, hard scale etc.).

In future, it will be necessary to measure jet cross sections corrected for detector effects extending to very low and very large x and Q^2 in order to find out where perturbative QCD is able to describe the data and where the mechanism governing jet production are more complicated. To achieve this goal the understanding of QCD models will play a major role.

7. Acknowledgements

I would like to thank my colleagues J. Dainton, G. Iacobucci, M. Kuhlen, E. Mirkes and M. Wobisch for their critical reading of the manuscript and helpful comments.

8. References

1. A. Mendez, Nucl. Phys. B145 (1978) 199; R. Peccei and R. Rückl, Nucl. Phys. B162 (1980) 125; C. Rumpf, G. Kramer and J. Willrodt, Z. Phys. C7 (1981) 337.
2. Y.L. Dokshitzer, Sov. Phys. JETP 46 (1977) 641; V.N. Gribov and L.N. Lipatov, Sov. J. Nucl. Phys. 15 (1972) 438 and 675; G. Altarelli and G. Parisi, Nucl. Phys. 126 (1977) 297.
3. A. Edin, G. Ingelman and J. Rathsman, Z. Phys. C75 (1997) 57.
4. K. Flamm, Phd thesis, University of Hamburg, DESY FH1-96-04 Hamburg, 1996.
5. D. Graudenz, Phys. Lett. B256 (1992) 518; Phys. Rev. D49 (1994) 3291; Comp. Phys. Comm. 92 (1995) 65.

6. T. Brodkorb and J. G. Körner, Z. Phys. C54 (1992) 519; T. Brodkorb and E. Mirkes, Z. Phys. C66 (1995) 141.
7. JADE Collab., W. Bartel et al., Z. Phys. C33 (1986) 23.
8. E. Mirkes and D. Zeppenfeld, Phys. Lett. B380 (1996) 205.
9. K. Fabricius, G. Kramer, G. Schierholz and I. Schmitt, Z. Phys. C11 (1981) 315; G. Kramer and B. Lampe, Fortschr. Phys. 37 (1989) 161; W.T. Giele, E.W.N. Glover, Phys. Rev. D46 (1992) 1980; W.T. Giele, E.W.N. Glover and D. A. Kosower, Nucl. Phys. B403 (1993) 663.
10. S. Catani and M. Seymour, Nucl. Phys. B485 (1997) 291.
11. R.K. Ellis, D.A. Ross and A.E. Terrano, Nucl. Phys. B178 (1981) 421.
12. S. Catani and M. Seymour, Phys. Lett. B378 (1996) 287.
13. S. Catani, Y.L. Dokshitzer and B.R. Webber, Phys. Lett. B285 (1992) 291.
14. E. Mirkes and D. Zeppenfeld, in: 5th Int. Workshop on DIS and QCD, eds. J. Repond and D. Krakauer, Chicago (USA) 1997.
15. L. Lönnblad, Comp. Phys. Comm. 71 (1992) 15.
16. G. Ingelman, in: Physics at HERA, eds. W. Buchmüller and G. Ingelman, Hamburg (Germany) 1991, Vol. 3, p. 1366; G. Ingelman, A. Edin and J. Rathsman, Comp. Phys. Comm. 101 (1997) 108.
17. G. Marchesini et al., Comp. Phys. Comm. 67 (1992) 465 .
18. T. Sjörstrand, Comp. Phys. Comm. 82 (1994) 74 .
19. H. Jung, Comp. Phys. Comm. 86 (1995) 147.
20. M. Bengtson and T. Sjöstrand, Z. Phys. C37 (1988) 465; M. Bengtson, G. Ingelman, T. Sjöstrand, in: HERA Workshop 1987, ed. R.D. Peccei, Hamburg (Germany) 1988, Vol. 1, p. 149.
21. G. Gustafson and U. Petterson, Nucl. Phys. B306 (1988); G. Gustafson, Phys. Lett. B175 (1986) 453; B. Andersson, G. Gustafson, L. Lönnblad and U. Petterson, Z. Phys. C43 (1989) 625.
22. B.R. Webber, Nucl. Phys. B238 (1984) 492.
23. B. Andersson et al., Phys. Rep. 97 (1983) 31.
24. T. Sjöstrand, Comp. Phys. Comm. 39 (1986) 347; T. Sjöstrand and M. Bengtsson, Comp. Phys. Comm. 43 (1987) 367, and for JETSET 7.3, T. Sjöstrand, CERN-TH-6488-92 (1992).
25. CDF Collab., F. Abe et al., Phys. Rev. D45 (1992) 1148.
26. OPAL Collab., R. Askers et al., Z. Phys. C63 (1994) 197.
27. CDF Collab., F. Abe et al., Phys. Rev. Lett. 70 (1993) 713.
28. D0 Collab., S. Abachi et al., Phys. Lett. B357 (1995) 500.
29. M. Weber, in: 5th Int. Workshop on DIS and QCD, eds. J. Repond and D. Krakauer, Chicago (USA) 1997.
30. V. Blobel, DESY 84-118, Hamburg 1984, and CERN Comp. School 1984.
31. H1 Collab., T. Ahmed et al., Phys. Lett. B346 (1995) 415.
32. ZEUS Collab., M. Derrick et al., Phys. Lett. B363 (1995) 201.
33. Particle Data Group, R.M. Barnett et al., Phys. Rev. D54 (1996).
34. K. Rosenbauer, in: Workshop on DIS and Related Phenomena, ed. G. D'Agostini and A. Nigro, Rome (Italy) 1996, p. 444.
35. E. Mirkes and D. Zeppenfeld, in: Workshop on DIS and Related Phenomena, ed. G. D'Agostini and A. Nigro, Rome (Italy) 1996, p. 428.
36. E. Mirkes and D. Zeppenfeld, Acta Phys. Polon. B27 (1996) 1393.
37. T. Trefzger, in: Workshop on DIS and Related Phenomena, ed. G. D'Agostini and A. Nigro, Rome (Italy) 1996, p. 434.
38. H1 Collab., S. Aid et al., Nucl. Phys. B470 (1996) 3.
39. ZEUS Collab., M. Derrick et al., Phys. Lett. B345 (1995) 576 .
40. R.C.E. Devenish and A.D. Martin, in: Workshop on DIS and Related Phenomena, ed. G. D'Agostini and A. Nigro, Rome (Italy) 1996, p. 193.

41. H1 Collab., S. Aid et al., Nucl. Phys. B449 (1995) 3.
42. D. Graudenz et al. Z. Phys. C70 (1996) 77.
43. F. Zomer, in: 5th Int. Workshop on DIS and QCD, eds. J. Repond and D. Krakauer, Chicago (USA) 1997.
44. G. Lobo, in: 'Future Physics at HERA', eds. G. Ingelmann, A. De Roeck and R. Klanner, Hamburg (Germany) 1996, p. 537.
45. Y.L. Dokshizer and B.R. Webber, Phys. Lett. B352 (1995); R. Akhoury and V.I. Zakharov, Phys.Lett. B357 (1995) 646.
46. H1 Collab., C. Adloff et al., DESY 97-098, Hamburg, 1997.
47. D. Mikunas, in: 5th Int. Workshop on DIS and QCD, eds. J. Repond and D. Krakauer, Chicago (USA) 1997.
48. J. Spiekermann, in: XXXIInd Rencontres de Moriond, ed. J. Tran Thanh Van, Les Arcs (France) 1997.
49. M. Wobisch, in: 5th Int. Workshop on DIS and QCD, eds. J. Repond and D. Krakauer, Chicago (USA) 1997.
50. L. Lönnblad, Z. Phys. C65 (1995) 285.
51. J. Rathsman, Phys. Lett. B393 (1997) 181.
52. L. Lönnblad, Z. Phys. C65 (1995) 285; A. H. Mueller, Nucl. Phys. B415 (1994) 373; J. Rathsman, Phys. Lett. B393 (1997) 181.
53. G. Schuler and T. Sjöstrand, Z. Phys. C68 (1995) 607 and Phys. Lett. B376 (1996) 193.
54. N. Brook et al., in: Future Physics at HERA, eds. G. Ingelman, A. DeRoeck and R. Klanner, Hamburg (Germany) 1996, Vol. 2, p. 613.
55. T. Carli, in: Workshop on Deep-Inelastic Scattering and Related Phenomena, ed. G. D'Agostini and A. Nigro Rome (Italy) 1996 p.415.

Deeply Inelastic Hadronic Final States: QCD Corrections

Dirk Graudenz*
Paul Scherrer Institut
5232 Villigen PSI, Switzerland
E-mail: Dirk.Graudenz@psi.ch

ABSTRACT

The status of the calculation of next-to-leading-order QCD corrections to hadronic final states in deeply inelastic scattering is reviewed, and an overview of the phenomenology (including the measurement of the strong coupling constant and the gluon density via jet rates) is given. We also describe a new universal program (DISASTER++) for the calculation of (2+1)-jet observables.

1. Introduction

The electron–proton collider HERA has started to accumulate large luminosity. This opens up the possibility to study the hadronic final state in deeply inelastic scattering to high precision. The motive for this is twofold: QCD can be tested at large scales in the spacelike regime, and an extraction of physical quantities, in particular the running strong coupling constant $\alpha_s(Q^2)$ and the gluon density $f_g(\xi, Q^2)$, is possible. The aim of this paper is to give an overview of the status of next-to-leading-order calculations both for jet-like quantities and for one-particle-inclusive processes. We will also discuss the phenomenology of processes with (2+1) jets[a] in the final state, and of inclusive particle spectra.

In the last few years, the technology for the calculation of QCD corrections in next-to-leading order of perturbation theory has developed considerably. There are explicit algorithms available which permit the calculation to be done in a "universal" way: the infrared singularities are subtracted in such a way that arbitrary infrared-safe observables can be calculated numerically. In principle, all existing algorithms are variations on a common theme, namely the interplay of the factorization theorems of perturbative QCD and the infrared-safety of the observables under consideration. This will be reviewed in Section 2.

The practical implementations of this principle can be quite different. In Section 3, we will discuss the two possible ways to do this: the phase-space-slicing method and the subtraction method. The extraction of the singular terms can be done in different ways, and we will describe in some detail a new calculation based on the subtraction formalism and on a general partial fractions formula.

An overview of applications for (2+1)-jet-like observables will be given in Sec-

*WWW URL: http://www.hep.psi.ch/graudenz/index.html
[a]Here "$(n+1)$" stands for n hard jets and the proton remnant jet.

tion 4, where we state some results for the cut dependence of jet cross sections and for transverse momentum spectra. One-particle-inclusive processes will be discussed in Section 5. The theoretical predictions in the framework of the fragmentation function picture fail to describe the experimental data for small values of the photon virtuality Q^2 and for small values of the energy E_h of the observed hadron, although there is excellent agreement at large Q^2 and large E_h. We will give a semi-quantitative estimate of the region where the fragmentation function picture should be applicable. The paper closes with a summary and an outlook.

2. Infrared-safe Observables and Factorization in QCD

Perturbative QCD permits the calculation of processes with partons in the final state. The gap to experimental data with final-state hadrons is bridged by means of infrared-safe observables $\mathcal{O}$. Examples are jet cross sections for various jet definitions, and event shape variables such as Thrust. The expectation value of an observable $\mathcal{O}$ for experimental data is given by

$$\langle \mathcal{O} \rangle_{\text{exp}} = \frac{1}{N} \sum_{I=1}^{N} \mathcal{O}^{(n_I)} \left(h_1^{(I)}, \ldots, h_{n_I}^{(I)} \right), \tag{1}$$

where N is the number of events, n_I is the number of final-state hadrons of event I, and the $h_i^{(I)}$ are the momentum vectors of the hadrons. This is to be compared with the expectation value for parton final states:

$$\langle \mathcal{O} \rangle_{\text{th}} = \sum_n \int \mathrm{dPS}^{(n)} \, \sigma^{(n)} \left(p_1, \ldots, p_n \right) \, \mathcal{O}^{(n)} \left(p_1, \ldots, p_n \right). \tag{2}$$

Here n is the number of final-state partons with momenta p_1, …, p_n, $\mathrm{dPS}^{(n)}$ is the phase space measure, and $\sigma^{(n)}$ is the hard scattering cross section for partons in the final state, calculated in perturbation theory[b]. For the next-to-leading-order calculation of the (2+1)-jet-production cross section to be described later on, the relevant values are $n = 2$ for the Born term and the virtual corrections, and $n = 3$ for the real corrections. Both virtual and real corrections exhibit infrared singularities. The soft singularities and the collinear singularities in the final state cancel, owing to the Kinoshita–Lee–Nauenberg mechanism. The collinear singularities from the initial state can be consistently absorbed into redefined parton densities. The resulting hard scattering cross section is finite.

The factorization theorems of perturbative QCD (see Ref. [1] and references therein) show that parton cross sections σ have a very simple behaviour in collinear and soft phase space regions:

---

[b] We have not indicated technical complications such as different types of final-state partons (quarks, gluons) and the convolution with parton densities if there are hadrons in the initial state.

- In the collinear limit, for the partons with labels j and k being collinear, the limit for σ is

$$\sigma \sim \frac{\alpha_s}{2\pi} \frac{1}{s_{jk}} \hat{P}_{kj\leftarrow i}(u)\, \sigma_{\mathrm{Born}}, \tag{3}$$

where u is the momentum fraction of the parent parton i carried by parton j, $s_{jk} = 2p_j p_k$ and σ_{Born} is the Born cross section corresponding to the real correction σ in the limit under consideration. The $\hat{P}_{kj\leftarrow i}(u)$ are the unsubtracted Altarelli–Parisi splitting functions (see, for example, Ref. [2]). There is a slight technical complication: factorization in this form holds only for polarized cross sections. The sum over the parton polarizations introduces residual azimuthal correlations, even in the collinear limit. These have to be taken into account in the construction of the subtraction terms, see below. After integration over the azimuthal angle, factorization of the form of Eq. (3) holds also in the unpolarized case. The splitting functions $\hat{P}_{kj\leftarrow i}(u)$ are universal, process-independent functions. This has the important consequence that physical parton densities and fragmentation functions can be defined in a process-independent way.

- In the soft limit, the situation is slightly more complicated. For a gluon[c] with label k becoming soft, the factorization formula now reads

$$\sigma \sim \frac{\alpha_s}{2\pi} \sum_{i,j\neq k} C_{ijk} \frac{p_i p_j}{(p_i p_k)(p_j p_k)}\, \sigma_{\mathrm{Born}}. \tag{4}$$

The sum runs over all pairs of partons $\{i,j\}$, and the C_{ijk} are constant coefficients. The structure of this formula can be easily understood in terms of an eikonal approximation, where the matrix element $\mathcal{M}$ factorizes as

$$\mathcal{M} \sim \frac{p_i \epsilon(\lambda)}{p_i p_k} \mathcal{M}_{\mathrm{Born}}. \tag{5}$$

The sum over the gluon polarization λ, $\epsilon(\lambda)$ being the gluon polarization vector, leads to the form given in Eq. (4).

In order not to spoil the cancellation of soft and collinear singularities from the real corrections against those of the virtual corrections, the observables $\mathcal{O}$ have to fulfill certain conditions. This is the topic we now turn to.

An observable $\mathcal{O}$ is called *infrared-safe*, if the functions $\mathcal{O}^{(n)}(p_1, \ldots, p_n)$ of the parton momenta respect soft and collinear limits:

$$\begin{aligned} \mathcal{O}^{(n)}(p_1, \ldots, p_i, \ldots, p_n) &\underset{p_i\to 0}{\longrightarrow} \mathcal{O}^{(n-1)}(p_1, \ldots, \hat{p}_i, \ldots, p_n), \\ \mathcal{O}^{(n)}(p_1, \ldots, p_i, \ldots, p_j, \ldots, p_n) &\underset{p_i \| p_j}{\longrightarrow} \mathcal{O}^{(n-1)}(p_1, \ldots, \hat{p}_i, \ldots, \hat{p}_j, \ldots, p_n, p_i + p_j). \end{aligned} \tag{6}$$

[c]Soft quarks do not lead to a soft singularity. The matrix element is singular as $1/E$, but this singularity is compensated by a factor E in the phase space volume.

Momenta denoted by $\hat{p}$ are to be omitted. The property of infrared safety for observables has the consequence that the factorization from Eqs. (3) and (4) also works if σ is replaced by $\sigma\,\mathcal{O}$. This ensures the cancellation of the infrared singularities for the case of a convolution of the parton cross section with an observable, as in Eq. (2).

3. Universal Calculations in Next-to-Leading Order for Jet Quantities

The factorization theorems of perturbative QCD make it possible to perform calculations for a particular process in such a way that arbitrary infrared-safe observables can be evaluated numerically. For this it is necessary to extract and cancel the infrared singularities in an observable-independent way. Technically, there are two different procedures to achieve this: the phase-space-slicing method [3] and the subtraction method [4]. These techniques can be illustrated by means of a simple example [5][d]. Assume that the integral

$$I = \int_0^A \mathrm{d}x\, x^{-\epsilon}\, \frac{1}{x}\, f(x) \tag{7}$$

is to be evaluated. The x-integration stands for the phase-space integral, the factor $x^{-\epsilon}$ is the regulator[e], the term $1/x$ is the infrared singularity (a "propagator term"), and the integrable function $f(x)$ contains regular parts of the cross section and of the observable.

- In the case of the *phase-space-slicing method*, the range of integration is split into two parts $[0, a]$ and $[a, A]$ by means of an arbitrary small technical cut-off parameter a. For the lower interval, the function $f(x)$ can be approximated by $f(0)$ up to terms $\mathcal{O}(x)$, which regulate the $1/x$ singularity, and, after integration, give rise to terms $\mathcal{O}(a)$. For the upper interval, the regulator ϵ can be set to zero. The final result for the integral is thus

  $$I = -\frac{1}{\epsilon} f(0) + f(0) \ln a + \int_a^A \mathrm{d}x\, \frac{1}{x}\, f(x) + \mathcal{O}(a) + \mathcal{O}(\epsilon). \tag{8}$$

  The first term is the singular part, which would cancel against a similar singular part of opposite sign from the virtual corrections. The remaining terms are finite. The integral in the third term can be evaluated numerically. The method is exact only in the limit $a \to 0$ (in practice, very small values for a are sufficient). The large logarithm in the second term is compensated by a similar logarithm from the numerical integration in the third term.

- The *subtraction method* requires the definition of a subtraction term, such that a splitting of the integral into a finite and a singular part is possible. As an

[d] Reference [5] actually contains the first fully general algorithm to calculate subtraction terms.
[e] The calculations are performed in $d = 4 - 2\epsilon$ space-time dimensions, so that UV and IR singularities are regulated and made explicit in the form of poles in ϵ. Infrared singularities are regulated for $\epsilon < 0$.

example we give a specific way to define the subtraction term:

$$I = \int_0^A \mathrm{d}x\, x^{-\epsilon}\, \frac{1}{x}\, (f(x) - f(0)) + \int_0^A \mathrm{d}x\, x^{-\epsilon}\, \frac{1}{x}\, f(0). \tag{9}$$

The subtraction ("local in phase space") in the first integral renders the integrand integrable for $\epsilon \to 0$, and the final result is thus

$$I = -\frac{1}{\epsilon} f(0) + f(0) \ln A + \int_0^A \mathrm{d}x\, \frac{1}{x}\, (f(x) - f(0)) + \mathcal{O}(\epsilon). \tag{10}$$

The integral in the third term can again be performed numerically. Because $f(0)$, the subtracted term in the integrand, is the residue of $f(x)/x$ at $x = 0$, this method is also called the *residue method.*

The advantage of the phase-space-slicing method is that it can be implemented in a straightforward way: all that is required is the integral of the real corrections over small cones in phase space (corresponding to the approximation $x \approx 0$) and a numerical implementation in the form of a Monte-Carlo program of the cross section itself. The disadvantage is that the cancellation of the term proportional to $\ln a$ is delicate; the statistical fluctuations require a large number of Monte-Carlo events. Moreover, the convergence of the final result in the limit $a \to 0$ has to be checked, in principle, for every calculated expectation value. These technical problems are not present in the case of the subtraction method, because no small cut-off has to be used[f]. However, the implementation of the method requires the construction of the subtraction term, which is a non-trivial task. If this can be afforded, the subtraction method is the method of choice.

For the particular case of (1+1)- and (2+1)-jet production in deeply inelastic scattering, there are by now several calculations in the form of weighted Monte-Carlo programs available:

- `PROJET` [6]: The jet definition is restricted to the modified JADE scheme; the program is based on the calculation in Refs. [7–9].
- `DISJET` [10]: Again the jet definition is restricted to the modified JADE scheme; the program is based on the calculation in Refs. [11, 12].
- `MEPJET` [13]: This is a program for the calculation of arbitrary observables which uses the phase-space-slicing method. The calculation [14] uses the Giele–Glover formalism [15] for the analytical calculation of the IR-singular integrals of the real corrections, and the crossing-function technique [16] to handle initial-state singularities. The latter requires the calculation of "crossing functions" for each set of parton densities.

[f]Strictly speaking, a tiny cut-off of the order of 10^{-8} to 10^{-10} has to be introduced in order to make the cancellation $(f(x) - f(0))/x$ work, because there is only a finite number of significant digits available.

- `DISENT` [17]: This program is based on the subtraction method. The subtraction term is defined by means of the dipole formalism[g] [18, 19].

- `DISASTER++` [20]: This is a `C++` class library[h]. The subtraction method is employed, and the construction of the subtraction term resembles the method of Ref. [5], i.e. it is obtained by the evaluation of the residues of the cross section in the soft and collinear limits. Double counting of soft and collinear singularities is avoided by means of a fully general partial fractions method.

Why a new calculation? There are two reasons: (a) The existing programs have the restriction that the number of flavours is fixed ($N_f = 5$ in the case of `MEPJET` and N_f fixed, but arbitrary for `DISENT`). For studies of the scale-dependence it is necessary to have a variable number of flavours, in order to be consistent with the scale evolution of the strong coupling constant and the parton densities. `DISASTER++` makes the N_f dependence explicit in the "user routine" on an event-by-event basis, and thus results for arbitrary renormalization and factorization scales can be binned simultaneously. (b) `DISASTER++` is already set up such that the extension to one-particle-inclusive processes will be possible without the necessity of re-coding the contributions which are already present for the jet-type observables.

In the following we will illustrate how the subtraction term for the real corrections is constructed. We will be concerned with infrared-safe observables that are zero if more than one parton is soft, or if more than two partons are collinear, or combinations thereof. These are the most relevant ones in practical applications. In principle, the procedure is simple. In energy and angle variables, soft singularities are of the form $1/E_i^2$, for parton i being soft, and collinear singularities are given by terms $1/v_{ij}$, for partons i, j being collinear, where $v_{ij} = (1 - \cos\vartheta_{ij})/2$, and ϑ_{ij} is the angle between the two partons. The singularities can be extracted by performing the limits $E_i \to 0$ and $v_{ij} \to 0$ of the terms $E_i^2\sigma$, $v_{ij}\sigma$ and $E_i^2 v_{ij}\sigma$, and the subtraction terms will consequently be given by

$$\frac{1}{E_i^2}\left[E_i^2\,\sigma\right]_{E_i=0}, \quad \frac{1}{v_{ij}}\left[v_{ij}\,\sigma\right]_{v_{ij}=0}, \quad \frac{1}{E_i^2 v_{ij}}\left[E_i^2\,v_{ij}\,\sigma\right]_{E_i=0,\,v_{ij}=0}. \tag{11}$$

There is, however, the problem of overlapping singularities, or double counting. Consider a term of the form $1/(E_1^2\,v_{12}\,v_{13})$. Here the problem is that if parton 1 is soft, then there are two regions which give rise to an additional collinear singularity: $v_{12} = 0$ and $v_{13} = 0$. In order to include every possible singular configuration only once in the

---

[g] The subtraction term is written as a sum over dipoles (an "emitter" formed from two of the original partons and a "spectator" parton). Besides the factorization theorems of perturbative QCD, the main ingredient is an exact factorization formula for the three-particle phase space, which allows for a smooth mapping of an arbitrary 3-parton configuration onto the various singular contributions.

[h] The acronym stands for "Deeply Inelastic Scattering: All Subtractions Through Evaluated Residues". Most of the program is written in `C++`. A `FORTRAN` interface is in preparation; thus there will not be any problem to interface the class library to existing `FORTRAN` code.

subtraction term, it is convenient to separate the singular terms. This can be done in two ways: (α) division of the phase space, or (β) separation of the singularities in the matrix element. In Ref. [5] a combination is used: $1/(v_{12}\,v_{13})$ is written in terms of partial fractions as $1/[v_{12}(v_{12}+v_{13})]+1/[v_{13}(v_{12}+v_{13})]$, and the energy of the E_1-integration is restricted to be smaller than $E_2/2$. In principle, phase-space cuts of this kind do not pose a problem in Monte-Carlo programs. However, it is preferable that the integrand be a smooth function. We achieve this by using the following general formula for partial fractions:

$$\frac{1}{x_1\,x_2\cdots x_n}=\sum_{\sigma\in S_n}\frac{1}{x_{\sigma 1}\,(x_{\sigma_1}+x_{\sigma_2})\cdots(x_{\sigma_1}+\ldots+x_{\sigma_n})}. \tag{12}$$

The sum runs over all $n!$ permutations of n objects[i]. The most straightforward way to apply this formula to the case at hand [(2+1)-jet production] would be to set the set of variables $\{x_1,\ldots,x_9\}$ to $\{E_1^2,E_2^2,E_3^2,v_{01},v_{02},v_{03},v_{12},v_{13},v_{23}\}$ (p_0 is the momentum of the incident parton), where the energies are rescaled such that they are dimensionless. The product of the cross section and the observable $\sigma\,\mathcal{O}$ is then rewritten as $[x_1\cdots x_n\,\sigma\,\mathcal{O}]/(x_1\cdots x_n)$, and the partial fractions identity (12) is applied to the denominator. The numerator is regular: the two-parton singularities are regulated by the x_i, and the remaining singularities are regulated by the observable (recall that we restrict the discussion to this kind of observables). The general partial fractions formula now gives us a "hierarchy" of singularities: the "leading" one $1/x_{\sigma_1}$, the "subleading" one $1/(x_{\sigma_1}+x_{\sigma_2})$, and so forth. It turns out that for the observables under consideration, we have to consider only the leading and subleading singularities. Let us write this singular part, for a specific term in the sum, as k_A, and the remaining terms from the product $\sigma\,\mathcal{O}$ as τ_A; the index A parametrizes the set of permutations. We then have $\sigma\,\mathcal{O}=\sum_A k_A\tau_A$. Each of the terms is singular only if the "leading variable" is zero. The subtraction term can therefore be constructed with respect to the leading and subleading variable in k_A. More precisely this is done in the following way. For each index A there is a particular parton label i_A related to it (for example, for $1/[E_1^2\,(E_1^2+v_{12})]$ or $1/[v_{12}\,(E_1^2+v_{12})]$ this would be $i_A=1$). The phase space $\mathrm{dPS}^{(n)}$ is factorized according to $\mathrm{dPS}_{i_A}\,\mathrm{dPS}^{(n-1)}$, i.e. the one-parton phase space corresponding to parton i_A is pulled out of the full phase space. Finally the phase space integration from Eqn. (2) is rewritten as

$$\begin{aligned}\int \mathrm{dPS}^{(n)}\,\sigma\,\mathcal{O} \;&=\; \sum_A\int \mathrm{dPS}_{i_A}\,k_A\left(\int \mathrm{dPS}^{(n-1)}\tau_A-\left[\int \mathrm{dPS}^{(n-1)}\tau_A\right]_{\text{soft/coll. limit}}\right)\\ &+\;\sum_A\int \mathrm{dPS}_{i_A}\,k_A\left[\int \mathrm{dPS}^{(n-1)}\tau_A\right]_{\text{soft/coll. limit}}. \end{aligned}\tag{13}$$

[i] This decomposition has a nice technical property. Holding σ_1 fixed, excluding $1/x_{\sigma_1}$ from the sum, and setting x_{σ_1} to zero in the remainder, the sum over the restricted set of permutations yields $1/(x_1\cdots\hat{x}_{\sigma_1}\cdots x_n)$. This property is useful to recombine terms after the soft and collinear limits have been performed.

The first integral is finite and can be calculated numerically. The second integral contains all infrared singularities. The term in the square bracket is simple because of the factorization theorems of QCD, and the one-particle integral over the kernel k_A and the factorization contribution from the term in the square brackets can be performed easily[j].

4. Numerical Results and Applications: α_s and the Gluon Density

We now turn to some applications of (2+1)-jet observables. Loosely speaking, observed jets of hadrons can be identified with partons at hard scales. The subsequent fragmentation process is assumed to be sufficiently soft in order to keep non-perturbative effects small. Jets cannot be defined in a canonical way, rather they are objects by definition. It is convenient to define jets in terms of iterative cluster algorithms[k]. There are three ingredients in the definition of a cluster algorithm: (i) a distant measure which determines the relative distance of two clusters in momentum space (for example their invariant mass $s_{ij} = (p_i + p_j)^2$, (ii) a mass scale M^2 that determines whether two clusters are to be combined into a single one (for example if $s_{ij} < M^2$), and (iii) a recombination procedure that prescribes how two clusters p_i and p_j are to be merged into a single cluster p_* (for instance $p_* = p_i + p_j$). There are various choices for (i), (ii) and (iii). The example given is the JADE algorithm [21] in the so-called E-scheme. This particular algorithm is known for large hadronization corrections. This problem can be reduced by explicitly forcing clusters to behave as if they were massless (JADE and P-schemes). The large corrections partly come from the fact that for JADE-type algorithms, soft partons are combined first, even if they differ considerably in their direction. Intuitively, such partons should possibly be combined with other partons nearby, even if they are hard. This can be achieved by means of a distance measure based on relative transverse momentum, leading to the k_T algorithm, defined in the Breit frame of reference. For deeply inelastic scattering, it has been introduced in Ref. [22][l].

---

[j] There are a few complications, though. The first one is that the required sum of 9! terms, which has to be done numerically, contains too many terms to be efficient (or even possible). The solution is to perform the partial fractions decomposition separately for the energy and angle terms. The second complication is the presence of correlations in azimuthal angle for collinear singularities, which have been mentioned above. They can be dealt with by introducing a fictitious azimuthal variable for the collinear configurations.

[k] "Non-cluster" algorithms of the cone-type suffer from the problem that the outcome may depend on the choice of a seed in the jet-finding process. This effect is particularly large at small transverse momentum. Cluster algorithms are well-defined without any ambiguity in the procedure.

[l] These algorithms have the property of factorization, which means that the $(n+1)$-jet cross section factorizes in exactly the same way into a product of a coefficient function and a parton density as a structure function does, without any residual explicit x_B-dependence. This kind of factorization is technically convenient for the calculation of resummed cross sections. We wish to stress that any acceptance cut on the final-state jets destroys this factorization property. What is important to have in a well-defined perturbative prediction is that the ***singular*** parts factorize in a universal way,

In the case of hadrons in the initial state, the corresponding remnants need a special treatment, because partons in the very forward direction lead to initial-state singularities. This problem is taken care of by either including the remnant jet in the clustering procedure (as in the case of the *modified* JADE algorithm) or by discarding partons which are too close to the remnant (as for the k_T algorithm). In any case a well-defined prediction for the (2+1)-jet cross section is obtained.

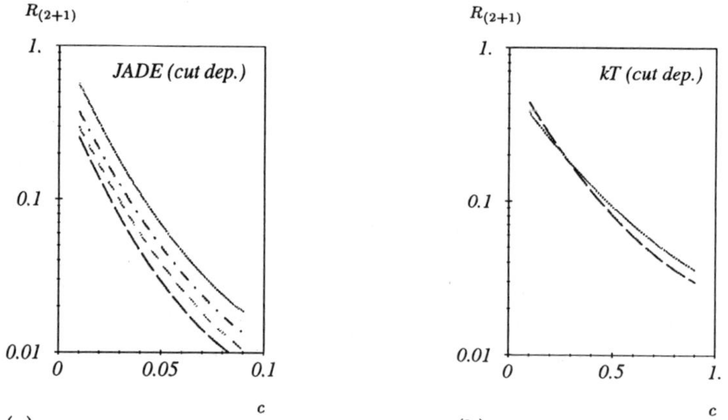

(a) (b)

Figure 1: *Cut-dependence of* $R_{(2+1)}$ *for the modified JADE cluster algorithm (a) and for the* k_T *algorithm (b). The jet resolution mass is* $M^2 = cW^2$ *for the JADE algorithm and* $M^2 = cQ^2$ *for the* k_T *algorithm. Leading order [— —], next-to-leading order: E-scheme [· · ·], P-scheme [- - -], JADE-scheme [- · -]. The recombination scheme dependence of the* k_T *algorithm is small, therefore only the E-scheme is shown in this case.*

In order to reduce systematic errors, it is convenient to consider the (2+1)-jet rate $R_{(2+1)} = \sigma_{2+1}/\sigma_{\rm tot}$. Figure 1 shows the cut-dependence of $R_{(2+1)}$ for two different jet algorithms[m]. The recombination scheme dependence of the modified JADE algorithm is large, as shown in Fig. 1a. For the E-scheme the QCD corrections may be as large as 100 %; they are considerably smaller for the "massless" recombination schemes[n].

which permits the definition of process-independent parton densities. This is, of course, the case for all infrared-safe observables, including the non-factorizing kind of jet algorithms.

[m]The centre-of-mass energy is 300 GeV, the photon virtuality is restricted by $(10\,{\rm GeV})^2 < Q^2 < (20\,{\rm GeV})^2$, the total hadronic energy W is assumed to be larger than 70 GeV and the lepton variable y is restricted to values smaller than 0.7. The parton densities are MRS A′ [23], and the renormalization and factorization scales are set to Q^2.

[n]The large recombination scheme dependence of the JADE algorithm sometimes leads to the conclusion that the algorithm is ill-defined. This is a misconception. Different recombination schemes define different observables, and there is no reason why different observables should not lead to different predictions. However, the strong dependence of the jet rate on the recombination scheme

The perturbative stability of the k_T algorithm is better, here the QCD corrections are of the order of a few per cent (Fig. 1b).

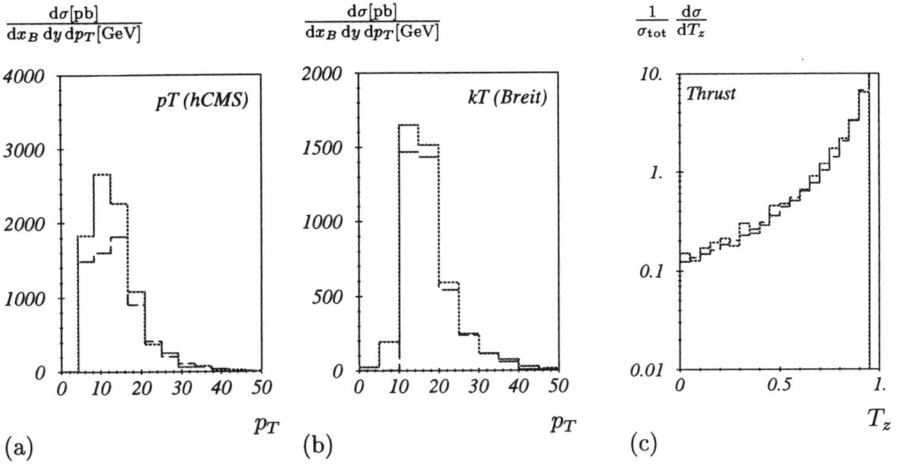

Figure 2: *Transverse momentum distribution in the hadronic CMS for the JADE algorithm (a), and in the Breit system for the k_T algorithm (b). Thrust distribution in the Breit frame (c). Shown are the leading order [— —] and the next-to-leading order [—].*

Deeply inelastic scattering is a multi-scale problem. Besides the photon virtuality Q^2 there are other hard scales related to the scattering process, for example the transverse momentum p_T of partons or jets. These hard scales can be very different from each other. The transverse momentum distribution of the jets for fixed lepton variables of $x_B = 0.01$ and $Q^2 = (20\,\mathrm{GeV})^2$ is shown in Fig. 2 for the JADE P-scheme with a recombination scale of $0.02\,W^2$ in the hadronic CMS (a) and for the k_T algorithm with a recombination scale of $0.5\,Q^2$ in the Breit frame (b). The transverse momentum distribution is quite broad, and therefore p_T^2 is not necessarily of the order of Q^2. This introduces a theoretical uncertainty, because it is not clear whether Q^2 or p_T^2 is the right scale to be used in the renormalized coupling $\alpha_s(\mu_r^2)$ and in the parton densities $f_i(\xi, \mu_f^2)$. As long as no resummed calculation is available, this uncertainty has to be considered as a contribution to the systematic error in case that physical quantities are extracted by using fixed-order next-to-leading-order calculations[o].

As an example for an event shape variable we show a Thrust distribution in Fig. 2c. The phase space region is the same as the one of the jet rates shown before. The

points to the fact that jet masses are actually important.

[o]Not shown here because of lack of space is a comparison of the scale dependence in leading and next-to-leading order. For the jet cross sections shown in Fig. 1 the scale dependence is reduced by about a factor of 2, if the renormalization and factorization scales are varied between $0.5\,Q$ and $2\,Q$.

Current Thrust variable is defined by $T_z = 2\sum_i p_{i\|}/Q$ [24], where the sum is over all partons in the current hemisphere (the hemisphere containing the virtual photon). The quantity $p_{i\|}$ is the longitudinal component of the parton momentum p_i along the virtual photon direction in the Breit frame of reference. The Thrust variable goes to one for a (1+1)-jet-like configuration. Values of $T_z < 1$ are therefore obtained from processes of $\mathcal{O}(\alpha_s)$. The QCD corrections are moderate. It turns out that the NLO calculations do not describe the experimental data well (see, for example, Ref. [25]). The inclusion of a power suppressed term of the form A/Q with a large coefficient A brings data and the theoretical prediction in good agreement. A quantitative derivation for the coefficient A in terms of an effective coupling constant at low scales is given in Ref. [26].

Applications of NLO calculations in deeply inelastic scattering include the measurement of the strong coupling constant [27, 28] and a direct fit of the gluon density [29]. The α_s measurements show a clear evidence for the running of the coupling constant as a function of the scale Q^2. To complement the measurement via jet rates, measurements of α_s based on event shape variables are currently under way [25][p]. The gluon density has been determined via jet rates by means of a Mellin transform method [30, 31] which makes the repeated evaluation of the NLO cross section for the purpose of the fitting procedure feasible[q]. The quark distributions are well known, and therefore the quark-initiated contribution can be subtracted. The obtained direct fit is in good agreement with gluon density parametrizations from global fits[r].

5. One-Particle-Inclusive Processes

One-particle-inclusive processes are another very promising field. The corresponding cross sections can be considered to be related to a special type of observable $\mathcal{O}$, the fragmentation functions $D_i\left(\xi, \mu_D^2\right)$. In the terminology used above, they are not infrared-safe, so that collinear singularities remain, which do not cancel. However, the singularities are of a universal form, and can be absorbed into the fragmentation functions. The redefined fragmentation functions are finite and universal, i.e. process-independent.

Figure 3 shows a comparison of recent preliminary experimental data for charged

---

[p] For the Thrust variable, the requirement of a power-suppressed term A/Q makes an α_s measurement conceptually difficult, because the coefficient A is parametrized by a universal effective coupling constant $\alpha_{s,\text{eff.}}$, whereas the perturbative contribution has the renormalized coupling constant as a coefficient. There are matching procedures to disentangle the two terms, but the size (about the same order of magnitude as the perturbative contribution) of the non-perturbative contribution is disturbing.

[q] This method is actually quite general and permits the calculation of arbitrary observables, the only limitation being that the factorization scale has to be fixed.

[r] For a specific value of the momentum fraction, the error bands show a cross-over. This phenomenon is well known from global fits [32] and is a consequence of the limited number of parameters in the *ansatz* for the gluon density.

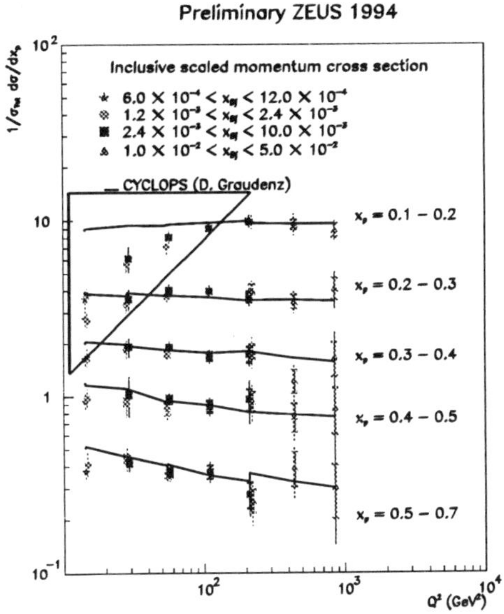

Figure 3: *Theoretical prediction vs. experimental data for the Q^2-dependence of $\sigma^h/\sigma_{\mathrm{tot}}$ for various bins of x_p. The kinks in the theoretical curves come from the overlay of several bins in x_B for the same bin in Q^2. The region within the triangle cannot be described by means of the fragmentation function picture; see text.*

hadron production from the ZEUS collaboration with a next-to-leading order calculation. The employed variable is $x_p = 2P_h/Q$, where P_h is the momentum of the observed charged hadron h in the current hemisphere of the Breit frame. The cross section σ^h is a convolution of parton densities, fragmentation functions and perturbatively calculated coefficient functions. The parton densities are from the MRS A′ set and the fragmentation functions are from Refs. [33, 34]. The coefficient functions have been calculated in Ref. [35]. The numerical results have been obtained by using the implementation in the program CYCLOPS of a recent recalculation [36]. The comparison of the theoretical prediction with experimental data shows a very good agreement, except for small values of Q^2 and small values of x_p.

What is the reason for this discrepancy? It is instructive to consider the situation in terms of the rapidity variable y_h of the observed hadron. We consider the leading-order process, with a single parton fragmenting into the hadronic final state. Under the assumption that the hadron has a typical transverse mass m_T of $\mathcal{O}(500\,\mathrm{MeV})$, the relation between x_p and y_h can be calculated. In the Breit frame, a positive value of y_h stands for production in the current direction, and a negative value for production in the target direction. In order to make the fragmentation function picture a valid description, the hadron should be produced closely in rapidity to the parent parton, which means that y_h should be larger than about $y_{\min} = 1$ units of rapidity. This can be translated into a lower bound on x_p:

$$x_{p,\min} = \frac{2m_T}{Q} \frac{1}{\sqrt{1 - (\tanh y_{\min})^2}}. \tag{14}$$

This excludes the region roughly indicated by the triangle in the figure. The theoretical prediction thus fails in this region because the fragmentation function picture is not applicable: the hadron is not produced sufficiently close to the parent parton. Another reason is that mass effects (which are not included in the fragmentation function formalism) become important at small Q^2 and small x_p for $x_p \approx 2m_\pi/Q$. This excludes a similar region in the plot. Outside of this critical region, in particular at large values of Q^2, the good agreement shows the universality of fragmentation functions.

A future application might be the determination of α_s via scaling violations of the fragmentation functions. This has been studied in some detail in Ref. [37]. It turns out that the main uncertainty is the dependence on a parametrization of the parton densities. This dependence is reduced at large values of Q^2, but then the available luminosity is the limiting factor. Another possibility might be the measurement of α_s from the hard scattering matrix element by means of p_T spectra of charged particles. For this the NLO calculation is not yet available; however, as indicated above, the implementation of the matrix elements in DISASTER++ is already such that it may become available in the near future. The advantage of such a measurement would be that the fragmentation process is no longer modelled by means of event generators as in the case of jet cross sections or event shape variables, but described by a universal parametrization of fragmentation functions with a QCD-predicted scale evolution.

Summary and Outlook

We have reviewed the status and the applications of next-to-leading-order calculations in deeply inelastic lepton–nucleon scattering. In the last few years, programs have become available that permit the calculation of arbitrary infrared-safe observables of the (2+1)-jet type. In next-to-leading order, these programs are still restricted to the case of photon exchange. A natural extension would be the case of Z exchange, and the inclusion of charged-current processes. Moreover, the next-to-leading order is not yet available for polarized initial states, and for the p_T spectra of identified hadrons.

Concerning the phenomenology of hadronic final states, the main problem is to understand the phase space region of small Q^2 (where, for example, the Thrust distribution is not well described in next-to-leading order), and the region of forward jets. With high-statistics data, the extraction of physical quantities can be improved by hard cuts to remove these dangerous regions, but a better theoretical understanding is certainly desirable. The physics of deeply inelastic hadronic final states will thus continue to be a very interesting topic in the future.

Acknowledgements

I would like to thank the organizers of the workshop for having prepared a very interesting and stimulating meeting. Moreover I gratefully acknowledge discussions with J. Bromley, N. Brook, J. Collins, T. Doyle, Z. Kunszt and T. Sjöstrand. The ZEUS Collaboration provided me with the plot from Fig. 3.

References

1. J.C. Collins, D.E. Soper and G. Sterman, in: *Perturbative Quantum Chromodynamics*, ed. A.H. Mueller, World Scientific (Singapore, 1989).
2. K. Konishi, A. Ukawa and G. Veneziano, Nucl. Phys. **B157** (1979) 45.
3. K. Fabricius, G. Kramer, G. Schierholz and I. Schmitt, Z. Phys. **C11** (1982) 315.
4. R.K. Ellis, D.A. Ross and A.E. Terrano, Nucl. Phys. **B178** (1981) 421.
5. Z. Kunszt and D.E. Soper, Phys. Rev. **D46** (1992) 192.
6. D. Graudenz, Comput. Phys. Commun. **92** (1995) 65.
7. D. Graudenz, *The three-jet cross section at order α_s^2 in deep-inelastic electron–proton scattering*, Doctoral thesis (September 1990).
8. D. Graudenz, Phys. Lett. **B256** (1991) 518.
9. D. Graudenz, Phys. Rev. **D49** (1994) 3291.
10. T. Brodkorb and E. Mirkes, `DISJET` *program manual*, preprint MAD/PH/821 (Madison 1994).
11. T. Brodkorb and J.G. Körner, Z. Phys. **C54** (1992) 519.
12. T. Brodkorb and E. Mirkes, Z. Phys. **C66** (1995) 141.
13. E. Mirkes and D. Zeppenfeld, `MEPJET 2.0` *program manual*, unpublished (1997).
14. E. Mirkes and D. Zeppenfeld, Phys. Lett. **B380** (1996) 205.
15. W.T. Giele and E.W.N. Glover, Phys. Rev. **D46** (1992) 1980.
16. W.T. Giele, E.W.N. Glover and D.A. Kosower, Nucl. Phys. **B403** (1993) 633.
17. S. Catani and M. Seymour, `DISENT 0.0` *program manual*, unpublished (1996).
18. S. Catani and M. Seymour, in: Proceedings of the workshop on future physics at HERA, eds. G. Ingelman, A. De Roeck, R. Klanner (Hamburg 1995/96).
19. S. Catani and M. Seymour, Nucl. Phys. **B485** (1997) 291.
20. D. Graudenz, `DISASTER++` *program manual*, in preparation.
21. JADE Collaboration, Z. Phys. **C33** (1986) 23.
22. S. Catani, Yu.L. Dokshitser and B.R. Webber, Phys. Lett. **B285** (1992) 291.

23. A.D. Martin, R.G. Roberts and W.J. Stirling, Phys. Rev. **D51** (1995) 4756.
24. K.H. Streng, T.F. Walsh and P.M. Zerwas, Z. Phys. **C2** (1979) 237.
25. K. Rabbertz, talk presented at the DIS 97 workshop in Chicago (April 1997).
26. M. Dasgupta and B.R. Webber, preprint Cavendish–HEP–96/5 (Univ. of Cambridge, April 1997).
27. H1 Collaboration, Phys. Lett. **B346** (1995) 415.
28. ZEUS Collaboration, Phys. Lett. **B363** (1995) 201.
29. M. Hampel, *Measurement of the Gluon Density from Jet Rates in Deep Inelastic Scattering*, Doctoral thesis, preprint PITHA 97/13 (Aachen Univ., February 1997).
30. D. Graudenz, M. Hampel, A. Vogt and Ch. Berger, Z. Phys. **C70** (1996) 77.
31. D. Graudenz, M. Hampel and A. Vogt, in: Proceedings of the workshop on future physics at HERA, eds. G. Ingelman, A. De Roeck, R. Klanner (Hamburg 1995/96).
32. A.D. Martin, R.G. Roberts and W.J. Stirling, Phys. Rev. **D43** (1991) 3648.
33. J. Binnewies, B.A. Kniehl and G. Kramer, Z. Phys. **C65** (1995) 471.
34. J. Binnewies, *Fragmentation Functions in Next-to-Leading Order QCD*, Doctoral thesis, preprint DESY–97–128 (July 1997).
35. G. Altarelli, R.K. Ellis, G. Martinelli and S.Y. Pi, Nucl. Phys. **B160** (1979) 301.
36. D. Graudenz, *Heavy-Quark Production in the Target Fragmentation Region*, Habilitationsschrift, preprint CERN–TH/96–52 (February 1996), to appear in "Fortschritte der Physik".
37. D. Graudenz, in: Proceedings of the workshop on future physics at HERA, eds. G. Ingelman, A. De Roeck, R. Klanner (Hamburg 1995/96).

HADRONIC FINAL STATES IN DIS AT HERA

JUTTA HARTMANN
McGill University, Department of Physics
3600 University St., Montréal PQ H3A 2T8, Québec, Canada
E-mail: Eisenmann@vxdesy.desy.de

On behalf of the ZEUS and H1 Collaboration

ABSTRACT

Hadronic final states in Deep Inelastic Scattering at HERA are investigated in order to test perturbative QCD and universality of QCD. Results on Bose Einstein correlations, on KNO scaling, on charged particle distributions, and on event shapes are presented and compared to results from e^+e^-, pp, $p\overline{p}$ and fixed target experiments. Perturbative QCD (pQCD) describes well all of these phenomena and thus supports the picture of universality of QCD.

1. Introduction

The HERA ep collider provides collisions of 820 GeV protons with 27.5 GeV positrons. Compared to the pre-HERA area the accessible kinematic range of Bjorken-x, x_{Bj}, and for the invariant mass of the hadronic final state, W, is extended significantly. x_{Bj} reaches three orders of magnitude lower values down to $\mathcal{O}(10^{-6})$ and W reaches a few hundred GeV which is an order of magnitude higher than before. The application of pQCD, in this new phase space and the comparison of HERA results to those from other experiments is an important test of the validity and universality of QCD. However, the basic particles of QCD are quarks and gluons, which are not observable as free particles. One needs the concept of local parton hadron duality (LPHD), which assumes that hadrons still have properties of the partons so that one can draw conclusions on the partons by measuring the kinematics of hadrons.

This review is structured as follows:
After a short overview of the kinematics and reference frames in the different experimental environments (e^+e^-, ep, pp, fixed target experiments) we first present studies on Bose-Einstein correlations and KNO scaling. Then several different studies on charged particle distributions and multiplicities, and on event shapes are reviewed and analysed in the view of universality of QCD.

2. Kinematics

The simplest reference frame to describe ep interactions at HERA is the laboratory frame. However, a direct comparison of results to other experiments is not possible. Physicists prefer a more natural frame which is independent of the experimental environment so that results on particle properties can be compared directly.

Comparison of *ep* scattering to $pp, p\overline{p}$ or fixed target experiments

In *ep* collisions the hadronic center of mass system (HCM) of the exchanged virtual photon, γ^*, and the proton is a natural frame, see Fig. 1. The photon is taken as the positive reference axis (z-direction). This frame is also the preferred one in pp, $p\overline{p}$ or fixed target experiments. The center of mass energy is denoted by W. Hadron momenta are normalised to the energy scale W, or to $W/2$ if the particle flow is separated in the forward ($+z$) and the backward ($-z$) hemisphere of the event. The investigation of the W-dependence of event properties in different experimental environments is thus a good test for the universality of QCD.

Comparison of diffractive *ep* scattering to $pp, p\overline{p}$ fixed target experiments

At HERA a special class of interactions occurs where the proton remains intact and there is a rapidity gap between the proton direction and the rest of the event (Fig. 1). These gaps suggest that there is a colorless particle exchanged which we call a pomeron, $I\!P$. The hadronic final state (excluding the p and the scattered e^+) can mostly be measured in the detector and its energy is denoted by M_X.

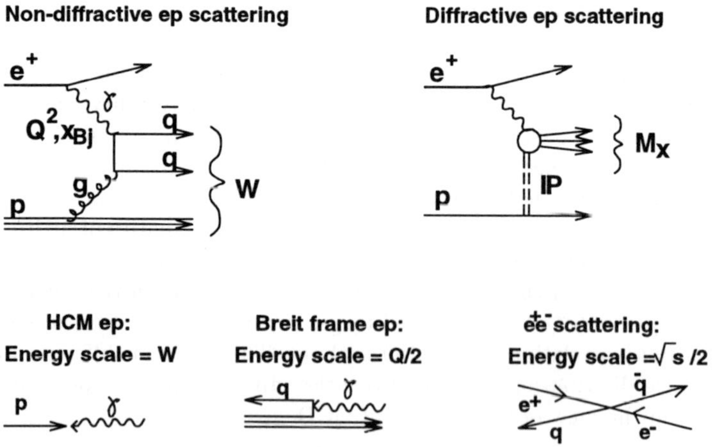

Fig. 1. Reference frames for different interactions: DIS *ep* collision, diffractive DIS in *ep* collisions, HCM in *ep* collisions, Breit frame in *ep* collisions, and e^+e^- collisions.

Comparison of *ep* scattering with e^+e^- experiments

The comparison of *ep* results to e^+e^- experiments is usually done in the Breit frame (Fig. 1). The fragments of the struck quark move to the current region of the Breit frame. The available energy in this hemisphere, i.e. the energy of the struck quark is

$Q/2$. All the rest of the event is found in the target region. Thus the current region corresponds to one hemisphere of e^+e^- collider events, where the available energy is $\sqrt{s}/2$ (Fig. 1).

3. Bose-Einstein Correlations

Bose–Einstein correlations[1] result in an enhanced probability to produce like–sign bosons close together in phase space compared to unlike–sign bosons. This can be understood from the Pauli principle, where identical bosons prefere to occupy the same quantum state.

Bose–Einstein correlations are studied[2] in normalised two–particle densities as function of a variable $T = -(p_1 - p_2)^2 = M^2 - 4m_\pi^2$. The particles under consideration are like–sign and unlike–sign pion pairs. Their 4–momenta are denoted by p_i and their invariant mass is M. T tends to zero if the particle masses m_1 and m_2 and the particle momenta are equal, i.e. if there are identical particles close together in phase space. Thus one expects Bose–Einstein correlations to appear at low T.

Theory predicts a functional form $RR(T) = R_o(1 + a * T)(1 + \lambda \exp^{r^2T^2})$ for the ratio of correlated to uncorrelated two–particle densities. Here λ is a parameter describing the coherence of the source, which emits the bosons, and r corresponds to the radius of the source. Figure 2 shows this distribution as a function of T. From the gaussian shape at low T the radius of the source, which emits the pions, is extracted to $0.61 \pm 0.04\ ^{+0.03}_{-0.05}$ fm. This is well consistent with the proton radius of about 1 fm. Similar studies with diffractive events result in a radius well consistent with that of non–diffractive events.

Many experiments have studied Bose–Einstein correlations and an increasing radius of the boson emitting source on the available energy in the interaction is found.

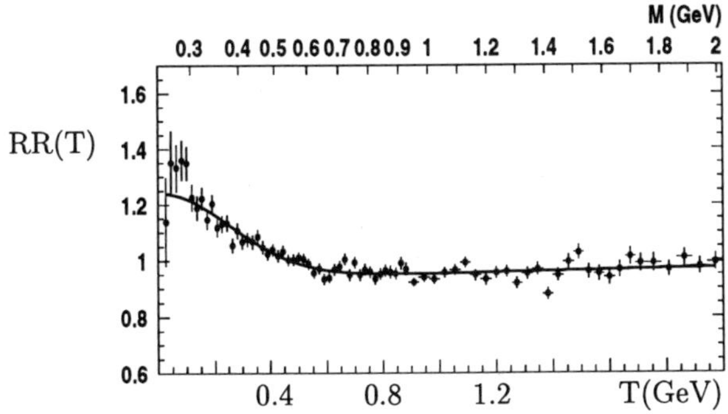

Fig. 2. Bose–Einstein correlations for non–diffractive events[2].

4. KNO Scaling

The KNO model[3] relates the mean number of charged particles $<n>$ and the probability of producing n charged particles in an event to a function $\Psi(z) = <n> \cdot P(n)$, where $z = n/<n>$. It was originally derived assuming Feynman scaling for the particle spectra and predicts in the limit, where $n, <n>$ and the energy scale become very large, that $\Psi(z)$ is independent of the energy. This is referred to as KNO scaling behavior. It was shown later that KNO scaling is an intrinsic property of the parton shower process.

KNO scaling was studied at HERA in the current region of the Breit frame[4,5] and in the hadronic center of mass system[6]. Figure 3 shows results from ZEUS in the Breit frame, where the energy scale Q is low. Here clear scaling violations are seen. In Fig. 4 similar studies by H1 in the HCM system are shown where the energy scale W is large. Here KNO scaling seems to be valid. In high energy e^+e^- experiments KNO scaling has been studied too[7] and at high energies scaling behavior is seen. The Monte Carlo prediction for e^+e^- collisions at high energy is indicated by the dotted line. It will be interesting to study whether KNO scaling at LEP2 at energy scales comparable to W will follow this prediction.

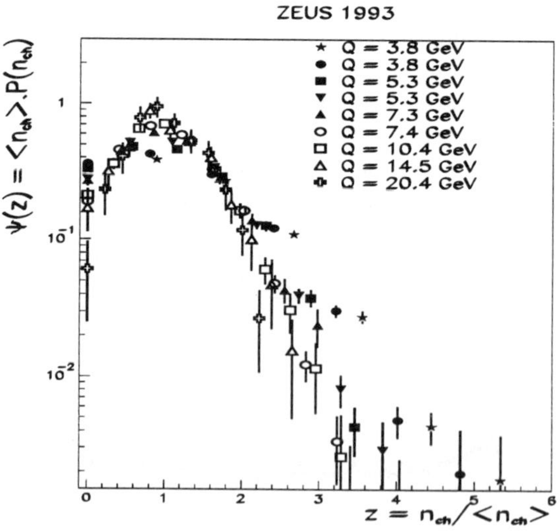

Fig. 3. KNO scaling in the current region of the Breit frame for different Q^2 bins[5].

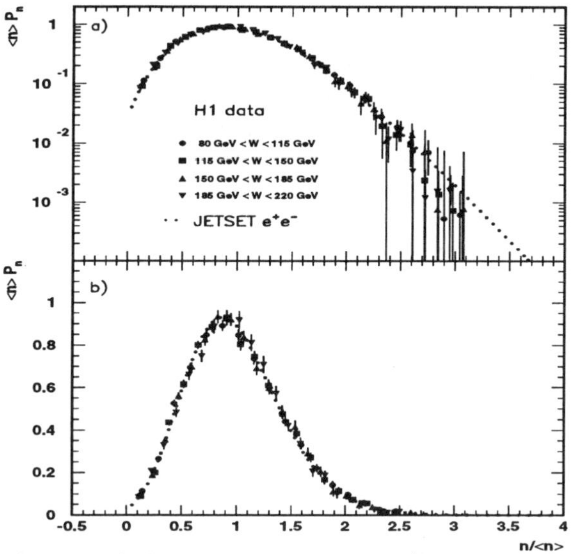

Fig. 4. KNO scaling in the hadron center of mass system, a) logarithmic and b) linear scale[4].

5. Charged Particle Multiplicities

Charged particle multiplicity distributions have been studied by many experiments. The mean number of charged particles increases about linearly with the logarithm of the available energy of the interaction. Figure 5 shows this for the current region of the Breit frame[4]. The full line is a fit to results from e^+e^- experiments. At high energies the results from ZEUS and H1 agree well with the ones from e^+e^- experiments, but at low energy the HERA experiments are systematically below the e^+e^- results. This can be understood as gluon emission from the struck quark when Q^2 is low, which results in migration of particles from the current region to the target region[8] and thus depopulates the particle density in the current region.

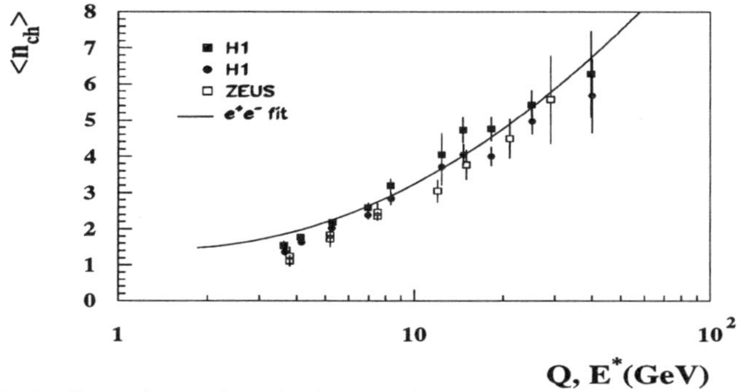

Fig. 5. Charged particle multiplicity in the current region of the Breit frame[4].

H1 has studied mean charged particle numbers $<n>$ and mean transverse energy flow $<E_T>$ as a function of W in the HCM[6], see Fig. 6. Results from other experiments are overlaied. $<n>$ is normalised to a certain rapidity interval. It is impressive, that although there are very different initial states in the experiments and the phase space is not comparable (different rapidity or pseudorapidity ranges) the results line up quite nicely for all the experiments. The prediction for e^+e^- results for the mean number of charged particles is shown as dotted line and is much steeper than for the hadron experiments. This can be understood from the smaller gluonic activity in e^+e^- experiments. The curves for $<E_T>$ and $<n>$ are about parallel, which means that the overall kinematic properties of the particles do not change much when W increases.

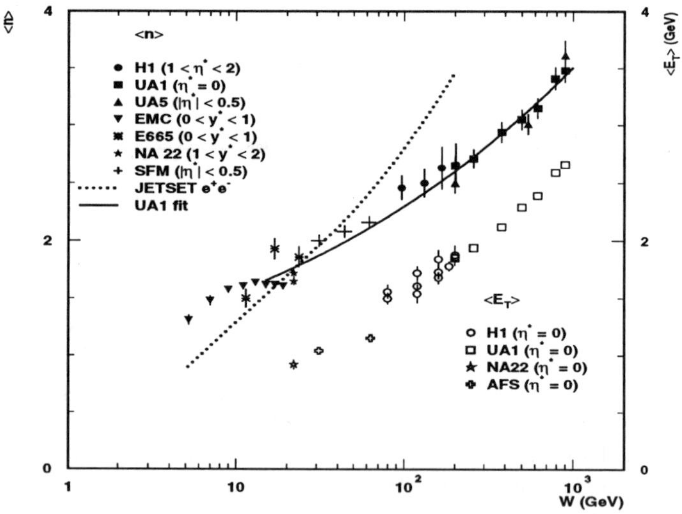

Fig. 6. Charged particle multiplicity and transverse energy flow[6] as function of W.

6. Scaled Momentum Distributions

Quarks and gluons fragment into observable hadrons. The momentum of the hadrons is usually studied as fraction x_p of the initial parton momentum. The latter is approximated by the energy in the relevant event hemisphere, i.e. in the current region of the Breit frame: $x_p = p_z/(Q/2)$. Perturbative QCD predicts an evolution of the scaled momentum with Q^2. Figure 7 shows[8] the peak position of the distribution $\ln 1/x_p$ as a function of Q^2. Results from e^+e^- experiments and different QCD models are overlaied. The assumption of incoherent parton branching processes

$(\ln(1/x_p)_{max} \sim \ln Q + const\ )$ is clearly excluded. The data follow the prediction of coherent parton branching $(\ln(1/x_p)_{max} \sim 0.5 \cdot \ln Q + const)$. The results from e^+e^- experiments are well consistent with the *ep* results in the current region as expected. Here is no effect of the depopulation of the current region visible at low Q^2 because events affected by this mainly are found in the tails instead of the peak position of the $\ln(1/x_p)$ distribution.

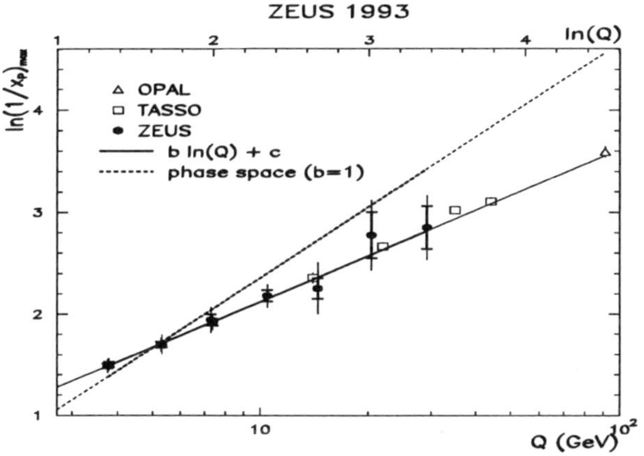

Fig. 7. Peak position of the $\ln(1/x_p)$ distribution[8] as a function of Q^2.

7. Charged Particle Fragmentation Functions

Fragmentation functions D^h describe the probability of partons to convert to a hadron h with a certain fractional momentum x_p: $D^h(x_p, Q^2) = 1/\sigma_{tot} \cdot d\sigma^h/dx_p$. They can not be predicted by QCD, but they can be evolved in Q^2 if they are measured at a certain Q^2 value. The fragmentation functions depend on the strong coupling constant and this offers an opportunity to measure α_S. Figure 8 shows[9] the fragmentation function measured in the current region of the Breit frame as function of Q^2 for several x_p bins. The scaling violations at high Q^2, i.e. the slope of the $D^h(x_p, Q^2)$ distribution, can be used for the determination of α_S. However, the scaled momentum cross section is a product of three factors: the matrix element, the proton structure function and the fragmentation function. A consistent NLO calculation in all three factors is needed for a meaningful determination of α_S. Today only calculations for one α_S are available, see full curve in Fig. 8. The slope of the fragmentation function at low Q^2 and low x_p is due to the depopulation of particles in the current region of the Breit frame as mentioned above. In the low Q^2 region the calculations are not applicable.

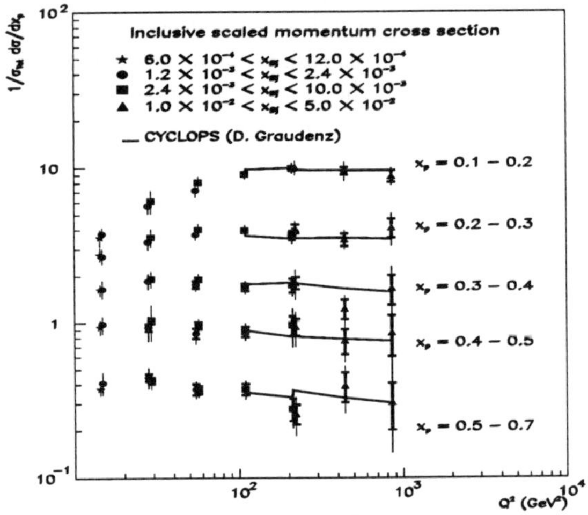

Fig. 8. Differential scaled momentum distribution as function of Q^2 with NLO curves[10]

DELPHI[11] has collected results on scaled momentum distributions from different high energy e^+e^- experiments up to $\sqrt{s} \sim 10^4$ GeV and has extracted α_S. The ZEUS results are well consistent in the overlap region of Q^2 (not shown), but also more statistics is needed before α_S can be extracted from scaling violations from one single experiment with this method.

8. Feynman–x and Mean Transverse Momentum Distributions in the HCM

In the HCM system the Feynman–x distribution $x_F = 2p_z/W$ and the mean transverse momentum $< p_t^{*2} >$ of charged particles were studied in the ZEUS and H1 experiments[5,12]. The longitudinal (p_z) and transverse momenta were measured with respect to the direction of the virtual photon. The x_F distribution of the high W non–diffractive ZEUS events are well consistent with results from high energy e^+e^- experiments where $\sqrt{s}$ is the energy scale, see Fig. 9a. In contrast to this, the same ZEUS data (high W) show a steeper x_F distribution than lower W data from fixed target experiments, see Fig. 9b. This is expected from the larger phase space for gluon emission at high W which results in a softer x_F distribution, i.e. more particles with lower mean momentum are produced. However, the mean $<p_t^{*2}>$ and x_F distribution of low M_X diffractive ZEUS events are well consistent with the results of similar low energy fixed target experiments, see Fig. 10a,b. M_X is obviously the relevant energy scale in diffractive events which corresponds to W in non–diffractive

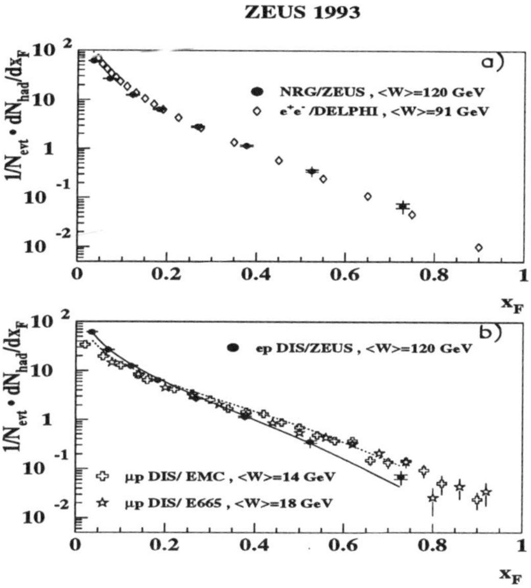

Fig. 9. Feynman–x distribution in the hadronic center of mass system. Shown are a) non–diffractive events with high W compared to e^+e^- data of about similar W and b) the same ZEUS data compared to lower energy data.

events. The energy scale rules uniquely the kinematics of the partons independently of the initial state and even if the event is initiated by a $I\!P$ exchange. This indicates that partons are involved in $I\!P$–initiated events as well as in non–diffractive events.

9. Event Shapes in Diffractive Events

The event topology can be used to study the influence of gluons. The apparence of gluon emission results in broader events. In Fig. 11 the sphericity and mean transverse momentum squared with respect to the event axis in diffractive events is shown[13] as a function of M_X. The result is compared to the corresponding distributions of e^+e^- experiments. The events are broader in diffractive ep scattering of similar energy scale. This is assumed to result from an additional graph of $I\!P \rightarrow gg$, which adds gluonic activity to the diffractive ep final state but does not exist in e^+e^- scattering. In e^+e^- experiments only $\gamma \rightarrow q\overline{q}$ and $\gamma \rightarrow q\overline{q}g$ occur. In diffractive ep scattering these two contributions correspond to the processes $\gamma I\!P \rightarrow q\overline{q}$ and $\gamma I\!P \rightarrow q\overline{q}g$. In ep collisions however there can be another contribution from the coupling $I\!P \rightarrow gg$: $\gamma I\!P \rightarrow \gamma gg \rightarrow q\overline{q}g$. This additional gluonic activity in the final state of ep collision results in the broader events compared to e^+e^- collision events.

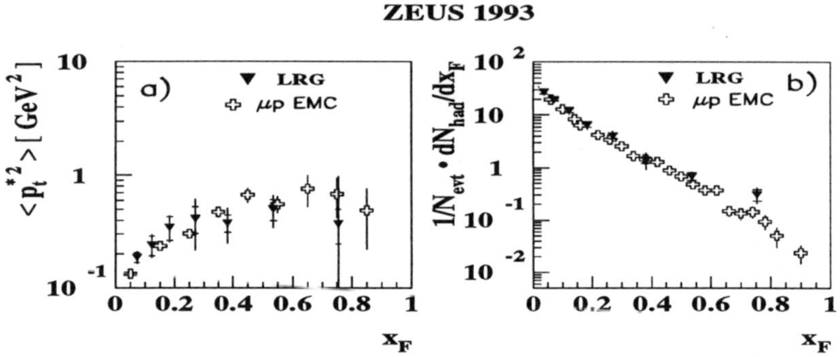

Fig. 10. Mean p_t^* distribution a) and differential x_F distribution b) as function of Feynman–x. Compared are ZEUS diffractive events[5] with low energy fixed target data.

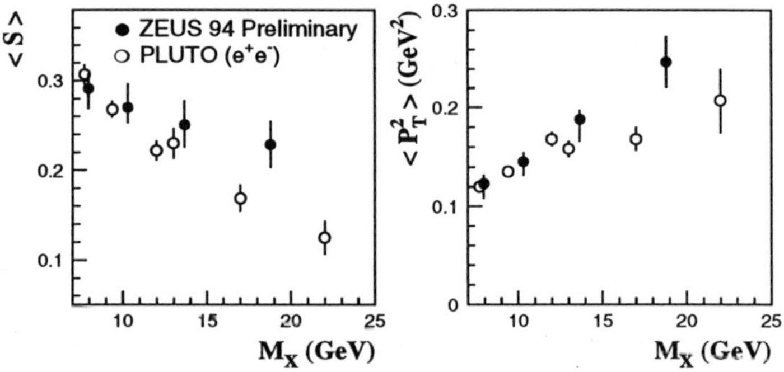

Fig. 11. Sphericity and $<p_t^2>$ as a function of M_X from[13].

10. α_s Determination from Event Shapes

The probability of gluon emission from a quark line is proportional to α_s. Thus the fraction of dijet events in DIS allows to determine α_s. This method was used by both HERA experiments[14,15] to determine α_s. The method requires to correct the detector level dijet rate to the parton level and to compare this to NLO calculations. However, it is known that parton level distributions depend significantly on the input parameters of the various models and thus this method suffers under large hadronisation uncertainties.

The hadron level is much less uncertain and is in principle only dependent on the quality of the detector simulation. Recently second order QCD calculations including non perturbative hadronisation corrections have become available which allow to compare data at hadron level to the theory. The calculations are done for mean values of event shape variables F like thrust, jet–broadening and jet–masses: $< F >=< F >^{pert} + < F >^{pow}$. They depend on α_s in both the perturbative and in the power part of the contribution. The term $< F >^{pow}$ is proportional to $1/Q$ and on a new universal effective coupling constant $\overline{\alpha}_{eff}$, which is independent of F. H1 has applied this approach to the data[16]. Figure 12a shows the hadron level differential distributions of the $1 - Thrust$ distribution in different Q bins. The curves are 2nd order parton level calculations. The parton level calculations converge to the measured hadron level distributions as Q becomes higher. In Fig. 12b the mean values of $1 - Thrust$ is plotted versus Q. The step between the full (fit to hadron level) and the dotted line (NLO partons) corresponds the size of the hadronisation correction. It decreases with increasing Q as expected. The fit to the data yields $\alpha_s = 0118 \pm 0.001 \pm\ ^{0.007}_{0.006}$. The new effective coupling constant $\overline{\alpha}_{eff}$ is determined to $\overline{\alpha}_{eff} = 0.491 \pm 0.003 \pm\ ^{0.079}_{0.042}$ and is valid for all four investigated event shape variables in[16].

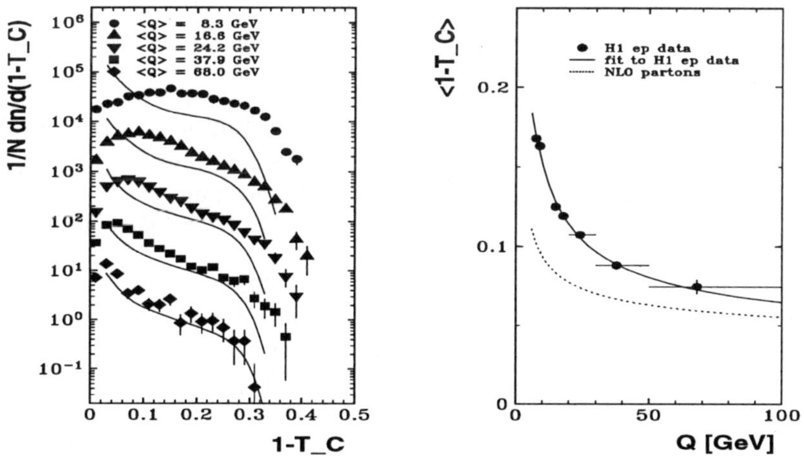

Fig. 12. Differential 1–*Thrust* distribution in Q bins and $<1 - Thrust>$ as function of Q from[16].

11. Summary

The predictions of perturbative QCD have been tested in a large Q^2 and x_{Bj} range and for many different initial state configurations. Particle properties in the different experimental environments have been transformed to comparable reference frames and have been compared to each other. After transformation the particle properties from experiments with different initial states usually are consistent with each other. Where small differences occur they can be understood in terms of pQCD. This supports the assumption that QCD is a universal theory.

1. G. Goldhaber et al., Phys. Rev. Lett. 3 (1959) 181;
   G. Goldhaber, S. Goldhaber, W. Lee and A. Pais, Phys. Rev. 120 (1960) 300.
2. H1 Collab., S. Aid et al., Z. Phys. C75 (1997) 437.
3. Z. Koba, H.B. Nielsen, P. Olesen, Nucl. Phys. B40 (1972) 317.
4. H1 Collab., S. Aid et al., Nucl. Phys. B445 (1995) 3.
5. ZEUS Collab., M.Derrick et al., Z. Phys. C70 (1996) 1.
6. H1 Collab., S. Aid et al., Z. Phys. C 72 (1996) 573.
7. OPAL Collab., P.D. Acton et al., Zeit. Phys. C53 (1992) 539;
   TASSO Collab., W. Braunschweig et al., Zeit. Phys. C45 (1989) 193;
   DELPHI Collab., P.Abreu et al., Zeit. Phys. C56 (1992) 63.
8. ZEUS Collab., M. Derrick et al., Z. Phys. C67 (1995) 93.
9. J. Bromley, Proceedings of the 5th International Workshop on Deep Inelastic Scattering and QDC, 11.-14. April 1997, Chicago, IL, USA.
10. D. Graudenz, CERN-TH/96-52 and CYCLOPS program.
11. DELPHI Collab., P. Abreu et al., CERN-PPE/96-185, submitted to Phys. Lett. B.
12. H1 Collab., I. Abt et al., Z. Phys. C 63 (1994) 377.
13. J. Hernandez, Proceedings of the 5th International Workshop on Deep Inelastic Scattering and QDC, 11.-14. April 1997, Chicago, IL, USA.
14. H1 Collab., S. Aid et al., Phys. Lett. B346 (1995) 415.
15. ZEUS Collab., M. Derrick et al., Phys. Lett. B363 (1995) 201.
16. H1 Collab., S. Aid et al., DESY-97-098, accepted for publication in Phys. Lett. B.

RECENT TESTS OF PARTON HADRON DUALITY IN MULTIPARTICLE FINAL STATES

WOLFGANG OCHS
Max-Planck-Institut für Physik (Werner-Heisenberg-Institut)
Föhringer Ring 6, D-80805 Munich, Germany
E-mail: wwo@mppmu.mpg.de

ABSTRACT

A large variety of jet properties can be described perturbatively by evolving the parton cascade down to small scales of order of a few 100 MeV. We discuss two recent applications of this approach: 1. the soft limit of the particle spectrum which is nearly energy independent as expected for the soft gluon bremsstrahlung; further tests of this interpretation in e^+e^- and DIS are presented; 2. the steep decrease of the distribution of rapidity gaps in e^+e^- annihilation can be explained by the Sudakov formfactor for quark jets.

1. Introduction

An important goal in the study of high energy hard collision processes is the detailed understanding of the hadronic final state and its characteristic jet structure. There are two mayor fields of interest, one concerning the predictions in QCD perturbation theory, mainly on the jet final states, the second concerning the colour confinement phenomena, how the partonic cascade evolves into the final state of hadrons.

In the popular models for particle production in hard collision processes it is assumed that first the primarely produced partons evolve by bremsstrahlung processes according to perturbation theory into jets of partons until a characteristic resolution scale of $Q_0 \sim 1$ GeV is reached. Thereafter, non-perturbative processes take over and the final hadronic particles, often through intermediate resonances, are produced, for example, by a string mechanism[1] or through cluster formation.[2] These models have been proven to be very successful in the description of the experimental data over the years. On the other hand, they are only accessible through a Monte Carlo code and involve quite a number of a priori unknown parameters in the description of the hadronization phase.

We consider here another approach, based on the concept of "Local Parton Hadron Duality" (LPHD).[3] The parton cascade is evolved further down to a scale of about $Q_0 \sim 250$ MeV where Q_0 is now the cut-off in the transverse momentum. Then one compares directly an observed quantity with the corresponding parton level calculation without any hadronization correction.

An immediate benefit of this approach are the analytical results in many cases which provide a deeper understanding of how the theory is actually tested by a particular measurement. For example, the scaling properties and the systematics of their violations can only be formulated in a meaningful way using analytical formulae. Some results can be expressed in compact form by simple QCD numbers.

Theoretical calculations can be carried out in the simplest case in the Double Log Approximation (DLA),[4,5] which provides the high energy limit, or in the Modified Leading Log Approximation (MLLA)[6] which includes finite energy corrections; they are usually essential to obtain quantitative agreement with experiment at present energies. The original success was in the description of the inclusive hadron distributions,[3] thereafter it has been applied to a large variety of different observables.[7] In recent years more applications have been worked out and the experimental results from LEP, HERA and TEVATRON gave further support to this approach.[8]

The duality picture is clearly rather bold and it is not clear a priori for which observables and in which kinematic ranges it applies. This has to be clarified ultimately by experiment. It is obvious that this model cannot compete with the standard hadronization models in the description of the various details of the final state like the production of different particle species or resonances. It has so far been applied successfully for suitably averaged inclusive quantities.

The main virtue of this approach is its intrinsic simplicity which involves only two essential parameters, namely the QCD scale Λ and the non-perturbative mass scale Q_0. In some cases the absolute normalization is taken as additional parameter; on the other hand, in a recent calculation of multiplicities beyond MLLA it was suggested that one parton corresponds to one hadron in this duality picture.[9]

The hope is that with increased phenomenological knowledge about successes and limitations of this approach our theoretical understanding of the confinement mechanism will improve. In this presentation we dicuss two recent applications, one concerning the soft limit of the energy spectrum[10] and the other one the rate for events with large rapidity gaps,[11] i.e. events in a quasi exclusive limit.

2. Main Ingredients of the Analytical Calculations

In this section we give a short overview of the analytical treatment of multiparticle processes. This involves the matrix elements for the primary hard subprocess and the evolution of the primary partons into jets by bremsstrahlung cascades. The gluon bremsstrahlung at small angles ϑ with energy E off the primary parton of type A ($A = q, g$) with jet momentum P_j is given by

$$dn_A = \frac{C_A}{N_C}\gamma_0^2(p_\perp)\frac{d\vartheta}{\vartheta}\frac{dE}{E}, \qquad \gamma_0^2(p_\perp) = \frac{2N_C\alpha_s(p_\perp)}{\pi} = \frac{\beta^2}{\ln(p_\perp/\Lambda)}, \qquad p_\perp \geq Q_0 \tag{1}$$

where $p_\perp \approx E\vartheta$, $\beta^2 = 4N_C/b$, $b \equiv (11N_C - 2n_f)/3$ with N_C, n_f the numbers of colours and flavours, also $C_g = N_C$, $C_q = 4/3$. Inside the cascade the soft gluons are coherently produced from all harder partons. For azimuthally averaged quantities the consequences of the coherence effect can be taken into account by the angular ordering prescription[12] which requires the angles of subsequent gluon emissions to be in decreasing order.

The multiparticle properties of the jet can be discussed conveniently by using the generating functional $Z_A(P_j, \Theta; u(p))$. Here P_j and Θ denote the initial parton momentum and opening angle of the jet, and $u(p)$ is a profile function for particle momentum p. The functional is constructed from all the exclusive final states. Then the inclusive densities can be obtained by functional differentiation with respect to the profile function $u(p)$, for example the one particle density by $\rho^{(1)}(p) = \delta Z\{u\}/\delta u(p)\,|_{u=1}$. The properties of these densities can be derived from the evolution equation for Z which relates the functional at scales P_j, Θ to the one at lower scales according to the "decay" $A \to BC$. In MLLA accuracy this evolution equation is given by[7]

$$\frac{d}{d \ln \Theta} Z_A(P_j, \Theta) = \frac{1}{2} \sum_{B,C} \int_0^1 dz \\ \times \frac{\alpha_s(p_\perp^2)}{2\pi} \Phi_A^{BC}(z)[Z_B(zP_j, \Theta)\, Z_C((1-z)P_j, \Theta) - Z_A(P_j, \Theta)] \quad (2)$$

where $\Phi_A^{BC}(z)$ denotes the DGLAP splitting functions. The initial condition for the parton evolution is given by

$$Z_A(P_j, \Theta; \{u\})|_{P_j\Theta = Q_0} = u_A(p = P_j), \quad (3)$$

i.e. at threshold there is only the primary parton.

From (2) one can obtain the evolution equations for particle densities by appropriate functional differentiation. Therefore this equation is the basic tool for deriving the multiparticle properties of a jet analytically. A more detailed discussion on the approximations can be found elsewhere.[7,8,13,14]

3. The Soft Limit of the Particle Spectrum

3.1. Theoretical Predictions Confronted with Experiment

An essential ingredient of the LPHD approach is the evolution of the parton cascade towards small scales. The normalization of all quantities is given by the initial condition (3). Therefore, one has to assume that the perturbative evolution can be applied at small *cms* energies as well. Alternatively, one could start the evolution from a larger *cms* energy whenever the perturbative QCD is considered trustworthy, but then one had to introduce unknown parameters for initialization instead of (3) and a lot of the predictive power of this approach would disappear. Indeed, applying the MLLA calculations[15] the moments of the energy spectra are well described down to such low energies as $\sqrt{s} = 3$ GeV using the initial condition (3).[16]

Of special interest is the behaviour of the soft end of the particle energy spectrum. The coherence of the soft gluon emission from all harder partons forbids the multiplication of the soft particles[3] and the spectrum in the variable $\xi = \ln(P_j/E)$ obtains a "hump-backed" shape.[4,17]

This problem has been studied recently in more detail.[10] The parton density in the variable $\xi = \ln(P_j/E)$ in the DLA is given in lowest order of α_s by

$$\frac{dn_A(E, P_j)}{d\ln E} = \frac{C_A}{N_C}\beta^2 \ln \frac{\ln(E/\Lambda)}{\ln(Q_0/\Lambda)} \tag{4}$$

This single bremsstrahlung contribution is independent of the jet energy P_j; the higher order contributions do depend on P_j but they are negligable in the soft limit $E \to Q_0$. Also the MLLA correction leaves this limit unaltered: the energy conservation effects and large z corrections from the splitting functions which make up the differences between DLA and MLLA can be neglected. If the perturbation theory and LPHD are valid towards such low particle energies E one expects then also that the hadron spectrum becomes independent of the *cms* energy in the soft limit. To exhibit the energy dependence it is convenient to consider the invariant density Edn/d^3p which remains finite for small momenta.[a] Then we expect an energy independent density I_0 of hadrons in the limit where the particle momentum p, or alternatively rapidity y and transverse momentum $k_\perp$, become small:

$$I_0 = \lim_{y\to 0, p_T\to 0} E\frac{dn}{d^3p} = \frac{1}{2}\lim_{p\to 0} E\frac{dn}{d^3p}. \tag{5}$$

The factor $\frac{1}{2}$ takes into account that both hemispheres are included in the limit $p \to 0$.

This prediction is a direct consequence of the coherence of the soft gluon emission: the emission rate for the gluon of large wavelength does not depend on the details of the jet evolution at smaller distances; it is essentially determined by the colour charge of the hard initial partons.

In Fig. 1 we show the experimental results on the invariant density of charged particles for *cms* energies from 3 to 130 GeV in e^+e^- annihilation. An approximate energy independence of the low momentum particle density (within about 20%) is indeed observed; the same is true for identified particles π, K and p.[10] The curves in Fig. 1 represent the MLLA results. At very low energies where $E \approx Q_0$ the QCD perturbative results have to be supplemented by a kinematical relation between parton and hadron spectra taking into account their different masses (for more details, see Ref. 10). The theoretical curves show the approach to the scaling limit and describe well the different slopes at larger particle energies. An important role here

[a] Another alternative is the non-invariant spectrum dn/d^3p^{19} which has the advantage of being independent of the particle mass.

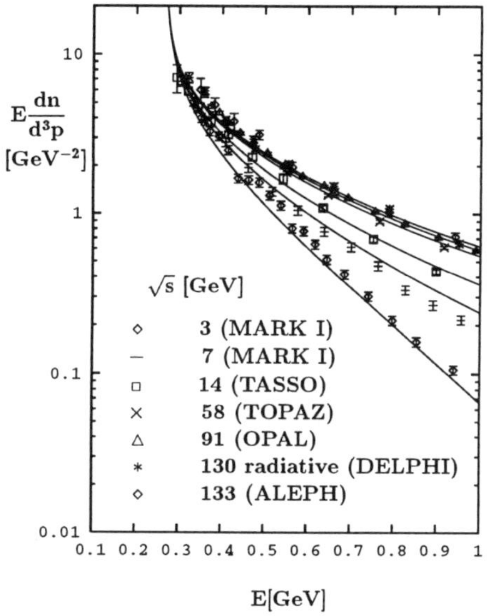

Fig. 1. Invariant density Edn/d^3p of charged particles in e^+e^- annihilation as a function of the particle energy $E = \sqrt{p^2 + Q_0^2}$ at $Q_0 = 270$ MeV. Experimental data at various *cms* energies from SLAC, TASSO and TOPAZ Collaborations, LEP-1 and LEP-1.5 are compared to MLLA predictions (taken from Ref. 10).

is played by the running α_s which provides the strong rise towards small energies for $E < 1$ GeV, for fixed α_s this rise would be much weaker.[16,10]

Recently inclusive spectra from HERA in the current fragmentation region in the Breit frame became available.[18] The data are found to be rather similar in shape to the observation in e^+e^-annihilation at comparable scales. Especially, the approximate energy independence of the soft part of the spectrum is verified in the region $12 < Q^2 < 100$ GeV^2. This supports the universality of the low energy phenomena.

3.2. Further Tests of the Perturbative Picture

One may suspect that the universal, approximately energy independent soft limit of the invariant spectrum could be some general hadronization phenomenon not related to the coherence properties of perturbative QCD. It is therefore very important to further investigate the perturbative origin of the effect. Whereas there is no parameter free prediction of the absolute size of the density I_0, the perturbative approach predicts its dependence on the colour charge of the primary parton as in (4). Especially, we expect for the ratio of soft particle densities in a gg system to the one in a

$q\bar{q}$ system

$$\frac{I_0^g}{I_0^q} = \frac{N_C}{C_F} = \frac{9}{4}. \tag{6}$$

Such a ratio is expected asymptotically for the total multiplicities of particles,[20] but at present energies there are large corrections. On the other hand, such a large ratio may nevertheless appear in case of the soft particles where the DLA asymptotic behaviour is valid to a good approximation. The preparation of gg final states is not easy; but there are two lines of approach which can test the hypothesis (6),[10,19] the first at a more qualitative, the second at a quantitative level.

Study of the central rapidity plateau at small $p_\perp$

Let's first consider e^+e^-annihilation. The rate I_0 in (5) is defined in the *cms*. If we perform a Lorentz transformation in direction of one primary parton, we arrive at a frame with unequal momenta of q and $\bar{q}$, say at rapidity y from the *cms*. As the soft radiation does not depend on the primary energy, the density $I_0(y)$ in the frame at y has not changed, so $I_0(y)$ should be constant away from the kinematical boundaries. Instead of the density at $p_\perp = 0$ we can also consider the density in an interval at small $p_\perp$. This picture is a bit idealistic, in that the primary parton direction is not known and should be replaced by a suitable jet-axis, then $I_0(y)$ will not be exactly constant.

Next we consider collisions with incoming hadrons or/and photons. The interaction may proceed through quark or gluon exchange where we have to assume some semihard momentum transfer of, say, at least 1 GeV—the soft processes may follow different rules. In case of a semihard quark or gluon exchange the outgoing partons will be either colour triplet or octet sources of radiation respectively, so in the soft limit we expect the ratio as in (6) for both cases.

Let's now consider specifically the DIS process studied at HERA and construct $I_0(y)$ as above. Analogous results are obtained for $\gamma\gamma$ collisions. In the Breit frame we have a quark in the current direction recoiling against another colour triplet system in proton direction. So in this frame we have $I_0(y_{Breit}) = I_0^q$, the density as in e^+e^-annihilation. However, the same DIS process can also be initiated by other subprocesses, for example, the photon gluon fusion. This process can be viewed best in a frame closer to the proton: a $q\bar{q}$ pair moves in current direction and the remainder in proton direction. This process is mediated by gluon exchange, so the colour octet sources yield a soft particle rate $I_0^g(y) = \frac{9}{4}I_0^q$ in such a frame. The overall picture then looks as in Fig. 2.

As already mentioned the exchanged parton should be sufficiently hard. The gluon virtuality cannot be simply enforced by the external conditions but it should become larger with increasing Q^2. We have no explicit calculation for this effect in the moment, so the conditions are formulated at a qualitative level. However, there

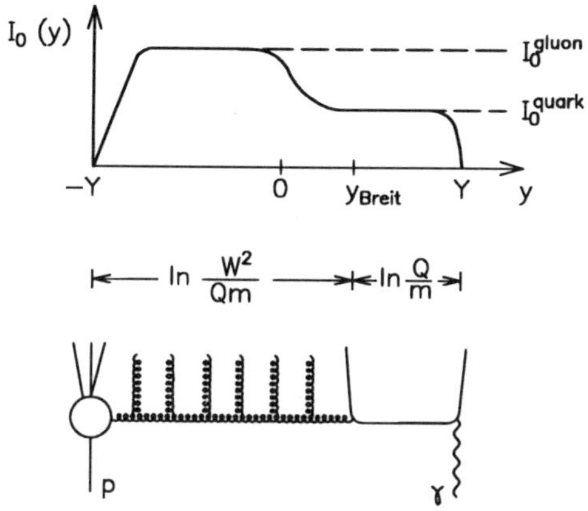

Fig. 2. Schematic view of particle density at small $p_\perp$ as a function of rapidity y in the *cms*: in frames at rapidity y with dominant gluon exchange the soft particle density $I_0(y)$ is 9/4 times larger than in the frames with dominant quark exchange (the Breit frame, for example).

are quantitative asymptotic expectations and predictions which could be falsified. We consider the distribution of particles in rapidity and at small $p_\perp$, at best in the Breit frame. An increase of Q^2 should make the small $k_\perp$ current plateau longer but not higher than I_0^q. At the same time the plateau in proton direction should be rising and approach for large Q^2 and W^2 a limit, the "gluon plateau" I_0^g, about two times larger than the "quark plateau" I_0^q.

Soft radiation perpendicular to the production plane in multi-jet events

The collinear parton configurations just discussed correspond to quark or gluon exchanges with the soft radiation expected in the ratio (6). We consider now the configurations not exactly collinear but with moderate transverse momentum, so that the corresponding parton jets can be resolved. In that case we consider the soft particle production in direction perpendicular to the production plane which can be analytically calculated in the soft region taking into account the cut-off Q_0.

The simplest case is the process $e^+e^- \to q\bar{q}g$. The soft radiation in this process has been analysed already some time ago[21] and applied to the case of multiplicity flow in the production plane. The spectrum perpendicular to the plane (say, in a small cone) is conveniently normalized to the corresponding flow in a 2 jet $q\bar{q}$ event by

$$R_\perp^a(p) \equiv \frac{dN_\perp^a/d\Omega_{\vec{n}}dp}{dN_\perp^{q\bar{q}}/d\Omega_{\vec{n}}dp} \tag{7}$$

for a general process a. For the case of $q\bar{q}g$ events one finds for the soft bremsstrahlung in order α_s[10]

$$R_{\perp}^{q\bar{q}g}(p) = \frac{N_C}{4C_F}\left[2 - \cos\Theta_{qg} - \cos\Theta_{\bar{q}g} - \frac{1}{N_C^2}(1 - \cos\Theta_{q\bar{q}})\right]. \tag{8}$$

where Θ_{ij} are the angles between the jets in the *cms*. This formula yields the two collinear limits $R_{\perp} = 1$ for collinear or soft gluons and $R_{\perp} = \frac{9}{4}$ for antiparallel quarks recoiling against the gluon. In this approximation the shape of the momentum spectrum does not depend on the inter-jet angles but the absolute magnitude does.

The analogous analysis can be carried out for photo-production of dijets in γp or $\gamma\gamma$ collisions and also in $p\bar{p}$ collisions.[19] One can distinguish between direct and resolved production.[22,23] The former involves the soft radiation from a $q\bar{q}g$ system, very much like in $e^+e^- \to q\bar{q}g$, and for the dominant small angle scattering corresponds to quark exchange, whereas the latter at small angles corresponds to gluon exchange. One expects again the ratio (6) of the soft particle production in both processes at small scattering angles.

It will be interesting to find out whether the particle density varies with the angles as predicted by (8) and occurs in the predicted ratios in different processes even down to low momenta of a few hundred MeV. This would provide a strong argument in favour of the perturbative QCD interpretation of the soft particle production.

4. Large Rapidity Gaps in e^+e^-Annihilation

In the previous example of inclusive particle spectra the partons and hadrons are typically close in phase space and the hadronization requires only small rearrangements of momenta. If during the evolution of the parton cascade in e^+e^-annihilation a large rapidity gap occurs, one may expect the colour confinement forces at large distances to fill the gap by hadrons in the hadronization phase. Then there is no close relation between parton and hadron final states. The study of large rapidity gap events can give us therefore further insight into the colour neutralization mechanism.

The interest in such events has been awakened recently by the findings of large rapidity gaps in dijet events at TEVATRON[24] and HERA[25] which are interpreted by the exchange of colour neutral objects.

In e^+e^-annihilation—if one wants to create a large gap without colour exchange such as to avoid final state rearrangements by confinement forces—one is lead to hard processes of the type $e^+e^- \to q\bar{q}q\bar{q}$ or $q\bar{q}gg$ where the hadrons are produced from colour singlet low mass $q\bar{q}$ or gg clusters.[26,27] As these processes in the perturbative analysis involve a highly virtual intermediate quark or gluon their rate is rather small though. One finds that the rate keeps decreasing with increasing gap size, contrary to the case of $p\bar{p}$ or ep collisions.

Indeed, the recent measurement of rapidity gaps by the SLD collaboration[28] has shown an unlimited decrease of the gap rate over five orders of magnitude (such a decrease has also been seen at lower energies some time ago[29]). However, the gap rate in absolute terms exceeds the expectations from the above calculations[26,27] by about two orders of magnitude and therefore, this type of hard colour neutralization cannot be the dominant process for the formation of large rapidity gap events.

An alternative approach,[11] in line with the previously discussed LPHD, considers a rather soft confinement mechanism, i.e. the hadron formation does not change the parton final state considerably, not even if a large rapidity gap is formed. Clearly, such an application of LPHD cannot be justified a priori. Rather one tries and looks how different the rates of large gaps really are in the parton and hadron cascades.

One has to calculate the probability that no parton is emitted into a certain angular interval, say between Θ_1 and Θ_2 ($\Theta_1 < \Theta_2$) where Θ is measured with respect to the initial parton direction. The rapidity for a massless parton is then obtained from $y = -\ln \mathrm{tg}\frac{\Theta}{2}$. This probability is given by the so-called Sudakov formfactor. For the actual problem of the angular ordered cascade with transverse momentum cut-off it has been calculated in connection with the multi-jet rates in the Durham/$p_\perp$ algorithm.[30,13]

The probability that no parton be emitted below the angle Θ in the cascade with cut-off $p_\perp > Q_0$ initiated by the primary parton A of momentum P_j is given by

$$\Delta_A(P_j,\Theta,Q_0) = \exp(-w_A(P_j,\Theta,Q_0)) \tag{9}$$

$$w_A(P_j,\Theta,Q_0) = \int d\omega' \int_{k_\perp > Q_0} d\Theta' \wp_A(\omega',\Theta'), \tag{10}$$

where $\wp_A(\omega',\Theta') = dn_A/d\omega' d\Theta'$ is the emission density of a gluon at energy ω' and angle Θ'. The probability for a gap between the two angles $\Theta_1 < \Theta_2 < \frac{\pi}{2}$ is then given by the ratio of two Sudakov formfactors

$$f_A(\Theta_1,\Theta_2) = \Delta_A(\Theta_2)/\Delta_A(\Theta_1) = \exp\left(-w_A(\Theta_2) + w_A(\Theta_1)\right). \tag{11}$$

In the simplest approximation, the DLA, one finds

$$w_A(Y,\lambda) = \frac{C_A}{N_C}\beta^2\{(Y+\lambda)\ln\frac{Y+\lambda}{\lambda} - Y\} \tag{12}$$

In this approximation the jets evolve independently in the two hemispheres and the respective gap probabilities factorize. In particular, for the symmetric gap in e^+e^-annihilation one obtains accordingly

$$f_A(\Theta_G) = e^{-2(w_A(\frac{\pi}{2}) - w_A(\Theta_G))}. \tag{13}$$

A result with improved accuracy is obtained by taking into account the exact form of the matrix element for $e^+e^- \to q\bar{q}g$ with numerical evaluation of the integral (10).[11] An analytical approximation in MLLA can be obtained by simplifying the integral over the nonsingular parts of the splitting functions.[13,11]

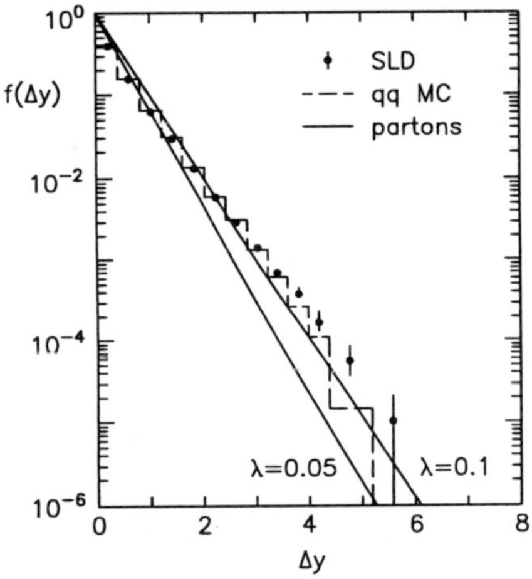

Fig. 3. Fraction of rapidity gap events in e^+e^- annihilation as a function of the full width of the symmetric gap. The data points refer to the measurement of gaps between charged particles (τ-lepton events included) by the SLD Collaboration.[28] The dashed histogram shows the expectation from the JETSET [1] Monte Carlo without τ-leptons. Also shown are the DLA predictions for the gaps of a parton cascade for two values of $\lambda = \ln(Q_0/\Lambda$ at $Q_0 = 0.244$ GeV.

As the result of these calculations one finds the probability for the gap in the symmetric rapidity interval $\Delta y \approx -2\ln\frac{\Theta_G}{2}$. It shows an almost exponential decrease very much like the experimental data. The slope is not very different for DLA and MLLA calculations but strongly depends on the parameter $\lambda = \ln(Q_0/\Lambda)$.

The SLD data[28] are shown in Fig. 3. They refer to rapidity gaps between charged particles and still include τ-lepton events. The analytical calculations[11] from the DLA are shown for two parameters of λ at $\Lambda = 0.244$. An upper limit $\lambda < 0.1$ has been derived from a moment analysis of the spectra.[16] These calculations should rather be compared to data on gaps empty of any particles, not only of charged ones. Their distribution is expected to have a steeper slope than shown by the data in the figure. As an example the prediction for $\lambda = 0.05$ is shown. A recent analysis of particle and jet multiplicities beyond MLLA[9] has given the small value $\lambda = 0.015 \pm 0.005$.

One can conclude that the distribution of gaps in a hadron cascade is well represented by the one in the parton cascade, provided the cut-off Q_0 is taken small enough, i.e. very close to the QCD scale Λ in the present approximation scheme. If one took the larger cut-off $Q_0 \sim 1$ GeV the gap probability at $\Delta y = 6$ would be larger by 4 orders of magnitude! This suggests that the hadronization phase in a conventional hadronization model with shorter parton cascade has a similar effect as the evolution of the parton cascade from $Q_0 = 1$ GeV down to $Q_0 = 0.25$ GeV. This is not too surprising as it is in this last phase where the running α_s increases most

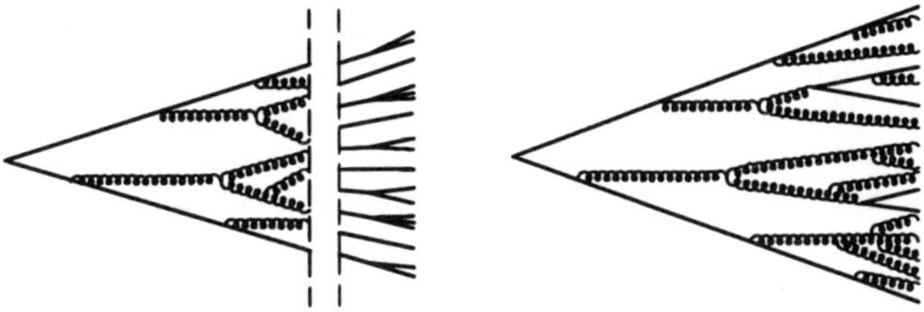

Fig. 4. In the standard hadronization model (left) the parton cascade is followed by a hadronization phase. In the dual approach the hadronization phase is assumed to be represented by the parton cascade in an average sense after evolution towards a small cut-off scale Q_0

strongly and makes the suppression of radiation increasingly difficult.

An interesting prediction from this approach concerns the distribution of rapidity gaps in single quark and gluon jets. As the exponent of the Sudakov formfactor is derived from the one-particle-density it is proportional again to the colour factor C_A, i.e. the slope of the gap distribution is correspondingly steeper (roughly twice) for gluon jets than for quark jets. This prediction could be checked in high $p_\perp$ jets at the TEVATRON or at HERA.

5. Conclusions

The description of detailed properties of the hadronic jets in hard processes in terms of a parton cascade assuming a duality between parton and hadron production continues to be successful. It is surprising that this duality works even for observables where a priori there is no good reason to believe in a perturbative approach.

It appears that the evolution of the parton cascade at very low scales below 1 GeV resembles in its main inclusive aspects the hadronization phase of the phenomenological models with all its resonance decays, so one can speak of a dual description of the hadronic phenomena after appropriate averaging (see Fig. 4). As the cut-off Q_0 is close to the QCD scale Λ within less than 10% the coupling α_s becomes large in the last phase of the perturbative cascade. So it may not be unreasonable that the strong coupling perturbative regime corresponds to the hadronic phase with resonance production.

The QCD expectation of the energy independence of the low momentum particle production from the soft gluon coherence is well met by the hadronic data in the full energy region explored, i.e. in e^+e^-annihilation from ADONE at 1.6 GeV up to LEP-1.5 and in DIS from $Q^2 = 12$ GeV^2 to $Q^2 = 100$ GeV^2. It will be interesting to test the predictions on the higher production rates of soft particles for other colour configurations than $q\bar{q}$ which can be obtained in various processes.

The parton cascade with low cut-off Q_0 and without hadronization also reproduces the approximately exponential decrease of the rapidity gap distribution in e^+e^-annihilation. This suggests in particular that the confinement of colour in the jet evolution is a soft process.

The correspondence between the parton and hadron final state according to these results is not like the conventional duality between a colour singlet $q\bar{q}$ and a hadronic cluster. Rather it is a correspondence between a coloured parton (typically a gluon) and a colourless hadron. These partons are taken at a particular scale $p_\perp > Q_0 \gtrsim \Lambda$, so that partons below that scale are not resolved. These soft non-perturbative sea-partons could take the important role in the confinement process without disturbing the momenta of the resolved perturbative partons.

Meanwhile it seems to be a challenge to test further this dual picture of hadron formation at the limit of what is accessible perturbatively.

6. Acknowledgements

It is a pleasure to thank V.A. Khoze, S. Lupia and T. Shimada for the collaboration on the topics discussed in this talk. Thanks also to B. Kniehl and R. Saffert for having the meeting so well organized.

7. References

1. T. Sjöstrand and M. Bengtsson, *Comp. Phys. Commun.* **43** (1987) 367.
2. G. Marchesini, B. R. Webber, G. Abbiendi, I. G. Knowles, M. H. Seymour, and L. Stanco, *Comp. Phys. Comm.* **67** (1992) 465.
3. Ya. I. Azimov, Yu. L. Dokshitzer, V. A. Khoze and S. I. Troyan, *Z. Phys.* **C27** (1985) 65 and **C31** (1986) 213.
4. Yu. L. Dokshitzer, V. S. Fadin and V. A. Khoze, *Phys. Lett.* **115B** (1982) 242; *Z. Phys.* **C15** (1982) 325.
5. A. Bassetto, M. Ciafaloni and G. Marchesini, *Phys. Rep.* **C100** (1983) 201.
6. A. H. Mueller, *Nucl. Phys.* **B213** (1983) 85; Erratum quoted ibid., **B241** (1984) 141;
   Yu. L. Dokshitzer and S. I. Troyan, *Proc. 19th Winter School of the LNPI*, Vol. 1, p.144; Leningrad preprint LNPI-922 (1984).
7. Yu. L. Dokshitzer, V. A. Khoze, A. H. Mueller and S. I. Troyan, *Basics of Perturbative QCD*, ed. J. Tran Thanh Van (Editions Frontiéres, Gif-sur-Yvette, 1991).
8. for a recent discussion of the phenomenological status, see V. A. Khoze and W. Ochs, *Int. J. Mod. Phys.* **A12** (1997) 2949.

9. S. Lupia and W. Ochs, "*Unified QCD Description of Hadron and Jet Multiplicities*", MPI-PhT/97-46, July 1997 (hep-ph/9707393).
10. V. A. Khoze, S. Lupia and W. Ochs, *Phys. Lett.* **B394** (1997) 179; *Proc. 7th Int. Workshop on Multiparticle Production*, Eds. R.C. Hwa, W. Kittel, W.J. Metzger and D.J. Schotanus, June 1966, Nijmegen, The Netherlands (World Scientific, Singapore, 1997), p. 358 (hep-ph/9610348).
11. W. Ochs and T. Shimada, *"Rapidity Gaps in e^+e^- Annihilation and Parton-Hadron Duality"*, *Proc. 33rd Eloisatron workshop*, Erice, Italy, Oct. 1996, to be publ. (hep-ph/9706485)
12. B. I. Ermolayev and V. S. Fadin, *JETP. Lett.* **33** (1981) 285; A. H. Mueller, *Phys. Lett.* **104B** (1981) 161.
13. Yu.L. Dokshitzer and M. Olsson, *Nucl. Phys.* **B396** (1993) 137.
14. W. Ochs, *Acta Phys. Pol.* **B27** (1996) 3505.
15. Yu. L. Dokshitzer, V. A. Khoze and S. I. Troyan, *Int. J. Mod. Phys.* **A7** (1992) 1875.
16. S. Lupia and W. Ochs, *Phys. Lett.* **B365** (1996) 339; MPI-PhT/97-26 (hep-ph/9704319) April 1997, *Z. Phys.* **C** in press.
17. A. Bassetto, M. Ciafaloni, G. Marchesini and A. H. Mueller, *Nucl. Phys.* **B207** (1982) 189.
18. H1 Collaboration: C. Adloff et al., *"Evolution of ep Fragmentation and Multiplicity Distributions in the Breit Frame"*, DESY 97-108 (hep-ex/9707005).
19. V.A. Khoze, S. Lupia and W. Ochs, paper in preparation.
20. S. J. Brodsky and J. F. Gunion, *Phys. Rev. Lett.* **37** (1976) 402.
21. Ya. I. Azimov, Yu. L. Dokshitzer, V. A. Khoze and S. I. Troyan, *Phys. Lett.* **B165** (1985) 147.
22. ZEUS Collaboration: M. Derrick et al., *Phys. Lett.* **B348** (1995) 665; **B384** (1996) 401. H1 Collaboration: S. Aid et al., *Z. Phys.* **C70** (1996) 17.
23. J.M. Butterworth, these proceedings.
24. D0 collaboration: S. Abachi et al., *Phys. Rev. Lett.* **72** (1994) 2332. CDF collaboration: F. Abe et al., *Phys. Rev. Lett.* 74 (1995) 855.
25. ZEUS collaboration: M. Derrick et al., *Phys. Lett.* **B332** (1994) 228. H1 collaboration: T. Ahmed et al., *Nucl. Phys.* **B429** (1994) 477.
26. J. D. Bjorken, S. J. Brodsky and H. J. Lu. *Phys. Lett.* **B286** (1992) 153.
27. J. Ellis and D.A. Ross. *Z. Phys.* **C70** (1996) 115.
28. SLD collaboration: *Phys. Rev. Lett.* **76** (1996) 115.
29. HRS collaboration: M. Derrick et al., *Z. Phys.* **C35** (1987) 323.
30. S. Catani, Yu.L. Dokshitzer, M. Olsson, G. Turnock and B.R. Webber, *Phys. Lett.* **B269** (1991) 432.

QCD INSTANTONS AT HERA

MICHAEL KUHLEN
Max-Planck-Institut für Physik, Werner-Heisenberg-Institut
Föhringer Ring 6, D-80805 München, Germany
E-mail: kuhlen@desy.de

ABSTRACT

Phenomenological aspects of chirality violating processes induced by QCD instantons in deep inelastic scattering are discussed. First instanton searches and the prospects for their experimental discovery at HERA are presented.

1. Introduction

Since a long time it has been recognized that the standard model contains processes which cannot be described by perturbation theory, and which violate classical conservation laws like baryon number (B) and lepton number (L) in the case of the electroweak interaction, and chirality (Q_5) in the case of the strong interaction[1]. Such anomalous processes are induced by so-called instantons[2]. The name indicates that these are non-perturbative fluctuations that are confined to "an instant" in space-time, with no corresponding free particle solutions for $t \to \pm\infty$. The interest in instantons remained somewhat academic, as observable effects were predicted to exist only at extremely high energies of $\mathcal{O}(10^5$ TeV), until it was discovered that their exponential suppression is much reduced by the emission of gauge bosons[3]. In electroweak theory with massive gauge bosons still rather high energies of $\mathcal{O}(\gtrsim 10$ TeV) would be required, but not so in QCD with massless gluons and strong coupling. Instanton effects could play a rôle in QCD reactions already at present day colliders. Deep inelastic ep scattering at HERA is particularly interesting, because the virtuality of the photon probe Q^2 provides a hard scale for the instanton subprocess, which is needed for theoretically sound predictions[4,5,6]. Instanton effects have not yet been observed in nature. Their experimental discovery would be of fundamental significance for particle physics.

In this report a short introduction to the basic theoretical ideas will be given. Instanton phenomenology in deep inelastic scattering (DIS) will be discussed, covering cross sections and event topologies. Finally, prospects for instanton searches and first results from the analysis of HERA data will be presented.

2. Instanton theory

Instantons originate from the non-trivial topological structure of the vacuum in non-Abelian gauge field theories, where the vacuum is degenerate in the Chern-Simons number N_{CS} . N_{CS} is defined as an integral over the gauge fields A^a_μ with coupling g,

$$N_{\mathrm{CS}} := \frac{g^2}{16\pi^2}\int \mathrm{d}^3x\epsilon_{ijk}\left(A^a_i\partial_j A^a_k - \frac{g}{3}\epsilon_{abc}A^a_i A^b_j A^c_k\right). \tag{1}$$

Neighbouring vacua have the same (minimal) potential energy, but differ in their topological winding number N_{CS} , and are separated by a potential barrier of height E_B (fig. 1).

The usual perturbative expansion of the scattering amplitudes in the coupling constant α around *one* minimum (fig. 1a), conveniently represented by a series of Feynman graphs, does not allow for transitions between neighbouring minima. They may however occur classically when the energy E is large enough $E > E_B$, or by quantum mechanical tunnelling when $E < E_B$, corresponding to so-called instanton solutions of the classical field equations. The transition amplitude for the instanton–induced tunnelling process is exponentially suppressed $\propto \exp(-4\pi/\alpha)$, a very small number.

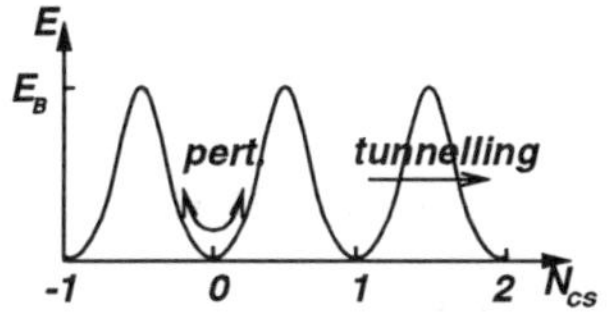

Fig. 1. The structure of the vacuum. Instanton solutions represent tunnelling transitions between topological inequivalent minima, which cannot be reached perturbatively.

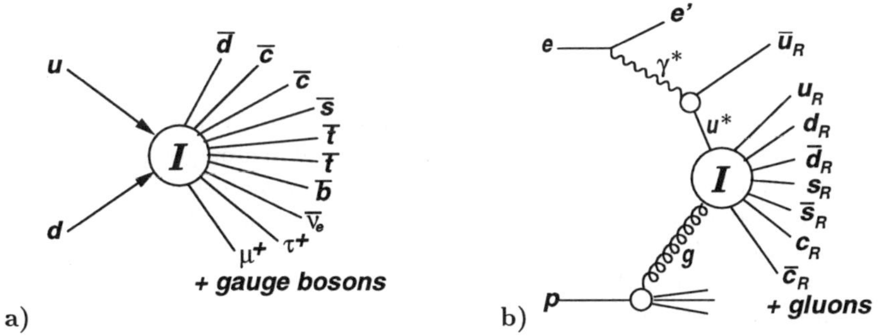

Fig. 2. **a)** the electroweak interaction with $\Delta(B+L) = -6$ and in **b)** the strong interaction with $\Delta Q_5 = 8$.

In the electroweak theory, the minimal barrier height is $E_B \approx m_W/\alpha_w = \mathcal{O}(10\mathrm{TeV})$. Instanton transitions between vacua separated by ΔN_{CS} (see fig. 2a for an example) would violate baryon (B) and lepton numbers ($L = L_e + L_\mu + L_\tau$) according to

$$\Delta(B+L) = -2\, n_{\mathrm{gener.}} \cdot \Delta N_{\mathrm{CS}} , \tag{2}$$

but respect

$$\Delta(B-L) = 0 \qquad\qquad \Delta L_e = \Delta L_\mu = \Delta L_\tau = \Delta B/3. \tag{3}$$

$n_{\text{gener.}} = 3$ is the number of fermion generations.

In instanton induced QCD reactions (see fig. 2b) chirality is violated. The chirality Q_5 is the difference between the number of left- and right-handed fermions, $Q_5 = \#L - \#R$. For n_f active quark flavours, the selection rule is

$$\Delta Q_5 = 2\, n_f \cdot \Delta N_{\text{CS}} \,. \tag{4}$$

The minimal barrier height is given by the hard scale of the process, e.g. $E_B = \mathcal{O}(Q)$ for DIS[5]. The exponential suppression is less severe than in the electroweak case, because $\alpha_s \gg \alpha_w$.

3. Instantons at HERA

In recent years, it has been realized[5,6,7,8,9,10] that quantitative calculations are possible for processes induced by QCD instantons in DIS due to the presence of a hard scale, Q^2. In DIS, events due to QCD instantons I (and anti-instantons $\overline{I}$) are predominantly produced in a photon-gluon fusion processes[a] (Fig. 3)

$$\gamma^* + g \xrightarrow{I} \sum_{n_f}(\overline{q}_R + q_R) + n_g g \qquad \gamma^* + g \xrightarrow{\overline{I}} \sum_{n_f}(\overline{q}_L + q_L) + n_g g. \tag{5}$$

In each event, quarks and antiquarks of all n_f active flavours are found, with n_g gluons in addition.

The kinematics is depicted in fig. 3. The DIS variables Bjorken x and Q^2 can be measured from the scattered electron, $q = e - e'$. A measurement of the other variables is more challenging. A measurement of the invariant mass of the hadronic system, excluding the remnant, would determine $\hat{s}$. If the outgoing "current jet" could be identified and measured, it's 4-momentum q'' would determine $q' = q - q''$, and thus the variables x' and Q'^2 which characterise the instanton subprocess. In practice, when not all of the five independent invariants (for example $x, Q^2, x', Q'^2, \hat{s}$) can be measured, they are being integrated over.

The instanton induced cross section is given by a convolution of the probability to find a gluon in the proton $P_{g/p}$, the probability that the virtual photon splits into a quark-antiquark pair in the instanton background $P^{(I)}_{q^*/\gamma^*}$, and the cross section $\sigma^{(I)}_{q^*g}(x', Q'^2)$ of the instanton subprocess[6,7]. Multi-gluon emission enhances the cross section[3]

$$\sigma^{(I)}_{q^*g;n_g} \propto \frac{1}{n_g!}\left(\frac{1}{\alpha_s}\right)^{n_g} \exp(-4\pi/\alpha_s). \tag{6}$$

---

[a]Quark initiated processes have not yet been considered. Due to the large gluon content of the proton in the HERA domain at small x, they are expected to be of minor importance. In addition, they are expected to be suppressed by $\mathcal{O}(\alpha_s^2)$ with respect to the gluon initiated processes.

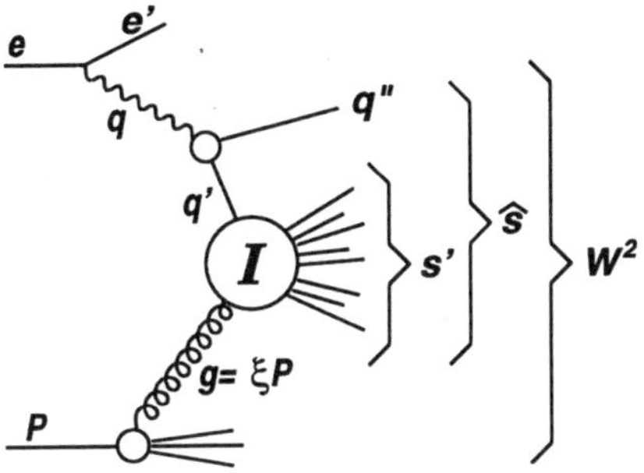

DIS variables:
$Q^2 := -q^2$
$x := Q^2 /(2P \cdot q)$
$W^2 := (q+P)^2 = Q^2 (1-x)/x$
$\hat{s} := (q+g)^2$
$\xi = x(1+\hat{s}/Q^2)$

Variables of instanton subprocess:
$Q'^2 := -q'^2$
$x' := Q'^2 /(2\, g \cdot q')$
$s' := (q'+g)^2 = Q'^2 (1-x')/x'$

Fig. 3. Kinematics of instanton induced processes in DIS. The labels denote the 4-vectors of the particles. A virtual photon γ^* (4-momentum $q = e - e'$) emitted from the incoming electron fuses with a gluon (4-momentum g) from the proton (4-momentum P). The gluon carries a fraction ξ of the proton momentum. The virtual quark q^* entering the instanton subprocess has 4-momentum q', and the outgoing quark from the $\gamma^* \to q\overline{q}$ splitting has 4-momentum q''. The invariant masses squared of the $\gamma^* g$ and $q^* g$ systems are $\hat{s}$ and s'. W is the invariant mass of the total hadronic system (the $\gamma^* p$ system). $0 \leq x \leq x/\xi \leq x' \leq 1$ holds. For completeness, we note $y := (Pq)/(Pe) = Q^2/(sx)$, where $s = (e+P)^2$ is the ep invariant mass squared.

The cross section of the instanton induced subprocess is then[6]:

$$\sigma^{(I)}_{q^* g}(x', Q'^2) = \sum_{n_g=0}^{\infty} \sigma^{(I)}_{q^* g; n_g} \approx \frac{\Sigma(x')}{Q'^2} \left(\frac{4\pi}{\alpha_s(\mu(Q'))} \right)^{\frac{21}{2}} \exp\left(\frac{-4\pi}{\alpha_s(\mu(Q'))} F(x') \right). \tag{7}$$

It depends critically on the functions $F(x')$ (called the "holy grail" function), which modifies the exponent in the suppression factor $\exp(-4\pi/\alpha_s)$, and on $\Sigma(x')$, which depends on $F(x')$. There exists also a scale dependence due to the choice of the renormalization scale $\mu(Q')$.

$F(x')$ can be estimated reasonably well (see fig. 4) for x' not too small, $x' \gtrsim 0.2$ [6]. The extrapolation to lower values of x' is unreliable due to inherent ambiguities. In addition, multi-instanton effects should be avoided by limiting the instanton size ρ_I (the spatial region occupied during the interaction) to $\rho_I < 2\ \mathrm{GeV}^{-1}$ with a cut-off $Q'^2 \gtrsim 25\ \mathrm{GeV}^2$ [6,7]. That requirement ensures also that $\alpha_s(\mu(Q'))$ stays small enough to apply instanton perturbation theory.

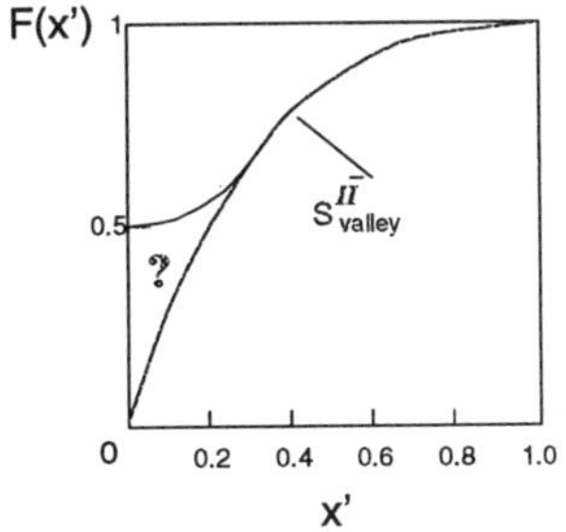

Fig. 4. The holy grail function $F(x')$ [6]. For small s' ($x' \approx 1$), instanton perturbation theory is applied. The calculation with the valley method matches smoothly with the perturbative result.

The resulting instanton induced subprocess cross section $\sigma^{(I)}_{q^* g}(x', Q'^2)$ (see fig. 5) is peaked at $Q' \approx 5$ GeV and exponentially grows with decreasing x'. The integrated instanton induced ep DIS cross section (see fig. 6) is sizeable; for $x > 0.001$ and $x' > 0.2$ it is of $\mathcal{O}(10 \text{ pb})$. The cross section is approximately scaling (depends only on x, not on Q^2 for large Q^2) [7]. It grows towards small x, and increases dramatically when the lower x' cut-off is relaxed. Eventually higher order instanton effects have to dampen the growth of the cross section.

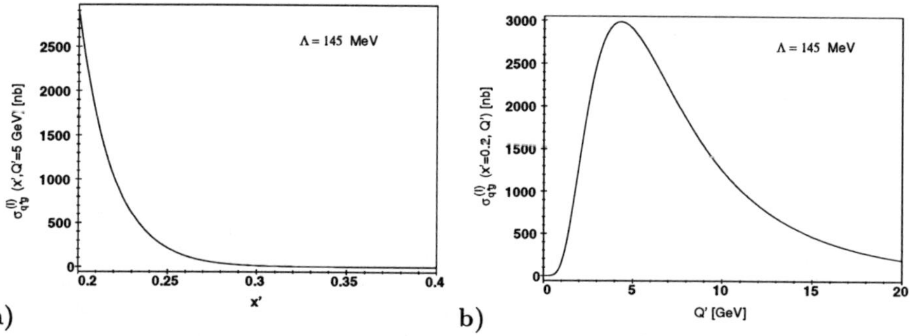

Fig. 5. The instanton subprocess cross section[9] for $q^*g \to$ hadrons as a function of **a)** x' for $Q' = 5$ GeV and **b)** Q' for $x' = 0.2$.

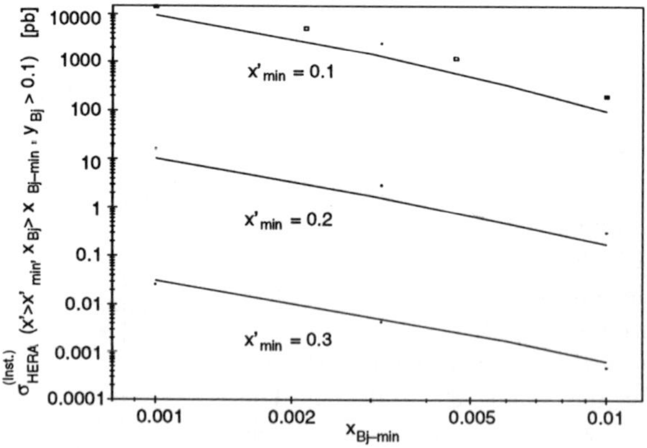

Fig. 6. The instanton induced DIS cross section[9] $ep \to e'X$ integrated over Bjorken $x > x_{\min}$, $y > 0.1$, and over the regions $Q' > 5$ GeV and $x' > x'_{\min}$ as indicated.

Two kinematic regions have to be distinguished. For $x' > 0.2$ the predictions are relatively safe, allowing the instanton theory to be tested. Either instantons are discovered at the predicted level – including the substantial theoretical uncertainties, which still need to be quantified –, or the theory has to be revised. For $x' < 0.2$ the cross section presumably continues to grow, but the extrapolation is extremely uncertain. For a discovery, this is the favourable region due to the large cross section. A negative result however cannot be turned against the theory, it would rather restrict the unknown behaviour of $F(x')$ at small x'. Most promising is the kinematic region of small Bjorken-x, because both the total DIS cross section and the predicted fraction of instanton induced events increase towards small x (see fig. 9b).

4. Experimental signatures

In the theoretically safe region, $x' > 0.2$, the expected fraction of instanton events in all DIS events is of $\mathcal{O}(10^{-3} - 10^{-4})$ (compare fig.9b), too small to be detected in inclusive cross section measurements (i.e. the structure function F_2). Instead, dedicated searches for the characteristic features of instanton events in the hadronic final state have to be performed. A Monte Carlo generator (QCDINS[11]) to simulate the hadronic final state of instanton events in DIS is available. In general, the event shape predictions are more stable than the rate predictions, because poorly known factors cancel. The instanton event properties can be contrasted with predictions from event generators for normal DIS events (ARIADNE[12], LEPTO[13] and HERWIG[14]) which give an overall satisfactory description of the DIS final state properties[15].

In the q^*g rest frame $2n_f - 1$ quark and antiquarks and n_g gluons are emitted isotropically from the instanton subprocess. n_g is Poisson distributed with[7,9]

$$\langle n_g \rangle \approx \frac{2\pi}{\alpha_s} x'(1-x') \frac{\mathrm{d}F(x')}{\mathrm{d}x'}. \tag{8}$$

After hadronization, this leads to a spherical system with a high multiplicity of hadrons, depending mainly on the available centre of mass energy $\sqrt{s'} = Q'\sqrt{1/x' - 1}$. For a typical situation ($x' = 0.2, Q' = 5$ GeV $\Rightarrow s' = 10$ GeV), $\langle n_g \rangle = \mathcal{O}(2)$. About $n_p = 10$ partons and $n = 20$ hadrons are expected. The expected parton momentum spectrum is semi-hard [5] with transverse momentum $\langle p_T \rangle \approx (\pi/4)(\sqrt{s'}/\langle n_p \rangle)$.

Hadronic final state properties are conveniently being studied in the centre of mass system (CMS) of the incoming proton and the virtual boson, i.e. the CMS of the hadronic final state. Longitudinal and transverse quantities are calculated with respect to the virtual boson direction (defining the $+z$ direction). The pseudorapidity η is defined as $\eta = -\ln\tan(\theta/2)$, where θ is the angle with respect to the virtual photon direction. When boosted to the CMS, the hadrons emerging from the instanton subprocess occupy a band in pseudorapidity of half width $\Delta\eta \approx 1$, which is homogeneously populated in azimuth [5].

The characteristics of instanton events by which they can be distinguished from normal DIS events are therefore: high multiplicity with large transverse energy; spherical event configuration (apart from the current jet); and the presence of all flavours (twice!) that are kinematically allowed in each event. One would therefore look for events which in addition to the other characteristics are rich in K^0s, charm decays, secondary vertices, muons etc.. In general, the strength of instanton signals in the hadronic final state increases somewhat towards low x' and large Q'^2 due to the increasing "instanton mass" $\sqrt{s'} = Q'\sqrt{1/x' - 1}$.

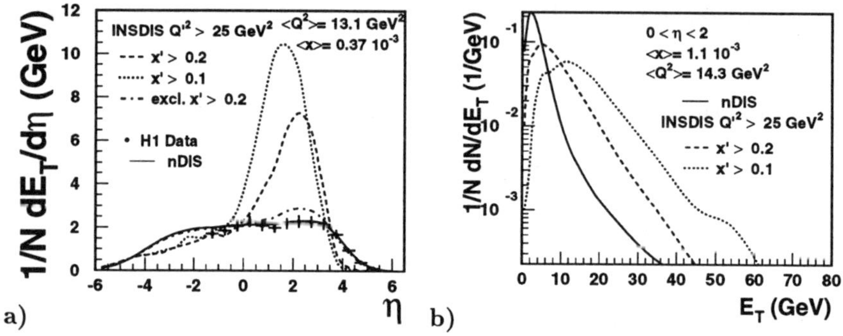

Fig. 7. Transverse energy E_T in the hadronic CMS [16]. **a)** The E_T flow vs. η. The proton remnant direction is to the left. The standard QCD model (nDIS=ARIADNE) and different instanton scenarios are confronted with the H1 data [17]. The excluded scenario [16] with an instanton fraction $f_I > 11.8\%$ for $x' > 0.2$ is indicated. **b)** The E_T distribution, where the transverse energy is measured in the CMS rapidity bin $0 < \eta < 2$, for two instanton scenarios, and the standard QCD model (nDIS=ARIADNE). The plots are normalized to the total number of events N that enter the distributions.

The "instanton band" shows up in the flow of hadronic transverse energy E_T as a function of η (fig. 7a). It's height and position depends on x' and Q'^2 (and also on x and Q^2). In normal DIS events on average an E_T of 2 GeV per η unit is observed. In instanton induced events, that number may go up to 10 GeV for low x'. A possible search strategy could involve the E_T distribution in a selected rapidity band (fig. 7b), looking for high E_T events in the tail of the distribution[18,16].

Further discrimination can be obtained[18] from the fact that for instanton events the E_T should be distributed isotropically, while normal DIS events are jet-like, in particular for large E_T . One defines

$$E_{\text{out}} := \min \sum_i \vec{p}_i \cdot \vec{n} \qquad\qquad E_{\text{in}} := \sum_i \vec{p}_i \cdot \vec{n}'. \tag{9}$$

The sum runs over all final state hadrons i with momentum $\vec{p}_i$. $\vec{n}$ is the unit vector perpendicular to the virtual photon axis which minimizes E_{out} and thus defines the event plane. $\vec{n}'$ lies in the event plane and is normal to both $\vec{n}$ and the virtual photon axis. It is easy to show that for an ideal isotropic "instanton decay", $E_{\text{out}} = \sqrt{s'}/2$ [18]. The "instanton mass" $\sqrt{s'}$ can thus be reconstructed experimentally (fig. 8a). Normal DIS events, either "1+1" or "2+1" jet events (the +1 refers to the unobserved proton remnant) are contained in the event plane, $E_{\text{out}} \ll E_{\text{in}}$, in contrast to instanton events with $E_{\text{out}} \approx E_{\text{in}}$ (see fig. 8b).

Instanton events are characterized by a large particle density localized in rapidity. In normal DIS events there are about 2 charged particles per unit of pseudorapidity[19],

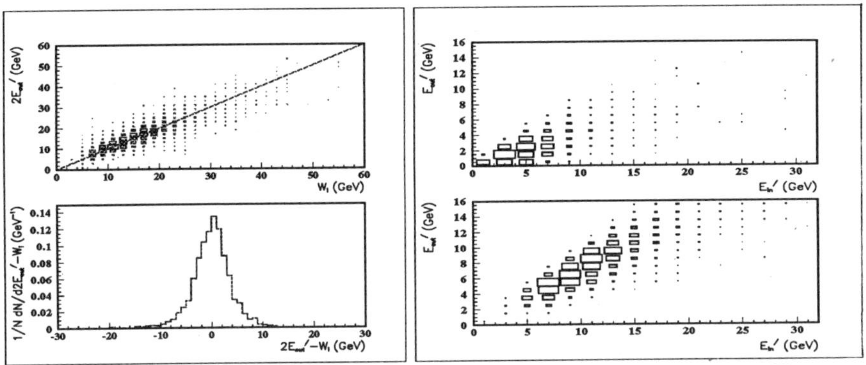

Fig. 8. **a)** The correlation between $2 \cdot E'_{\rm out}$ and the "instanton mass", $W_I = \sqrt{s'}$ (top), and the resolution for $\sqrt{s'}$ that can be achieved (bottom)[18]. The primes indicate additional cuts in η to minimize higher order QCD radiation which may wash out the relation between $E_{\rm out}$ and $\sqrt{s'}$. **b)** $E'_{\rm out}$ vs. $E'_{\rm in}$ for normal (top, HERWIG) and instanton induced events (bottom, QCDINS)[18]. Both distributions are taken in the hadronic CMS for events with $0.001 < x < 0.01$, $0.1 < y < 0.6$ and $20\ {\rm GeV}^2 < Q^2 < 70\ {\rm GeV}^2$.

rather uniformly distributed in η. For a low x' cut-off, that number goes up to 10 in the peak of the instanton band[16]. Very sensitive to instanton events is the charged particle multiplicity distribution[16], see fig. 9a. A significant fraction of the instanton events would lead to charged multiplicities which are very unlikely to be found in normal DIS events. Furthermore, particle-particle correlation functions should be influenced by instanton effects[20].

5. Searches for instanton processes

The fact that instanton events look very different from the expectation for standard QCD events can be exploited to search for instanton signals in the HERA data. One strategy is to compare the shape of hadronic final state distributions to the expectation from standard QCD events (nDIS) with an admixture of instanton events (INSDIS) of fraction f_I . In case the measured distribution agrees with the standard QCD expectation, a limit on the fraction of instanton induced events in DIS $f_I < f_{\rm lim}$ can be set. The caveat of this method is that one has to make an assumption on what standard QCD looks like. In particular at small x that issue is under debate [22,23,24]. There exists a danger that an instanton effect is tuned or explained away by stretching the standard QCD predictions by generator tuning, introducing for example BFKL effects etc.. A good understanding of standard QCD will be crucial for the positive identification of instanton effects.

In the first search for instanton events[25] an anomalous K^0 yield has been looked for. For $x > 10^{-3}$ about 0.12 K^0 mesons (including $\overline{K^0}$) have been measured per

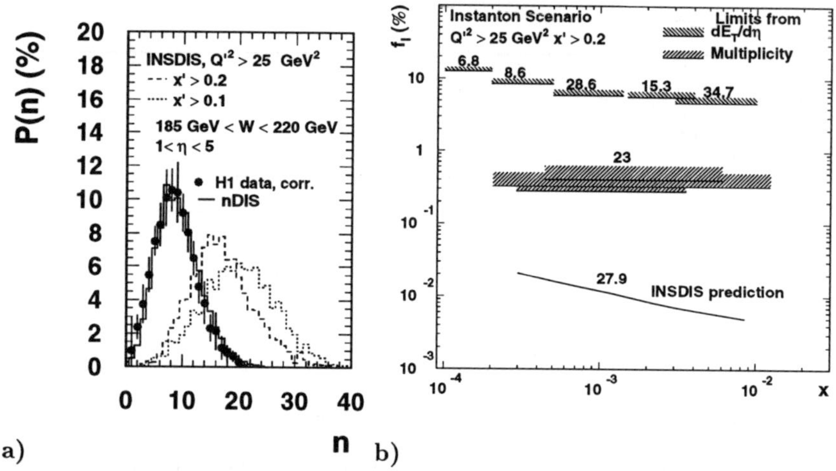

Fig. 9. **a)** The probability distribution $P(n)$ of the charged particle multiplicity n from the CMS pseudorapidity range $1 < \eta < 5$ for events with 185 GeV $< W <$ 220 GeV. Shown are the unfolded H1 data [21], the expectation from a standard DIS model (nDIS=ARIADNE)), and the predictions for instanton events with different cut-off scenarios [16]. **b)** The maximally allowed fraction f_{lim} of instanton induced events in DIS for $Q'^2 > 25$ GeV^2 and $x' > 0.2$ from transverse energy flows and the multiplicity distribution as function of x [16]. Regions above the lines are excluded at 95% C.L.. The numbers give the average Q^2 values in GeV^2 for the x bins. The theory prediction, calculated with QCDINS [11], for 10 GeV^2 $< Q^2 < 80$ GeV^2 is superimposed (full line, label INSDIS).

event and unit pseudorapidity, with a relatively flat η distribution. For instanton events with $x' > 0.2$, a peaked distribution with about 0.55 K^0 per event and unit η is expected. From the comparison with standard QCD event generators[12,13,14] and the instanton generator[11] an upper limit of $f_I < f_{\text{lim}} = 6\%$ at 95% C.L. is obtained[25].

The charged particle multiplicity distribution $P(n)$ in high energy reactions can often by described by a negative binominal distribution (NBD). Also the DIS data are relatively well described by NBDs [21]. The multiplicity distribution from the CMS interval $1 < \eta < 5$ for events with $W = 80 - 115$ GeV (corresponding to $x > 0.0007$) can be parametrized with an NBD of mean $\langle n \rangle = 6.90 \pm 0.33$. Possible deviations from an NBD allow for an instanton fraction of at most f_I =2.7% at 95% C.L. [21].

Other measured event shapes have been systematically analysed in terms of their sensitivity to instanton events[16], and their dependence on the kinematic variables x, Q^2, x', Q'^2. The most sensitive distributions were the transverse energy flows[17], the pseudorapidity distribution of charged particles and their p_T spectra[19]. For example, the E_T flow has been measured over a wide range of x and Q^2, allowing to extend the search region down to $x = 0.0001$. From a shape analysis[16] (see fig. 7), instanton fractions f_I between 5 and 13 % can be excluded for $x' > 0.2$ (see figs. 9b, 10). For lower x' the signal is more prominent, and somewhat better limits are obtained.

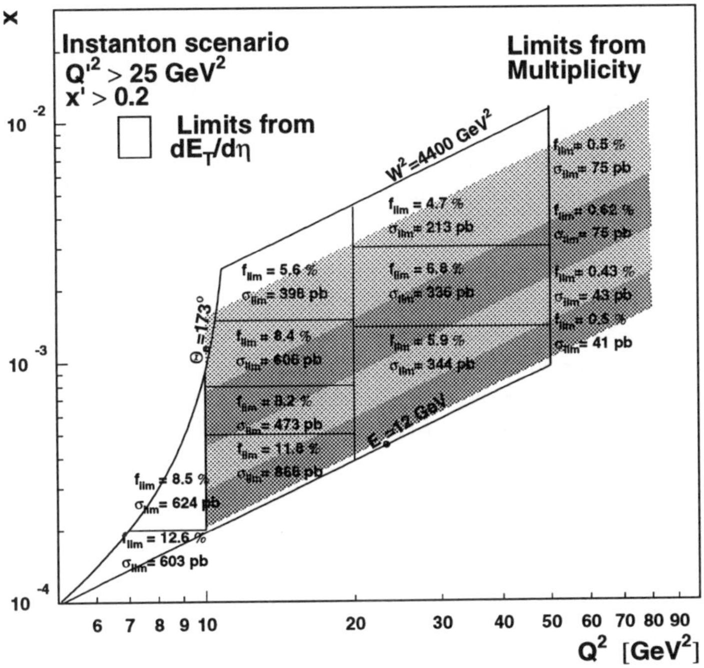

Fig. 10. 95% C.L. limits on instanton production with $Q'^2 > 25\ \mathrm{GeV}^2$ and $x' > 0.2$ [16]. The cross-section limits (σ_{lim}) together with the maximally allowed instanton fraction f_{lim} are shown in the x $-Q^2$ plane. They are obtained from the E_T flow analysis (open fields), and from the multiplicity analysis (shaded fields) with their numbers at the right edge. The boundaries implied by the analysis cuts of the energy flow analysis in the angle and energy of the scattered electron, $\theta_e < 173°$ and $E_e > 12$ GeV , and by the requirement $W^2 > 4400\ \mathrm{GeV}^2$ are indicated.

The fact that H1[21] did not observe any events above a certain multiplicity $n_{\max}$ has been exploited[16] to place more stringent limits on instanton production[b]. A significant fraction of instanton induced events would have multiplicities $n > n_{\max}$. Instanton fractions $f_I > 0.4 - 0.6\%$ can therefore be excluded for $x' > 0.2$ (see figs. 9b, 10), and somewhat lower f_I values for a lower cut-off $x' > 0.1$ [16]. This search method has the advantage that, in contrast to the previous shape comparisons, it does not rely on assumptions for standard QCD event topologies, since no background needs to be subtracted. Unavoidable of course is the dependence on the expected instanton event shape, which may be even more uncertain than the standard QCD event shapes.

[b]The previous limits from the H1 multiplicity analysis[21] were derived from the shape of the multiplicity distribution for $n < n_{\max}$.

Table 1. Limits on QCD instantons in DIS. A fraction $f_I > f_{\rm lim}$ of instanton induced events in DIS is excluded at 95% C.L..

analysis	DIS kinematics covered			instanton scenario		limit
	Q^2 (GeV^2)	x	W (GeV)	Q'^2(GeV^2)	x'	$f_{\rm lim}$
K^0 [25]	10 – 70	0.001 – 0.01	95 – 230	$\gtrsim 1$	$\gtrsim 0.2$	6 %
multipl. [21]	10 – 80	0.0007 – 0.012	80 – 115	$\gtrsim 1$	$\gtrsim 0.2$	2.7 %
E_T flows [16]	5 – 50	0.0001 – 0.01	65 – 230	> 25	> 0.2	5 – 13 %
multipl. [16]	10 – 80	0.0001 – 0.01	80 – 220	> 25	> 0.2	0.4 – 0.6 %

The available bounds on instanton production are summarised in tab. 1. The most stringent limits for the theoretically "safe" scenario $x' > 0.2$ are still a factor 20 higher than what is predicted from the instanton theory, see fig. 9b. Limits for other scenarios can be found in[16]. For $x' > 0.1$ they are already below the naive extrapolation into the theoretically uncertain region, providing a constraint for the theory and the holy grail function $F(x')$.

6. Conclusion

Instanton transitions, a yet unexplored facette of non-abelian gauge field theories, have been discussed. While in the electroweak theory the $B + L$ violating effects induced by instantons are expected only at energies $\gtrsim 10$ TeV , their chirality violating pendant in QCD could lead to striking signatures already at present day colliders. In DIS at HERA, these are a high particle multiplicity with large transverse energy localized in rapidity, and s, c and possibly b quarks in the final state. The expected contribution to DIS events is of $\mathcal{O}(10^{-3} - 10^{-4})$, with substantial theoretical uncertainties. First analysis of HERA data taken in the years ≤ 1994, corresponding to an integrated luminosity of $\mathcal{O}(1.3\ \mathrm{pb}^{-1})$, are still a factor ≈ 20 above the prediction. With higher statistics data samples ($\mathcal{O}(25\ \mathrm{pb}^{-1})$ up to summer 1997) and improved search strategies, a fundamental discovery at HERA appears to be in reach. It might be possible to exploit also other reactions than DIS, such as photoproduction, where the hard scale needed for reliable instanton calculations could be provided by the p_T of a jet.

7. Acknowledgements

I would like to thank B. Kniehl and G. Kramer for their invitation to this beautifully set workshop, and I thank T. Carli, A. Ringwald and F. Schrempp for the exciting time we are having with instantons, and for their critical reading of the manuscript.

8. References

1. G. 't Hooft, Phys. Rev. Lett. 37 (1976) 8; Phys. Rev. D14 (1976) 3432.
2. A. Belavin, A. Polyakov, A. Schwarz and Yu. Tyupkin, Phys. Lett. B59 (1975) 85.
3. A. Ringwald, Nucl. Phys. B330 (1990) 1; O. Espinosa, Nucl. Phys. B343 (1990) 310.
4. I.I. Balitskii and V.M. Braun, Phys. Lett. B314 (1993) 237.
5. A. Ringwald and F. Schrempp, DESY 94-197, hep-ph/9411217, Proc. of Int. Sem. "Quarks 94", Vladimir, Russia, 1994, p. 170.
6. A. Ringwald and F. Schrempp, DESY 96-203, hep-ph/9610213, to appear in Proc. IXth Int. Sem. "Quarks 96", Yaroslavl, Russia, 1996.
7. S. Moch, A. Ringwald and F. Schrempp, DESY 96-202 hep-ph/9609445; and in prep.
8. A. Ringwald and F. Schrempp, DESY 96-125, hep-ph/9607238, Proc. of the Workshop DIS 96 on "Deep Inelastic Scattering and Related Phenomena", Rome 1996, eds. G. D'Agostini and A. Nigro, p. 481.
9. A. Ringwald and F. Schrempp, DESY 97-115, hep-ph/9706399, to appear in Proc. of the DIS97 workshop, Chicago 1997, eds. J. Repond and D. Krakauer.
10. S. Moch, A. Ringwald and F. Schrempp, DESY 97-114, hep-ph/9706400, to appear in the Proc. of the DIS97 workshop, Chicago 1997, eds. J. Repond and D. Krakauer.
11. T. Carli, M. Gibbs, M. Kuhlen, A. Ringwald and F. Schrempp, in preparation (version used: QCDINS 1.4.1).
12. L. Lönnblad, Comp. Phys. Comm. 71 (1992) 15 (version used: 4.08).
13. G. Ingelman, Proc. of the Workshop on Physics at HERA, Hamburg 1991, eds. W. Buchmüller and G. Ingelman, vol. 3, p. 1366;
    G. Ingelman, A. Edin and J. Rathsman, DESY 96-057.
14. G. Marchesini et al., Comp. Phys. Comm. 67 (1992) 465 .
15. N. Brook et al., Proc. of the Workshop on Future Physics at HERA, Hamburg 1996, eds. G. Ingelman, A. DeRoeck and R. Klanner, vol. 2, p. 613.
16. T. Carli and M. Kuhlen, DESY 97-151, MPI-PhE/97-18, hep-ex/9708008.
17. H1 Collab., S. Aid et al., Phys. Lett. B356 (1995) 118.
18. M. Gibbs, T. Greenshaw, D. Milstead, A. Ringwald and F. Schrempp, Proc. of the Workshop on "Future Physics at HERA", Hamburg 1996, eds. G. Ingelman, A. DeRoeck and R. Klanner, vol. 1, p. 509.
19. H1 Collab., C. Adloff et al., Nucl. Phys. B485 (1997) 3.
20. V. Kuvshinov and R. Shulyakovsky, Acta Physica Polonica B28 (1997) 1629.
21. H1 Collab., S. Aid et al., Z. Phys. C72 (1996) 573.
22. T. Carli, to appear in Proc. of the Ringberg Workshop "New Trends in HERA Physics", 1997.
23. G. Grindhammer, to appear in Proc. of the Ringberg Workshop "New Trends in HERA Physics", 1997.
24. M. Kuhlen, to appear in the Proc. of the Madrid Workshop on low x Physics, 1997.
25. H1 Collab., S. Aid et al., Nucl. Phys. B480 (1996) 3.

Chapter 5

Photon Structure

PHOTON STRUCTURE FUNCTION AND PHOTOPRODUCTION OF LARGE $P_\perp$ JETS AT HERA

Michel FONTANNAZ
Laboratoire de Physique Théorique,* *Bâtiment 211 Université Paris XI*
91405 ORSAY Cedex, France
E-mail: fontanna@qcd.th.u-psud.fr

ABSTRACT

We present a theoretical discussion of the photon structure function and of the fragmentation functions of partons into photons. The fragmentation functions are compared to LEP data, and we study the sensitivity of photoproduction reactions at HERA to the parton distributions in the photon.

1. Introduction

The distribution functions of partons in (almost) real photons, and the parton into photon fragmentation functions are interesting theoretical objects, which, at asymptotically large Q^2, can be calculated exactly in perturbative QCD. This result is very different from the hadronic case, in which a nonperturbative input is always required; this comes from the pointlike coupling of photon to quark. A full perturbative calculation can be performed as long as Q^2 is large enough. However this statement is not true at non asymptotic values of Q^2 and a nonperturbative component must be added to the distributions. I will discuss this point in detail in Section 2.

The first QCD calculation of the parton distributions, performed by Witten [1], generated a considerable theoretical and experimental activity [2].More recently, TRISTAN, LEP and HERA experimental results have reactivated the interest for these distributions and triggered new theoretical works [3,4,5]. The possibility offered by photoproduction experiments at HERA to measure the gluon distribution is particularly interesting, and many efforts have been devoted to the calculation of jet photoproduction cross sections. This point is discussed in Section 4 (see also M. Klasen's talk, these proceedings).

Here I am going to discuss both the quark and gluon distributions in the photon $q_\gamma(x,Q^2), g_\gamma(x,Q^2)$ (Section 2) and the quark and gluon fragmentation functions $D_q^\gamma(z,Q^2)$ and $D_g^\gamma(z,Q^2)$ (Section 3). These functions can be measured in reactions related by crossing, namely $\gamma^* + \gamma \to X$ and $\gamma^* \to \gamma X$, and the theoretical approaches are very similar. As already stated, the parton distributions in the photon are basic ingredients of photoproduction cross sections. The fragmentation functions are needed in the calculation of direct photon cross sections, and experiments now begin

*Laboratoire associé au CNRS, URA D0063

to measure them at LEP [6,7].

2. QCD calculation of the photon structure function

The parton contents of the photon can be measured in deep inelastic scattering experiments in which the virtual photon γ^* of momentum q ($Q^2 = -q^2 \gg \Lambda^2$) probes the short distance behavior of the real photon γ of momentum p. The structure function F_2^γ of this reaction is proportional, in the Leading Logarithm (LL) approximation, to the quark distributions in the real photon

$$F_2^\gamma(x,Q^2) = x \sum_{f=1}^{n_f} e_f^2 \left(q_\gamma^f(x,Q^2) + \bar{q}_\gamma^f(x,Q^2) \right) \quad . \tag{1}$$

The sum in (1) run over the quark flavors and $x = Q^2/2p.q$.

In the LL approximation the perturbative quark distribution is given by the sum of ladder diagrams [8] (Fig.1) (for simplicity we forget the gluons and consider only the non singlet quark distribution)

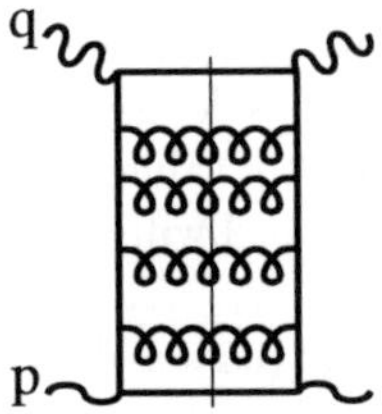

Fig.1: Ladder diagram contribution to F_2^γ (the thin line cuts final partons)

The sum of the ladder diagrams can be written in a very compact form (AN is for Anomalous, a designation introduced in ref.[1])

$$q_\gamma^{AN}(n,Q^2) = \frac{\alpha}{2\pi} \int_{\alpha_s(Q_0^2)}^{\alpha_s(Q^2)} \frac{d\alpha_s'}{\beta(\alpha_s')} k^{(0)}(n) e^{\int_{\alpha_s'}^{\alpha_s(Q^2)} \frac{d\alpha_s''}{\beta(\alpha_s'')} \frac{\alpha_s''}{2\pi} P^{(0)}(n)} \tag{2}$$

in terms of moments of the Altarelli-Parisi kernels $k^{(0)}(n) = \int_0^1 dx \; x^{n-1} k^{(0)}(x)$ and $P^{(0)}(n)$ describing the splitting of a photon into a $q\bar{q}$ pair (the bottom rung of the ladder) and the splitting of a quark into a quark and a gluon (the other rungs of the ladder). The beta function has the usual definition $\partial \alpha_s(Q^2)/\partial \ell n \; Q^2 = \beta(\alpha_s) = -\alpha_s(\frac{\alpha_s}{4\pi}\beta_0 + (\frac{\alpha_s}{4\pi})^2 \beta_1 + \cdots)$. This quark distribution verifies the inhomogeneous equation

$$Q^2 \frac{\partial q^{AN}\gamma(n,Q^2)}{\partial Q^2} = \frac{\alpha}{2\pi} k^{(0)}(n) + \frac{\alpha_s(Q^2)}{2\pi} P^{(0)}(n) q_\gamma^{AN}(n,Q^2) \quad . \tag{3}$$

If we forget the effects of the gluons (the exponential in formula (2)), we find

$$q_\gamma^{AN}(n,Q^2) = \frac{\alpha}{2\pi}\log\frac{Q^2}{Q_0^2}k^{(0)}(n) \tag{4}$$

which is the logarithmic contribution of the "Box" cross section (Fig.1 without gluon). We notice a factorization between the Q^2-behavior and the n-behavior which is conserved when QCD corrections are switched on. This peculiar behavior of the quark distributions in the photon, which grow with Q^2 whatever n may be, lead Witten to coin the term Anomalous for these distributions.

Q_0^2 must be chosen large enough to allow a perturbative calculation. It corresponds to the lowest virtuality of quarks in the ladder. The process is no longer perturbative if the quark virtualities are smaller than Q_0^2, and we expect a strong reinteraction of the quark and antiquark emitted by the real photon. In this case, the photon ceases to couple in a pointlike manner to the quark.

Therefore the full distribution consists of two contributions: the Anomalous one (2) and a non perturbative contribution:

$$q_\gamma(n,Q^2) = q_\gamma^{AN}(n,Q^2,Q_0^2) + q_\gamma^{NP}(n,Q^2) \tag{5}$$

with $q_\gamma^{AN}(n,Q_0^2,Q_0^2) = 0$.

The modifications of these LL results due to HO QCD corrections [9,10,11] are obtained by replacing the LL kernels of (2) by kernels involving HO contributions[12,13,14]

$$\begin{aligned} k(n) &= \tfrac{\alpha}{2\pi}k^{(0)}(n) + \tfrac{\alpha}{2\pi}\tfrac{\alpha_s}{2\pi}k^{(1)}(n) + \cdots \\ P(n) &= \tfrac{\alpha_s}{2\pi}P^{(0)}(n) + \left(\tfrac{\alpha_s}{2\pi}\right)^2 P^{(1)}(n) + \cdots \quad , \end{aligned} \tag{6}$$

and by a modification of the expression of F_2^γ in terms of parton distributions (the gluon contribution is now explicitly written)

$$F_2^\gamma(n-1,Q^2) = \sum e_f^2 C_q(n,Q^2)\left(q_\gamma(n,Q^2) + \bar{q}_\gamma(n,Q^2)\right) + C_g(n,Q^2)g_\gamma(n,Q^2) + C_\gamma(n) \tag{7}$$

where C_γ is the "direct term". C_q and C_g are the well-known Wilson coefficients which are identical to those found in the case of a hadronic target [13,14].

A subtle point when working beyond the LL approximation is that of the factorization scheme. A change in the factorization scheme is translated into a change in $k^{(1)}$ and C_γ but in such a way that the physical quantity F_2^γ remains unmodified (at order α_s^0). On the other hand q^γ is not an invariant with respect to the factorization scheme and a change in $k^{(1)}$ causes modifications in q_γ^{AN} and q_γ^{NP}. Therefore the separation (2) in a perturbative and a non perturbative part is not scheme invariant [4].

A simple model, based on the Box diagram, allows us to find the part of q_γ^{NP} which is scheme dependent, and thus to isolate the non perturbative scheme independent contribution. We then describe this contribution by using the Vector meson Dominance Model (VDM): the photon couples to a vector meson which is probed by the virtual photon. Therefore this non perturbative component of the quark distribution corresponds to parton distributions in vector mesons.

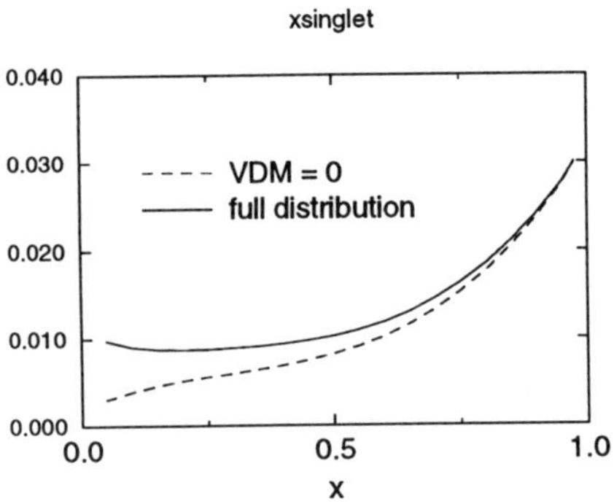

Fig.2

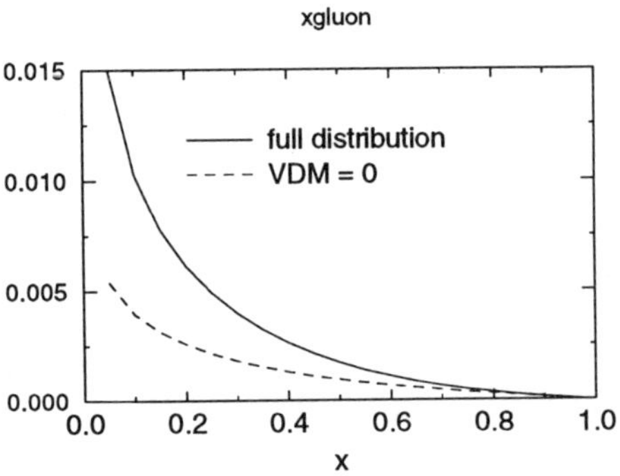

Fig.3

Another approach, elaborated in ref.[3], consists in choosing a factorization scheme (the definition of $k_i^{(1)}(n)$) in which $C_\gamma(n)$ vanishes. Then it is assumed that, in this scheme called DIS_γ, the non perturbative part of (5) is given by the VDM contribution. It turns out that the approaches of ref.[3] and ref.[4] lead to similar predictions.

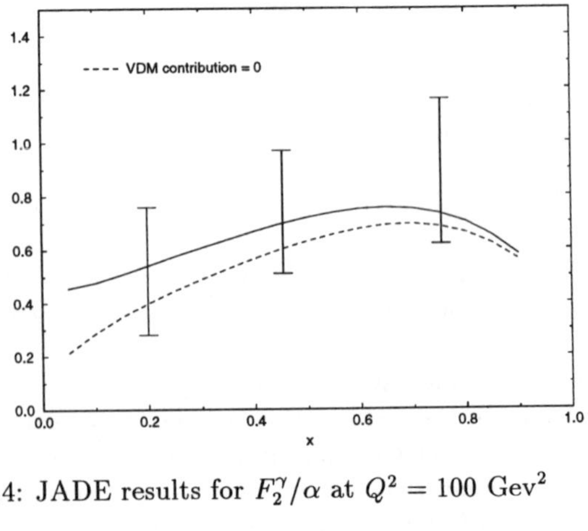

Fig.4: JADE results for F_2^γ/α at $Q^2 = 100$ Gev^2

We show a few results obtained in ref.[4] (with $Q_0^2 = .5$ GeV^2 and $Q^2 = 45.$Gev^2) for the quark and gluon distributions and for $F_2^\gamma(x, Q^2)$. In Fig. 2 and 3 the singlet quark distribution (multiplied by x) and the gluon distribution (multiplied by x) are displayed. The full lines correspond to the full distributions and the dashed lines to distributions with a VDM component set equal to zero. One can see that the VDM contribution is large at small values of x, especially in the gluon case. The structure function $F_2^\gamma(x, Q^2)$ (7) is compared with JADE [15] data in Fig.4. One notices that the VDM contribution is non negligible, even at $Q^2 = 100$ GeV^2.

3. Parton into photon fragmentation functions

The study of the fragmentation functions follows a similar theoretical approach [16]. The inclusive photon cross-section $d\sigma^\gamma/dz$ with $z = 2p.q/q^2$ observed in e^+e^- annihilation reactions is given by an expression which has the same structure as expression (7), obtained in the space-like case. However the direct term and the Wilson coefficients are different; they have been calculated in ref.[17]. The HO corrections to the Altarelli-Parisi kernels are also different and can be found in ref.[12,18].

As in the space-like case, a VDM component must be added to the perturbative component. It is obtained from parton into ρ -meson fragmentation functions [16]. In the time-like case also there exist different approaches [16,19] to specify the non perturbative contribution. They lead to similar results.

At present, there is no data with which to compare [20]. ALEPH data could seem to be a basis for comparison, but it is produced by an analysis in jets. In Ref.[21], this Collaboration defines the fragmentation into a photon within a jet. The fragmentation function D_{jet}^γ defined in that way [22] does not correspond to the functions calculated in this paper which are fully inclusive ; we do not put any limitation on the phase-space of the hadrons which accompany the photon.

In order to better understand the difference between ALEPH results and our predictions, let us consider the decay of a Z-boson (of momentum q) into a photon (p_1), a quark (p_2) and an anti-quark (p_3). We define $z_i = 2p_i.q/q^2$ and $y_{ij} = 2p_i.p_j/q^2$, where z_1 is the inclusive photon fragmentation variable. We have $1 - z_i = y_{jk}$ (i,j,k different) and $\sum_{i<j} y_{ij} = 1$. The variable used by ALEPH to describe the photon in the jet (here the jet is made of the photon and the quark) is $z_\gamma = z_1/(z_1 + z_2) = z_1/(1 + y_{12})$. However an integration is performed on y_{12} within the jet, so that the effective value z_1^{eff} at which one should compare our results is larger than z_γ. But if we assume that the largest contribution to the integral comes from the collinear region $y_{12} \approx 0$, we obtain $z_1 \approx z_\gamma$. Hereafter we use this assumption.

One must also notice that the ALEPH Collaboration uses the Durham algorithm to define a jet. According to this algorithm, $y_{12}^{max} = (1 - z_\gamma)/(1 + z_\gamma)$; therefore the scale in the fragmentation function is no longer q^2, but $(1 - z_\gamma)/(1 + z_\gamma)q^2/z_\gamma$ (the extra $1/z_\gamma$ comes from the fact that in the inclusive case $y_{12}^{max} = z_1 q^2$ and this factor $z_1 \approx z_\gamma$ is already included in our calculation).

Finally, one must keep in mind that higher order QCD corrections to the $q\bar{q}\gamma$ process can generate logarithms of the jets parameter (e.g. y_{cut}) coming from the limited phase space integration, which are not present in the fully inclusive case. However for z_γ large enough, y_{cut} does no longer constrain the phase space and the comparison between our predictions and ALEPH results is not spoiled by this effect.

This comparison is shown in fig.5 for 2-jets events and $y_{cut} = 0.1$. We display our predictions for two scales in order to exhibit their sensitivity to the latter. The agreement is quite satisfactory. It is interesting to notice that, in this z_γ-region, one essentially tests the anomalous component of the fragmentation functions; once Q_0^2 is chosen, these parts are a pure prediction of the perturbative QCD. Q_0^2, which is of the order of m_ρ^2, caracterizes the border between the perturbative and non perturbative regions (cf eq.(5)).

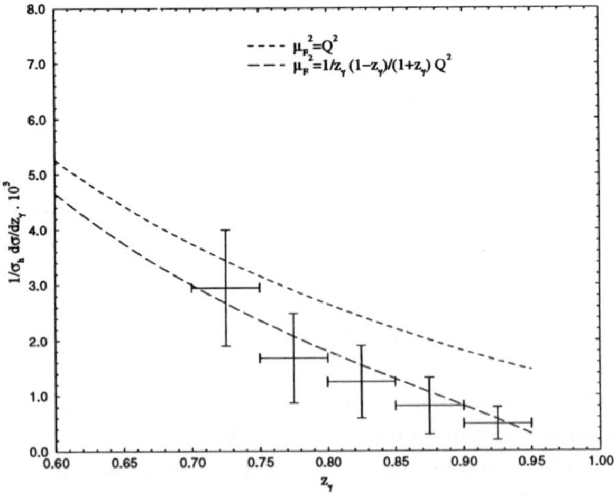

Fig.5: Comparison between ALEPH data and the predictions of ref.[16]

This good agreement between theory and experiment puts the calculation of the bremsstrahlung contribution to the direct photon cross section on a firmer ground. This contribution which, may be non negligible, is now constrained by LEP results.

4. The quark and gluon distributions in photoproduction

The resolved component of the photoproduction cross section of large $p_\perp$ hadrons and jets is important in the HERA kinematical domain; for instance for a jet of rapidity $\eta = 1$ and transverse momentum $p_\perp = 10$ GeV/c, the direct cross section is negligible (for a factorisation scale $M^2 = p_\perp^2$). Therefore accurate HERA results should allow a good measure of the quark and gluon distribution in the photon. Moreover in photoproduction the gluon couples to the hard subprocess at the lowest order in α_s (contrarily to the DIS case where the gluon contribution is a higher order contribution) and the cross section is sensitive to $g(x, Q^2)$.

To illustrate this point, let us consider the dijet photoproduction cross section $d\sigma/d\vec{p}_1 d\eta_1 d\eta_2$ which has been calculated, including HO corrections, by several groups [23,24,25]. This cross section depends on the transverse momentum $p_{\perp_1}$ and rapidity η_1 of the first jet, and on the rapidity η_2 of the second jet (fixing $p_{\perp_2}$ would be too great a constraint on the phase space and the cross section would be a distribution in $p_{\perp_2}$). From this variables one can define

$$x_{LL} = \frac{p_{\perp_1}(e^{-\eta_1} + e^{-\eta_2})}{2E_\gamma} \tag{8}$$

which is equal to the parton momentum x in the photon (E_γ is the photon energy) when the cross section is calculated at the Leading Logarithm (LL) approximation. This is not the case in a beyond LL expression in which these variables no longer fix the parton momentum. However one can show that the cross section (calculated in the BLL approximation)

$$d\sigma/dx_{LL} = \int_{p_{\perp min}} d\vec{p}_{\perp_1} d\eta_1 d\eta_2 \delta(x_{LL} - \frac{p_{\perp_1}(e^{-\eta_1} + e^{-\eta_2})}{2E_\gamma}) \frac{d\sigma}{d\vec{p}_{\perp_1} d\eta_1 d\eta_2} \tag{9}$$

is still very sensitive to the x-shape of the distribution function [26].

The ZEUS variable $x_\gamma^{obs} = (p_{\perp_1} e^{-\eta_1} + p_{\perp_2} e^{-\eta_2})/2E_\gamma$ requires the measure of the second jet momentum $p_{\perp_2}$. This may lead to non Infra Red Stable results when, for instance, a more inclusive cross-section is defined

$$d\sigma/dx_\gamma^{obs} = \int_{p_{\perp 1 min}} d\vec{p}_{\perp_1} d\eta_1 \int_{p_{\perp 2 min}} d\vec{p}_{\perp_2} d\eta_2 \delta(x_\gamma^{obs} - (p_{\perp_1} e^{-\eta_1} + p_{\perp_2} e^{-\eta_2})/2E_\gamma) \frac{d\sigma}{d\vec{p}_{\perp_1} d\eta_1 d\vec{p}_{\perp_2} d\eta_2} \tag{10}$$

with $p_{\perp_1 min} = p_{\perp_2 min}$.

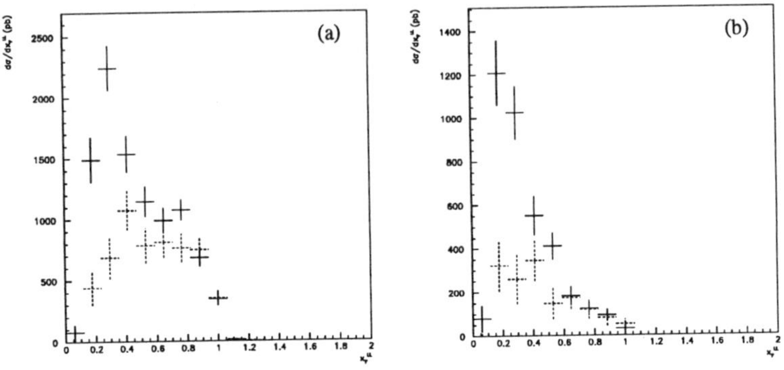

Fig.6: Cross sections $d\sigma/dx_{LL}$ for the resolved contribution without the gluon (dashed lines) and with it (full line).The cuts on the rapidity of the two high p_T jets are $0.5 < \eta < 2$ (a) and $1 < \eta < 2$ (b).

Let us now consider the sensitivity of $d\sigma/dx_{LL}$ on the gluon in the photon distribution function. Fig.6 shows the effect of artificially setting equal to zero the gluon distribution. We see that the importance of the gluon depends stongly on the rapidity cuts. When η_1 and η_2 become positive and large (a configuration which corresponds to a small value of x), the contribution of the gluon becomes large, and should be measurable in future accurate experiments at HERA.

1. E. Witten, *Nucl. Phys.* **B210** (1977) 189.
2. Reviews on the photon structure function may be found in
   C. Berger and W. Wagner, *Phys. Rev.* **146** (1987) 1;
   H. Kolanoski and P. Zerwas, in High Energy e^+e^- physics, World Scientific, Singapore, 1988, eds. A. Ali and P. Söding;
   J.H. Da Luz Vieira and J.K. Storrow, *Z. Phys.* **C51** (1991) 241.
   A. Vogt, Proceedings of "Two-Photon Physics at LEP and HERA", Lund, May 1994.
3. M. Glück, E. Reya, A. Vogt, *Phys. Rev.* **D45** (1992) 3986 and **D46** (1992) 1973.
4. P. Aurenche, M. Fontannaz and J. Ph. Guillet, *Z. Phys.* **C64** (1994) 62.
5. L.E. Gordon and J.K. Storrow, *Nucl. Phys.* **B489** (1997) 405.
6. D. Buskulic et al., ALEPH coll., *Z. Phys.* **C69** (1996) 365.
7. K. Ackerstaff et al., OPAL coll., CERN-PPE/97-086
8. Y.L. Dokshitzer, D.I. Dyakonov and S.I. Troyan, Phys. Rep. **58** (1980) 265.
9. W.A. Bardeen and A.J. Buras, *Phys. Rev.* **D20** (1979) 166, **21** (1980) 2041

(E).
10. M. Glück and E. Reya, *Phys. Rev.* **D28** (1983) 2749.
11. M. Fontannaz and E. Pilon, *Phys. Rev.* **D45** (1992) 382.
12. W. Furmanski and R. Petronzio, Phys. Lett. **B97** (1980) 435.
13. G. Altarelli, R.K. Ellis, G. Martinelli, Nucl. Phys. **157** (1979) 461.
14. E. Laenen, S. Riemersma, J. Smith and W.L. van Neerven, *Phys. Rev.* **D49** (1994) 5753.
15. JADE coll., W. Bartel et all. Z. Phys. **C24** (1984) 231.
16. L. Bourhis, M. Fontannaz, J.Ph. Guillet, (to appear in Z. Phys. C).
17. G. Altarelli, R.K. Ellis, G. Martinelli, S.Y. Pi, Nucl. Phys. **160** (1979) 301.
18. P. Aurenche, P. Chiappetta, M. Fontannaz, J.Ph. Guillet et E. Pilon, Nucl. Phys. **B399** (1993) 34.
19. M. Glück, E. Reya and A. Vogt, Phys. Rev. **D58** (1993) 116.
20. See however : OPAL coll. paper submitted to the 97 Lepton-photon conference, Hamburg, July 1997.
21. D. Buskulic et al., ALEPH Coll., Z. Phys. C69 (1996) 365.
22. E.W.H. Gloven, A.G. Morgan, Z. Phys. **C62** (1994) 311
23. P. Aurenche,Luc Bourhis, M. Fontannaz, J.Ph. Guillet, Proceedings of the workshop "Future Physics at HERA", G. Ingelman, A. De Roeck, R. Klanner eds.
24. M. Klasen and G. Kramer, DESY 96-246
25. B.W. Harris and J.F. Owens, FSO-HEP-970411
26. P. Aurenche,Luc Bourhis, M. Fontannaz, J.Ph. Guillet, in preparation.

PHOTON STRUCTURE AT LEP

STEFAN SÖLDNER-REMBOLD
Universität Freiburg
Hermann-Herder-Str. 3, D-79104 Freiburg i. Br., Germany
E-mail: soldner@ruhpb.physik.uni-freiburg.de

ABSTRACT

The structure of the photon is studied in photon interactions at high energies using photons radiated by the electron and positron beams at LEP. The current status of these measurements is reviewed.

1. Introduction

The photon is one of the fundamental gauge bosons of the Standard Model. At high energies photon interactions are dominated by quantum fluctuations of the photons into fermion-antifermion pairs and into vector mesons which have the same spin-parity ($J^{PC} = 1^{--}$) as the photon. This is called photon structure. Electron-Positron collisions at LEP are an ideal laboratory for studying photon structure in interactions of quasi-real and virtual photons, testing predictions of both QED and QCD. These results are complementary to the results obtained in γp scattering at HERA.

2. Electron–Photon Scattering

If one of the scattered electrons[a] is detected (tagged), the process $\mathrm{e^+e^-} \to \mathrm{e^+e^-} +$ hadrons can be regarded as deep-inelastic scattering of an electron on a quasi-real photon with the cross-section

$$\frac{\mathrm{d}^2\sigma_{\mathrm{e}\gamma\to\mathrm{e+hadrons}}}{\mathrm{d}x\mathrm{d}Q^2} = \frac{2\pi\alpha}{x\,Q^4}\left[\left(1+(1-y)^2\right)F_2^\gamma(x,Q^2) - y^2 F_\mathrm{L}(x,Q^2)\right], \tag{1}$$

where $Q^2 = -q^2$ is the negative four-momentum squared of the virtual photon and x and y are the usual dimensionless variables. The structure function F_2^γ is related to the sum over the parton densities of the photon. In order to identify an electron in the detector, a large tag energy E_tag has to be required, i.e only low values of y are accessible ($y^2 \ll 1$). The contribution of the term proportional to the longitudinal structure function F_L^γ is therefore negligible.

For measuring $F_2^\gamma(x,Q^2)$ the value of Q^2 can be well reconstructed from the energy, E_tag, and the angle, θ_tag, of the tagged electron via the relation $Q^2 \approx 2E_\mathrm{beam}E_\mathrm{tag}(1-\cos\theta_\mathrm{tag})$. The reconstruction of $x = Q^2/(Q^2+W^2)$, however, relies heavily on the

[a]In this paper positrons are also referred to as electrons

measurement of the invariant mass W from the hadronic final state.

2.1. Hadronic Energy Flows

OPAL [1], ALEPH [2] and DELPHI [3] have therefore studied the hadronic energy flow $1/N \cdot dE/d\eta$ as a function of the pseudorapidity $\eta = -\ln \tan\theta/2$, where the sign of η is chosen in such way that the tag-electron is always at negative η. OPAL has measured

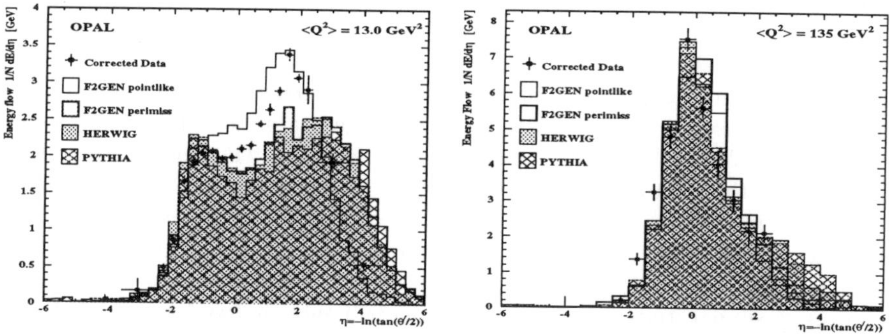

Figure 1: The energy flow per event as a function of the pseudorapidity η, compared to the HERWIG, PYTHIA and F2GEN Monte Carlo models. The data have been corrected for detector effects.

the energy flow at medium Q^2 ($\langle Q^2 \rangle = 13\,\mathrm{GeV}^2$) and at high Q^2 ($\langle Q^2 \rangle = 135\,\mathrm{GeV}^2$) [1]. In Fig. 1 the energy flows are compared to the two QCD based Monte Carlo generators HERWIG [4] and PYTHIA [5]. The data distributions have been corrected for detector effects. The generator F2GEN is used to model the unphysically extreme case of a two-quark state in the $\gamma^*\gamma$ centre-of-mass system with an angular distribution as in lepton pair production from two real photons ("pointlike"). The "perimiss" sample is a physics motivated mixture of pointlike and peripheral interactions, where peripheral means that the transverse momentum of the outgoing quarks is given by an exponential distribution as if all the photons interacted as pre-existing hadrons.

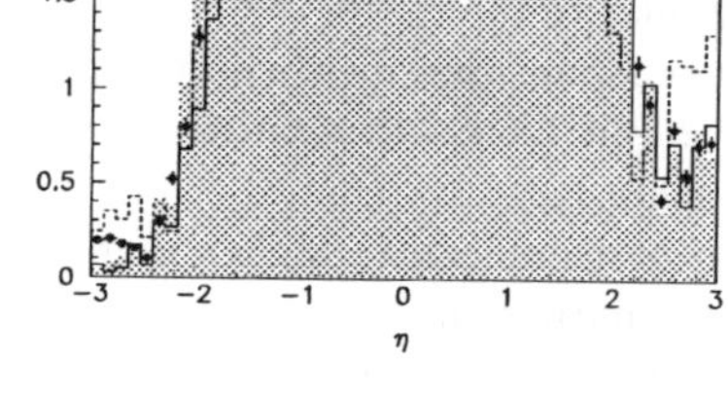

Figure 2: The energy flow per event as a function of η, compared to a QPM+VMD model, the standard and the tuned version of HERWIG.

Significant discrepancies exist between the data and all of the Monte Carlo models. The agreement improves at higher Q^2. Since x and Q^2 are correlated, the discrepancies at low Q^2 are observed also at low x. These discrepancies between the data and the Monte Carlo model for the hadronic final state are the dominant source of systematic uncertainty in the unfolding of $F_2^\gamma(x, Q^2)$ [1].

Tuning of the Monte Carlo to improve the modelling of the hadronic final state is complicated and must be done with care in order to avoid a bias of the result of the unfolding towards the parametrisation of the parton distribution functions used in the tuned Monte Carlo. ALEPH has measured the energy flow in tagged event for $\langle Q^2 \rangle = 14.2$ GeV^2 [2]. The distributions have not been corrected for detector effects. The energy flow shown in Fig. 2 is compared to the HERWIG generator [4] and a Monte Carlo which consists of a mixture of Quark Parton Model (QPM) and Vector Meson Dominance (VMD) similar to the F2GEN generator. In addition, a modified version of HERWIG ("HERWIG+power law p_T") was used. The modification is based on studies of the photon remnant by ZEUS [6]. In standard HERWIG a Gaussian distribution is used to describe the limitation of the transverse momentum of the outgoing partons with respect to the initial target photon. In the modified HERWIG the Gaussian is replaced by a power law spectrum. The agreement with the data improves. A similar study was performed earlier by Lauber in Ref. 7 using OPAL data. It is expected that such improvements of the Monte Carlo models will significantly reduce the systematic error of the structure function measurements.

2.2. Photon Structure Function

The photon structure function $F_2^\gamma(x, Q^2)$ can be measured at LEP in the range $x > 10^{-3}$ and $1 < Q^2 < 10^3$ GeV^2. This will make it possible to study the QCD evolution of F_2^γ in a wide range of x and Q^2. All currently available measurements of the photon structure function are shown in Fig. 3 [8]. The data are compared to the next-to-leading order (NLO) GRV parametrisation [9] and the leading order (LO) SaS-1D parametrisation [10].

For low x and not too small Q^2 a rise of the photon structure function is expected, similar to the rise of the proton structure function observed at HERA. An interesting new F_2^γ measurement is presented by OPAL in the x and Q^2 ranges $2.5 \times 10^{-3} < x < 0.2$ and $1.1 < Q^2 < 6.6$ GeV^2. The measurements is consistent with a possible rise within large systematic errors. It should also be noted that the OPAL points tend to be significantly higher than the previous measurement by TPC/2γ in a similar kinematic range.

For large x and asymptotically large Q^2 the value of F_2^γ can be calculated from perturbative QCD due to the pointlike coupling of the photon to $q\bar{q}$ pairs [11]. The next-to-leading order (NLO) result [12] can be written as

$$\frac{F_2^\gamma}{\alpha} = \frac{a(x)}{\alpha_s(Q^2)} + b(x), \tag{2}$$

where $a(x)$ and $b(x)$ are calculable functions which diverge for $x \to 0$. The first term corresponds to the LO result by Witten [11]. The measurement of F_2^γ could be a direct measurement of Λ_{QCD} if it were not for the large non-perturbative contributions due to bound states. These contributions are large at all experimentally accessible values of Q^2.

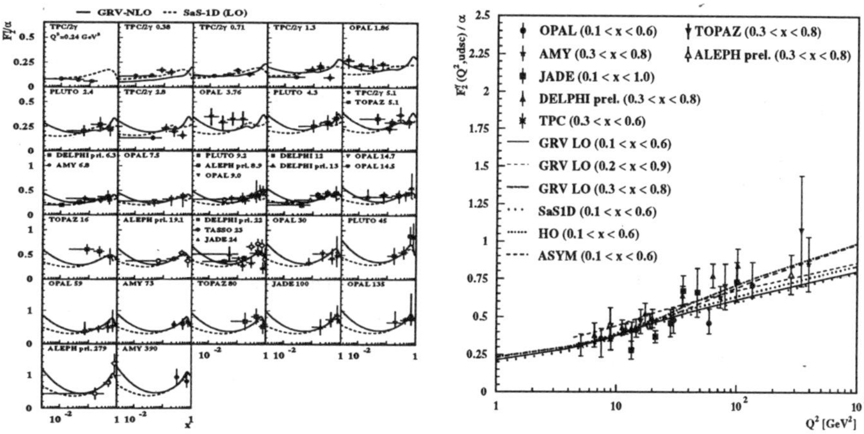

Figure 3: Measurements of the photon structure function F_2^γ in bins of x and Q^2.

The evolution of F_2^γ with $\ln Q^2$ is shown in Fig. 3 using the currently available F_2^γ measurements for 4 active flavours. The data are compared to the LO GRV and the SaS-1D parametrisations, and to a higher order (HO) prediction based on the HO GRV parametrisation for light quarks and on the NLO charm contribution calculated in Ref. 13. The data are measured in different x ranges. The comparison of the LO GRV curves for these x ranges shows that for $Q^2 > 100$ GeV^2 significant differences are expected. An augmented asymptotic prediction for F_2^γ is also shown. The contribution to F_2^γ from the three light flavours is approximated by Witten's leading order asymptotic form [11]. This has been augmented by adding a charm contribution evaluated from the Bethe-Heitler formula [14], and an estimate of the hadronic part of F_2^γ, which essentially corresponds to the hadronic part of the GRV (LO) parametrisation. In the region of medium x values studied here this asymptotic prediction in general lies higher than the GRV and SaS predictions but it is still in agreement with the data. The importance of the hadronic contribution to F_2^γ decreases with increasing x and Q^2, and it accounts for only 15 % of F_2^γ at $Q^2 = 59$ GeV^2 and $x = 0.5$.

As predicted by QCD the evolution of F_2^γ leads to a logarithmic rise with Q^2, but theoretical and experimental uncertainties are currently too large for a precision test of perturbative QCD.

2.3. Azimuthal Correlations

Only the structure function F_2^γ has so far been determined directly from measurements of double-differential cross-sections for $e\gamma$ events with hadronic or leptonic final states. It has been pointed out [15] that azimuthal correlations in the final-state particles from two-photon collisions are sensitive to additional structure functions. Azimuthal correlations can thus supplement the direct measurement of structure functions. ALEPH [16], L3 [17] and OPAL [18] have measured azimuthal correlations using single-tag $e^+e^- \to e^+e^-\mu^+\mu^-$ events.

For single-tag events two independent angles can be defined in the $\gamma\gamma^*$ centre-of-mass system assuming that the target photon direction is parallel to the beam axis: The azimuthal angle χ is the angle between the planes defined by the $\gamma\gamma^*$ axis and the two-body final state and the $\gamma\gamma^*$ axis and the tagged electron. The variable $\eta = \cos\theta^*$ is defined by the angle θ^* between the μ^- and the $\gamma\gamma^*$ axis.

Neglecting the longitudinal component of the target photon and setting $(1-y)$ to one, the cross-section can be written as ($F_2^\gamma = 2xF_T^\gamma + F_L^\gamma$):

$$\frac{d\sigma(e\gamma \to e\mu^+\mu^-)}{dx dy d\eta d\chi/2\pi} = \frac{2\pi\alpha^2}{Q^2}\left(\frac{1+(1-y)^2}{xy}\right)\left[2x\tilde{F}_T^\gamma + \tilde{F}_L^\gamma - \tilde{F}_A^\gamma \cos\chi + \frac{\tilde{F}_B^\gamma}{2}\cos 2\chi\right]. \quad (3)$$

The conventional structure functions are recovered by integration over η and χ: $F_i^\gamma = \int_{-1}^{1}\int_0^{2\pi} \frac{d\chi d\eta}{2\pi}\tilde{F}_i^\gamma (i = 2, A, B)$. The structure functions F_T and F_L are related to the scattering of transverse and longitudinally polarized virtual photons, respectively. F_A is related to the interference terms between longitudinal and transverse photons and F_B to the interference term between purely transverse photons. The longitudinal structure function F_L^γ has been shown to be equal to the structure function F_B^γ in leading order and for massless muons, although coming from different helicity states of the photons.

The variation of F_A^γ and F_B^γ with x is in general consistent with QED (Figs. 4 and 5 and Ref. 18). The measured values are significantly different from zero. Apart from being an interesting test of QED, these results are especially important as a first step towards measuring the additional structure function for hadronic events using azimuthal

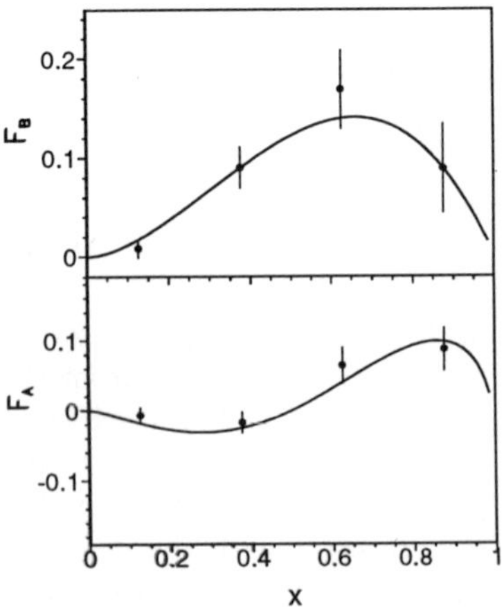

Figure 4: The structure functions F_B^γ and F_A^γ for the process $e^+e^- \to e^+e^-\mu^+\mu^-$ (L3). The lines show the QED expectation.

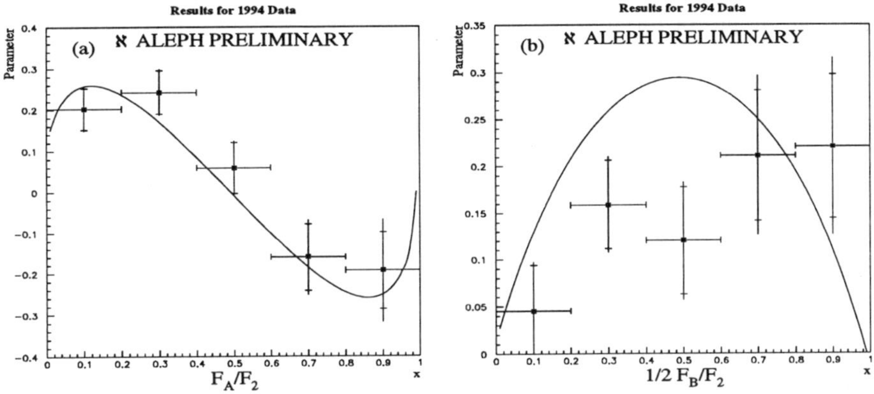

Figure 5: The ratio of structure functions, F_A^γ/F_2^γ and $1/2 \cdot F_B^\gamma/F_2^\gamma$, for the process $e^+e^- \to e^+e^-\mu^+\mu^-$ (ALEPH). The lines show the QED expectation.

correlations. Such a measurement will be much more difficult due to the problems related to the jet finding in hadronic events.

3. Photon–Photon Scattering

If both scattered electrons remain undetected, the Q^2 of the two interacting photons is very small and both photons can be considered to be quasi-real. At high $\gamma\gamma$ centre-of-mass energies $W = \sqrt{s_{\gamma\gamma}}$ the total cross-section for the production of hadrons in the interaction of two real photons is expected to be dominated by interactions where the photon has fluctuated into an hadronic state.

In LO QCD the $\gamma\gamma$ interactions can be classified by three different processes: direct, single- and double-resolved. The bare photons interact in the direct process, whereas in resolved events the partons (quarks or gluons) inside the hadronic state of the photon take part in the hard interaction. The probability to find partons in the photon is parametrised by the parton density functions.

3.1. Total Cross Sections

Measuring the $\sqrt{s_{\gamma\gamma}}$ dependence of the total $\gamma\gamma$ cross-section should improve our understanding of the hadronic nature of the photon and the universal high energy behaviour of total hadronic cross-sections.

The total cross-sections σ for hadron-hadron and γp collisions are well described by a Regge parametrisation of the form $\sigma = Xs^\epsilon + Ys^{-\eta}$, where $\sqrt{s}$ is the centre-of-mass energy of the hadron-hadron or γp interaction. The first term in the equation is due to Pomeron exchange and the second term is due to Reggeon exchange [19].

Assuming factorisation of the Pomeron term X, the total $\gamma\gamma$ cross-section can be related to the pp (or $\overline{\rm p}$p) and γp total cross-sections at high centre-of-mass energies $\sqrt{s}_{\gamma\gamma}$ where the Pomeron trajectory should dominate:

$$\sigma_{\gamma\gamma} = \frac{\sigma^2_{\gamma \rm p}}{\sigma_{\rm pp}}. \tag{4}$$

A slow rise of the total cross-section with energy is predicted, corresponding to $\epsilon \approx 0.08$. This rise can also be attributed to an increasing cross-section for parton interactions leading to mini-jets in the final state [20].

Before LEP the total hadronic $\gamma\gamma$ cross-section has been measured by PLUTO [21], TPC/2γ [22] and the MD1 experiment [23]. These experiments have measured at centre-of-mass energies W below 10 GeV before the onset of the high energy rise of the total cross-section. Using LEP data taken at $\rm e^+e^-$ centre-of-mass energies $\sqrt{s}_{\rm ee} = 130 - 161$ GeV L3 [24] has demonstrated that the total hadronic $\gamma\gamma$ cross-section in the range $5 \leq W \leq 75$ GeV is consistent with the universal Regge behaviour of total cross-sections which was also observed in γp scattering at HERA. The L3 measurement is shown in Fig. 6 together with a preliminary OPAL measurement [25] in the range $10 < W < 110$ GeV using data taken at $\sqrt{s}_{\rm ee} = 161 - 172$ GeV. The observed energy dependence of the cross-section is similar, but the values for $\sigma_{\gamma\gamma}$ are about 20 % higher. It should be noted that the errors are strongly correlated between the W bins in both experiments. Furthermore, L3 has used the Monte Carlo generator PHOJET [28] for the unfolding, whereas for the OPAL measurement the unfolding results of PHOJET and PYTHIA are averaged. The unfolded cross-section using PHOJET is about 5 % lower than the central value. In both experiments the cross-sections obtained using PHOJET are lower than the cross-section obtained with PYTHIA.

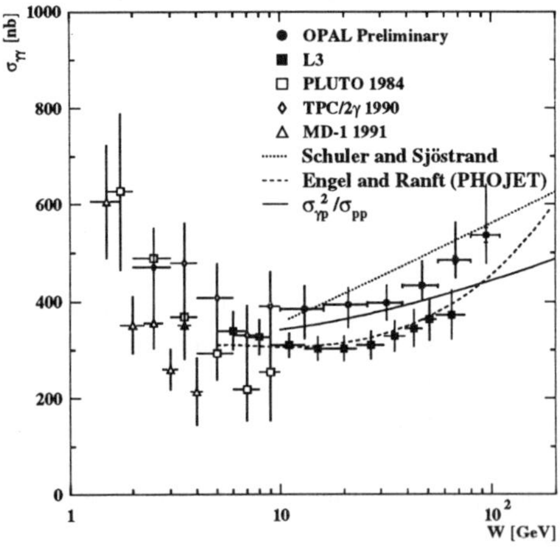

Figure 6: Total cross-section of the process $\gamma\gamma \rightarrow$ hadrons

The total cross-section is compared to several theoretical models. Based on the Donnachie-Landshoff model [26], the assumption of a universal high energy behaviour of $\gamma\gamma$, γp and pp cross-sections is tested. The parameters X and Y are fitted to the total $\gamma\gamma$, γp and pp cross-sections in order to predict $\sigma_{\gamma\gamma}$ via Eq. 4. This is done assuming that the cross-sections can be related at $\sqrt{s}_{\gamma\gamma} = \sqrt{s}_{\gamma \rm p} = \sqrt{s}_{\rm pp}$. The process dependent fit values for X and Y are taken from Ref. 19 together with the values of the universal parameters $\epsilon = 0.0790 \pm 0.0011$ and $\eta = 0.4678 \pm 0.0059$. This simple ansatz gives a reasonable description of the total $\gamma\gamma$ cross-section $\sigma_{\gamma\gamma}$. Schuler and Sjöstrand [27] give a total cross-section for the sum of all possible event classes in their model of $\gamma\gamma$ scattering where the photon has a direct, an anomalous and a VMD component. They consider the spread between this prediction and the simple factorisation ansatz as conservative estimate of the theoretical band of uncertainty. The prediction of Engel and Ranft [28] is also plotted which is implemented in PHOJET. It is in good agreement with the L3 measurement and significantly lower than the OPAL measurement. The steeper rise predicted by Engel and Ranft is in agreement with both measurements.

A large part of the cross-section (about 20 % in both MC models) is due to diffractive and elastic events (e.g. $\gamma\gamma \to \rho\rho$). At high W the detectors have only little acceptance for these events and the correction procedure has to rely heavily on the MC model. In the future it will therefore be very important to gain a better understanding of these processes in $\gamma\gamma$ interactions.

3.2. Jet Production

The measurement of inclusive jet cross-sections and the comparison with NLO QCD calculations [29,30] and LO QCD Monte Carlo simulations using different parametrisations of the parton distributions of the photons can constrain the relatively unknown gluonic content of the photon.

In contrast to deep inelastic electron-photon scattering, which in leading order is only sensitive to the quark content of the photon, the gluon content of the photon can be tested, especially in the case of double-resolved processes, in the interaction of two almost real photons. Inclusive one-jet and two-jet cross-sections have been measured at an e^+e^- centre-of-mass energy of $\sqrt{s}_{ee} = 58$ GeV at TRISTAN [31,32] and at $\sqrt{s}_{ee} = 130 - 172$ GeV by OPAL [33,34]. In all cases the cone jet finding algorithm has been used.

In Fig. 7 the $E_T^{\rm jet}$ distribution measured by OPAL [34] at $\sqrt{s}_{ee} = 161 - 172$ GeV is compared to a NLO perturbative QCD calculation of the inclusive two-jet cross-section by Kleinwort and

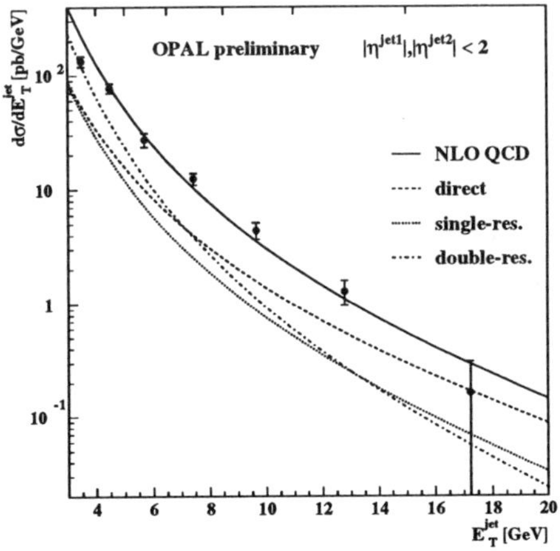

Figure 7: The inclusive two-jet cross-section as a function of $E_T^{\rm jet}$ for jets with $|\eta^{\rm jet}| < 2$ using a cone size $R = 1$.

Kramer [29] who use the NLO GRV parametrisation of the photon structure function [9]. The direct, single- and double-resolved parts and their sum are shown separately. The data points are in good agreement with the calculation except in the first bin where the calculation predicts a much higher cross-section. The resolved cross-sections dominate in the region $E_T^{\text{jet}} \lesssim 8$ GeV, whereas, at high E_T^{jet} the direct cross-section is largest.

The NLO QCD calculation gives the jet cross-section for massless partons, whereas the experimental jet cross-sections are measured for hadrons. The difference due to fragmentation mainly contributes at low E_T^{jet} values. A correction to the parton level would increase the cross-section by about a factor of 1.2 to 1.3.

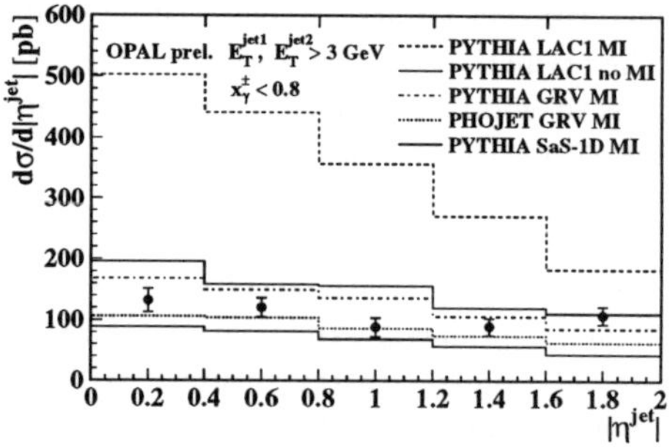

Figure 8: The inclusive two-jet cross-section as a function of η^{jet} for jets in mainly resolved event with $E_T^{\text{jet}} > 3$ GeV using a cone size $R = 1$.

The inclusive two-jet cross-section as a function of $|\eta^{\text{jet}}|$ is shown in Fig. 8 for events with a large double-resolved contribution after requiring $x_\gamma^\pm < 0.8$. The variable $x_\gamma^\pm$ specifies the fraction of the photon energy participating in the hard scattering:

$$x_\gamma^+ = \frac{\sum\limits_{\text{jets}=1,2} (E + p_z)}{\sum\limits_{\text{hadrons}} (E + p_z)} \quad \text{and} \quad x_\gamma^- = \frac{\sum\limits_{\text{jets}=1,2} (E - p_z)}{\sum\limits_{\text{hadrons}} (E - p_z)}, \tag{5}$$

where p_z is the momentum component along the z axis of the detector and E is the energy of the jets or hadrons. Ideally, for direct events without remnant jets $x_\gamma^+ = 1$ and $x_\gamma^- = 1$, whereas for double-resolved events both values x_γ^+ and x_γ^- are expected to be much smaller than 1.

The inclusive two-jet cross-section predicted by the two LO QCD Monte Carlo models, PYTHIA and PHOJET differ significantly even if the same photon structure function (here GRV LO) is used. This model dependence reduces the sensitivity to the parametrisation of the photon structure function. Different parametrisations were used as input to the PYTHIA simulation. The GRV-LO [9] and SaS-1D parametrisations [10] describe the data equally well, but LAC1 [35] which increases the cross-section for gluon-initiated processes overestimates the inclusive two-jet cross-section significantly. Turning off the simulation of multiple interactions (MI) within PYTHIA reduces the predicted cross-section using LAC1 by more than a factor of two.

For a more quantitative interpretation of these results in terms of parton distribution functions, it will be very important to understand the influence of multiple interactions on the jet cross-sections and to use jet definitions which will allow to compare theory (partons) and experiment (hadrons) directly. This is very similar to the problems of measuring jet cross-sections in photoproduction at HERA discussed in these proceedings [36,37].

3.3. Charm Production

Similar to jet production, open charm production in photon-photon collisions can also be used to constrain the gluon content of the photon. The charm production cross-sections have been calculated in NLO [38,39]. The mass m_c of the charm quark sets the scale for the perturbative QCD calculation. The cross-section is factorized into the matrix elements for the production of heavy quarks and the parton densities for light quarks (q) and gluons. This 'massive' approach is expected to be valid if the transverse momenta p_T of the charm quarks are of the same order, $p_T \approx m_c$, which is true for the kinematic range probed at LEP. At LEP energies only the direct process $\gamma\gamma \to c\bar{c}$ and the single-resolved process $g\mathrm{q} \to c\bar{c}$ are relevant.

The cleanest method to tag open charm is the reconstruction of $D^{*+} \to D^0\pi^+$ decays. Due to the small branching ratios of the D^0 into charged pions and kaons, this method is statistics limited. ALEPH has measured the charm cross-section using 33 ± 8 $D^{*\pm}$ mesons reconstructed in the LEP1 data. L3 has measured the charm cross-section in $\gamma\gamma$ interaction at LEP1 and LEP2 by tagging muons from semi-leptonic charm decays in the momentum range $2 < p_\mu < 7$ GeV/c at LEP1 and $2 < p_\mu < 15$ GeV/c at LEP2 [41]. The efficiency to tag muons is less than 10^{-3} leading to large systematic and statistical uncertainties.

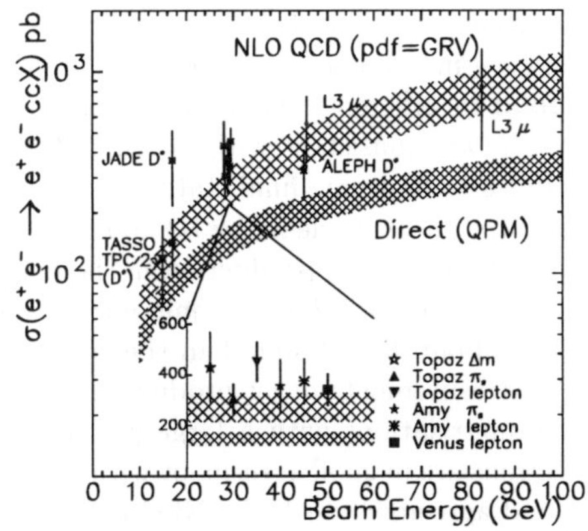

Figure 9: Cross-section for the process $e^+e^- \to e^+e^-c\bar{c}$ as a function of the electron beam energy.

The cross-section for the process $e^+e^- \to e^+e^-c\bar{c}$ as a function of the beam energy is shown in Fig. 9. The experimental results for various charm tagging methods used by pre-LEP experiments have been extrapolated to obtain a total charm cross-section [42]. The upper band shows the full NLO charm cross-section calculated by Drees et al [38] and the lower band the contribution from the Born term direct process (Quark Parton Model, QPM). The upper edge of the band is obtained by setting $m_c = 1.3$ GeV with a scale $\mu = m_c$ and the lower edge by setting $m_c = 1.7$ GeV with $\mu = \sqrt{2}m_c$. The data points obtained from the TRISTAN and JADE measurements cluster around the higher edge of the the 'massive' NLO calculation which uses the GRV parametrisation. At LEP energies the extrapolated measurement are in good agreement with the NLO calculation within the large errors.

4. Conclusions

Measurements of the photon structure function $F_2^\gamma(x, Q^2)$ from double-differential cross-section in eγ scattering mainly constrain the quark distribution in the photon. At low x LEP will be able to study the region where the onset of the rise of F_2^γ is expected from the HERA data on the proton structure function. LEP measurements will cover the kinematic range $x > 10^{-3}$ and $1 < Q^2 < 10^3$ GeV^2. A first measurement in the low x region has been presented by OPAL.

The logarithmic rise of F_2^γ with Q^2 for medium x and large Q^2 is observed as predicted, but theoretical and experimental uncertainties are currently too large for a precision test of perturbative QCD. The current systematic limitation of the F_2^γ measurements comes from the discrepancies between data distributions and Monte Carlo models for the hadronic final state. Work has started to improve the Monte Carlo models.

Azimuthal correlations have been used to determine the structure functions F_A and F_B in single-tagged $e^+e^- \to e^+e^-\mu^+\mu^-$ events. These structure functions correspond to different helicity states of the virtual photon and the target photon. The results are consistent with QED. In the future it will be interesting to extend these studies to hadronic final states.

First measurements by L3 and OPAL of the energy dependence of the total cross-section for hadron production in the interactions of quasi-real photons show the slow rise characteristic for hadronic interactions. However, the cross-section measured by L3 is about 20 % lower than the cross-section measured by OPAL.

Measurements of jet production in $\gamma\gamma$ interactions presented by OPAL disfavour parametrisations with a large gluon content in the photon like LAC1. The cross-sections are in good agreement with NLO calculations using the GRV parametrisation. Good agreement with the NLO calculation is also found for the measurement of the total charm production cross-section in $\gamma\gamma$ interactions by ALEPH and L3.

The experimental methods and the theoretical framework necessary for studying photon structure in the interactions of quasi-real photons at LEP and in photoproduction at HERA are very similar. This workshop has shown very clearly that both communities can learn a lot from each other.

5. Acknowledgements

I want to thank the Max-Planck-Institut für Physik and DESY, especially Bernd Kniehl, Gustav Kramer and Albrecht Wagner, for the excellent organisation of this interesting workshop. Thanks also to the LEP collaborations for providing the plots.

6. References

1. OPAL Coll., K. Ackerstaff et al, *Z. Phys.* **C74** (1997) 33.
2. ALEPH Coll., paper submitted to Lepton-Photon 1997, LP-253.
3. I. Tyapkin, *proceedings of PHOTON '97*, Egmond aan Zee, The Netherlands (1997).
4. G. Marchesini et al., *Comp. Phys. Comm.* **67** (1992) 465.
5. T. Sjöstrand, *Comp. Phys. Commun.* **82** (1994) 74; T. Sjöstrand, LUND University Report, LU-TP-95-20 (1995).
6. ZEUS Coll., M. Derrick et al., *Phys. Lett.* **B354** (1995) 163.
7. J. Lauber et al., *proceedings of PHOTON '97*, Egmond aan Zee, The Netherlands (1997).
8. A compilation can be found in D. Morgan et al, *J. Phys. G, Nucl. Part. Phys.* **20** (1994) A1: PLUTO Coll., Ch. Berger et al., *Phys. Lett.* **B142** (1984) 111; PLUTO Coll., Ch. Berger et al., *Nucl. Phys.* **B281** (1984) 365; JADE Coll., W. Bartel et al., *Z. Phys.* **C24** (1984) 231; TASSO Coll., M. Althoff et al., *Z. Phys.* **C31** (1986) 527; TPC/2γ Coll., H. Aihara et al., *Z. Phys.* **C34** (1987) 1; AMY Coll., Y. Sasaki et al., *Phys. Lett.* **B252** (1990) 491; TOPAZ Coll., K. Muramatsu et al., *Phys. Lett.* **B332** (1994) 477; DELPHI Coll., P. Abreu et al., *Z. Phys.* **C69** (1996) 223 ($Q^2 = 12$ GeV^2); AMY Coll., T. Kojima et al., *Phys. Lett.* **B400** (1997) 395; ($Q^2 = 6.8$ GeV^2); OPAL Coll., K. Ackerstaff et al, *Z. Phys.* **C74** (1997) 33 ($Q^2 = 7.5, 14.7, 135$ GeV^2); OPAL Coll., K. Ackerstaff et al, CERN-PPE/97-87 ($Q^2 = 9, 14.5, 30, 59$ GeV^2); OPAL Coll., K. Ackerstaff et al, CERN-PPE/97-103 ($Q^2 = 1.86, 3.76$ GeV^2); ALEPH Coll., paper submitted to Lepton-Photon 1997, LP-315; DELPHI Coll., submitted to Lepton-Photon 1997 ($Q^2 = 6.3, 13, 22$ GeV^2).
9. M. Glück, E. Reya and A. Vogt, *Phys. Rev.* **D46** (1992) 1973; Phys. Rev. **D45** (1992) 3986.
10. G. A. Schuler and T. Sjöstrand, *Z. Phys.* **C68** (1995) 607.
11. E. Witten, *Nucl. Phys.* **B120** (1977) 189.
12. W. A. Bardeen and A. J. Buras, *Phys. Rev.* **D20** (1979) 166; *Phys. Rev.* **D21** (1980) 2041;
13. E. Laenen, J. Smith, S. Riemersma and W. L. van Neerven, *Phys. Rev.* **D49** (1994) 5753.
14. E. Witten, *Nucl. Phys.* **B104** (1976) 445.
15. C. Peterson et al., *Nucl. Phys.* **B229** (1983) 301; CELLO Coll., H.-J. Behrend et al., *Z. f. Phys.* **C43** (1989) 1; J. H. Field, *proceedings of Photon '95*, Sheffield, U.K., eds. D. J. Miller, S. L. Cartwright and V. Khoze, World Scientific, Singapore, 490 (1995); N. Arteaga et al., *proceedings of Photon '95*,

Sheffield, U.K.; Phys. Rev. **D52** (1995) 4920; *Phys. Rev.* **D53** (1996) 2854; P. Aurenche et al., in *Physics at LEP2*, CERN 96-01, p.301 (1996).
16. C. Brew et al., *proceedings of PHOTON '97*, Egmond aan Zee, The Netherlands (1997).
17. L3 Coll., paper submitted to EPS97, Jerusalem.
18. OPAL Coll., K. Ackerstaff et al, *Z. Phys.* **C74** (1997) 49.
19. R. M. Barnett et al., *Review of Particle Physics, Phys. Rev.* **D54** (1996) 1.
20. M. Drees and R. M. Godbole, *Nucl. Phys.* **B339** (1990) 355; J. R. Forshaw and J. K. Storrow, *Phys. Rev.* **D46** (1992) 4955; M. Drees and R. M. Godbole, *Z. Phys.* **C59** (1993) 591; A. Corsetti, R. M. Godbole, G. Panchieri, *proceedings of PHOTON '97*, Egmond aan Zee, The Netherlands (1997).
21. PLUTO Coll., Ch. Berger et al., *Phys. Lett.* **B149** (1984) 421.
22. TPC/2γ Coll., H. Aihara et al., *Phys. Rev.* **D41** (1990) 2667.
23. S. E. Baru et al., *Z. Phys.* **C53** (1992) 219.
24. L3 Coll., M. Acciarri et al., CERN-PPE/97-48.
25. F. Wäckerle, *proceedings of the XXVIII Int. Symposium on Multiparticle Dynamics*, Frascati, Italy (1997).
26. A. Donnachie and P. V. Landshoff, *Phys. Lett.* **B296** (1992) 227.
27. G. A. Schuler and T. Sjöstrand, *Z. Phys.* **C73** (1997) 677.
28. R. Engel and J. Ranft, *Phys. Rev.* **D54** (1996) 4244; R. Engel, *Z. Phys.* **C66** (1995) 203.
29. T. Kleinwort, G. Kramer, *Z. Phys.* **C75** (1997) 489; *Nucl. Phys.* **B477** (1996) 3; *Phys. Lett.* **B370** (1996) 141.
30. P. Aurenche et al., *Progr. Theor. Phys.* **92** (1994) 175.
31. AMY Coll., B. J. Kim et al., *Phys. Lett.* **B325** (1994) 248.
32. TOPAZ Coll., H. Hayashii et al., *Phys. Lett.* **B314** (1993) 149.
33. OPAL Coll., K. Ackerstaff et al., *Z. Phys.* **C73** (1997) 433.
34. OPAL Coll., paper submitted to Lepton-Photon 1997, LP-201;
35. H. Abramowicz, K. Charchula and A. Levy, *Phys. Lett.* **B269** (1991) 458.
36. J. Butterworth, hep-ex/9707001, these proceedings.
37. M. Klasen, hep-ph/9706292, these proceedings.
38. M. Drees, M. Krämer, J. Zunft, P. M. Zerwas, *Phys. Lett.* **B306** (1993) 371.
39. M. Cacciari, M. Greco, B. A. Kniehl, M. Krämer, G. Kramer, M. Spira *Nucl. Phys.* **B466** (1996) 173.
40. ALEPH Coll., *Phys. Lett.* **B355** (1995) 595.
41. L3 Coll., paper submitted to Lepton-Photon 1997, LP-092.
42. P. Aurenche et al., in *Physics at LEP2*, CERN 96-01, p.328 (1996) and references therein.

Chapter 6

Photoproduction of Jets and Hadrons

PHOTOPRODUCTION OF JETS

J. M. BUTTERWORTH
Department of Physics and Astronomy, University College London, Gower St. London, WC1E 6BT, UK
E-mail: jmb@hep.ucl.ac.uk

On behalf of the H1 and ZEUS collaborations.

ABSTRACT

The status of HERA data on jet photoproduction is reviewed, and some suggestions and prospects for further work are given.

1. Introduction

The photoproduction of jets at HERA is proving to be a very fruitful process in which to study strong interactions. Aspects of QCD which are being investigated include the partonic structure of both the proton and the photon, the internal structure of jets, and the dynamics of jet production. I will omit jet production in association with prompt photons, charm and rapidity gaps - these reactions are covered in other contributions.

2. Starting Simply

In leading order QCD, jet photoproduction processes are divided into two classes: 'direct' and 'resolved'. Example diagrams for these processes are shown in Fig.1. Direct processes are apparently simple - the photon couples directly into the hard scattering (via high virtuality quarks), and thus the fraction (x_γ) of its momentum transferred to the high E_T partons is one. However, the photon may couple to a $q\bar{q}$ pair with a relative p_T much less than the E_T of the hard jets. In this case large logarithms of E_T/p_T enter into calculations of the cross section and for the lowest p_T's the splitting of the photon is non-perturbative. These features lead to the introduction of a partonic structure for the photon, to describe the 'cascade' of partons derived from such low p_T splittings. Reactions involving partons from this cascade are called resolved photon interactions, and they enter the jet production cross section at the same order as direct processes. In this case, $x_\gamma < 1$.

An obvious first question to address at HERA then is - How well does this leading order language correspond to the actual situation in experiment? To answer it one must decide carefully what to attempt to measure. The goal should not be to somehow try to extract 'leading order' cross sections or variables, but to define sensible observables which have natural interpretations in the LO picture but which are well defined independently of it.

The first such observable is a jet. At leading order this corresponds to a parton,

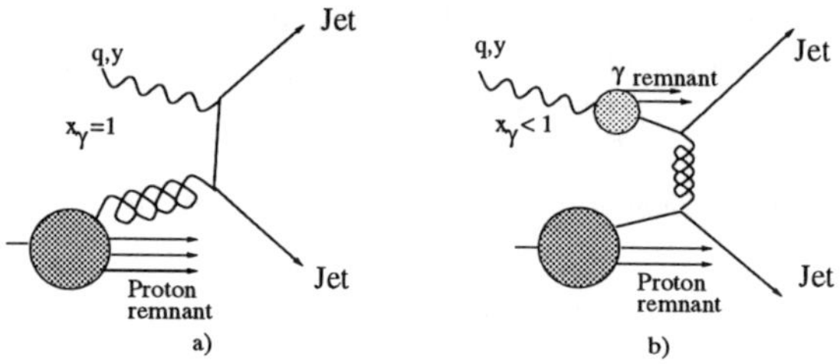

Fig. 1. Leading order diagrams for direct (a) and resolved (b) jet photoproduction.

but in experiment (as well as in more sophisticated calculations) it is defined by a jet algorithm. More of this later. For now, it is enough to remark that jets have well defined pseudorapidity[a](η^{jet}) and transverse energy (E_T^{jet}).

In photoproduction events at HERA the positron generally escapes down the beampipe. This constrains the negative of the four-momentum squared of the photon (Q^2) to be below about 4 GeV^2, and the median value is $Q^2 \approx 10^{-3}$ GeV^2. This is *not* the hard scale of the interaction (which is provided by E_T^{jet}), and is more usually referred to as P^2 in photon physics. Another important variable is the inelasticity y. This is defined in the same way as in deep inelastic scattering (DIS), but at low Q^2 it reduces to $y \approx E_\gamma / E_e$.

Using the jets and y, the variable $x_\gamma^{OBS} = \sum_{jets}(E_T^{jet} e^{-\eta^{jet}})/2yE_e$ is defined. It is the fraction of the photon's momentum appearing in the high E_T^{jet} jets. In dijet cross sections the sum runs over the two highest E_T^{jet} events. The superscript 'OBS' is meant to indicate that this is an observable, unlike the LO x_γ. However, in a two parton final state, x_γ^{OBS} reduces to the LO variable.

The (uncorrected) x_γ^{OBS} distribution from ZEUS[1] is shown in Fig.2. There is a clear two-component structure, with a peak at high values of x_γ^{OBS} and a rising tail to low values, which is cut off eventually by detector acceptance. Also shown are the distributions obtained from Monte Carlo (MC) simulations which include parton showers and hadronisation models as well as the LO diagrams. The direct MC events lie at high x_γ^{OBS}.

Thus the LO QCD picture has passed its first test (at least qualitatively - there are discrepancies at low x_γ^{OBS} between the data and the MC). There is a further test which can be made quite simply. The dominant LO diagrams for direct processes involve fermion (i.e. quark) exchange whereas those in resolved processes involve boson (i.e. gluon) exchange (see Fig.1). This leads to different predictions for the dijet angular distribution in the jet-jet centre-of-mass frame. The direct (high x_γ^{OBS}) should be distributed according to $|1 - cos\theta^*|^{-1}$, whereas the resolved (low x_γ^{OBS}) should be $\approx |1 - cos\theta^*|^{-2}$. ZEUS has measured these distributions[2] for dijet invariant mass above 23 GeV. The results are shown in Fig.2. The data agree well with the

[a] $\eta - \ln \tan \theta/2$ where θ is the polar angle of the jet w.r.t. the proton direction.

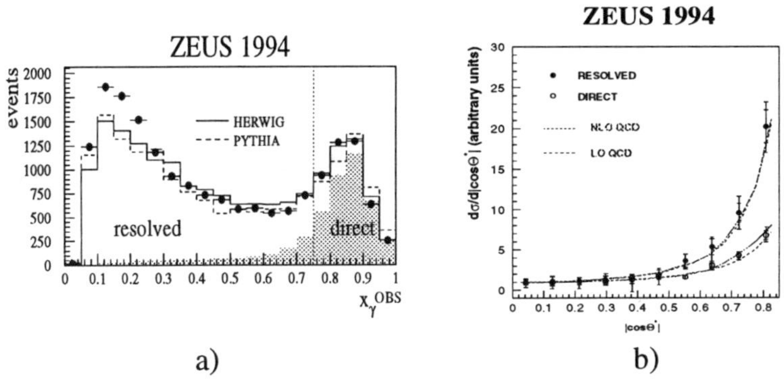

Fig. 2. a) Uncorrected x_γ^{OBS} distribution b) dijet angular distributions.

predictions. Also shown are NLO calculations from Harris and Owens[3], which agree well with both the data and with the LO curves.

3. Jet Cross Sections

So far the simple LO picture of these processes is in pretty good shape. What else can we learn? As well as being sensitive to the QCD dynamics, jet cross sections are sensitive to the parton distributions in the photon and the proton. Thus in principle they can give information about the quark and gluon distributions inside the photon and proton. H1 and ZEUS have measured inclusive and dijet cross sections and the statistics are now becoming sufficiently high for measurements of multijet jet cross sections to begin.

3.1. Inclusive Jet Cross Sections

Both ZEUS and H1 measure inclusive jet cross sections differential in η^{jet} and integrated above a given E_T^{jet}, and cross sections differential in E_T^{jet} integrated within a range of η^{jet}. All the cross sections are *ep* cross sections integrated within specified y and Q^2 ranges.

Some examples are shown in Fig.3. The H1 data[4] are compared to the expectations of LO MC simulations. For the standard PYTHIA there is in general reasonable agreement in the shape at high E_T^{jet} values. However, at low E_T^{jet} and in the forward region the data lie above the MC (whether GRV or LAC1 parton distributions are used for the photon). The other MC models shown contain multiparton interactions and give higher cross sections - more of this later. An example of the ZEUS inclusive jet data is also shown. This time, the difference between data and 'theory' is plotted, where in the first case the theory is PYTHIA again, and in the second case it is a NLO

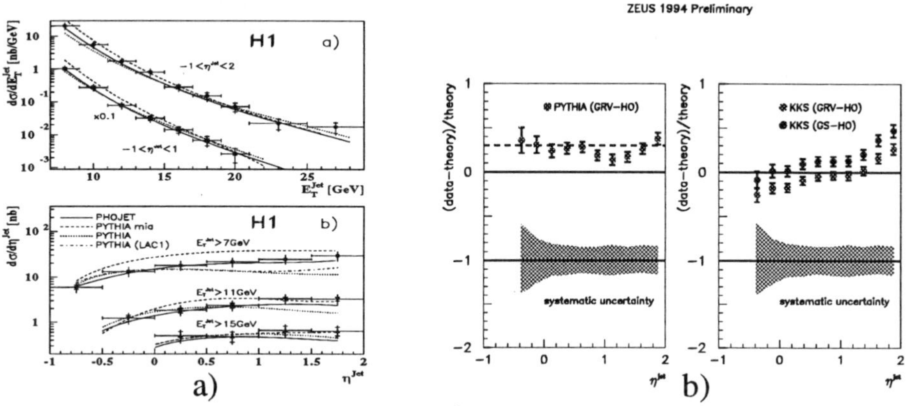

Fig. 3. Inclusive jet cross sections. The H1 data[4] have a normalisation uncertainty of 26% which is not shown. The uncertainty in the ZEUS data arising from the energy scale of the calorimeter is correlated between points and is indicated by the shaded band. The ZEUS data show the difference between data and theory for the cross section $d\sigma/d\eta^{jet}$ for $E_T^{jet} > 17$ GeV[6].

QCD calculation from Klasen and Kramer[5]. For these higher E_T^{jet} values PYTHIA lies below the data over the whole range. The NLO QCD calculations describe the normalisation of the data better, but lie below the data in the forward region. The sensitivity to the parton distribution in the photon is similar in size to the systematic uncertainties in the measurement.

3.2. Dijet Cross Sections

Once two (or more) jets per event are measured in the detector, many possible cross sections can be measured. The angular distributions shown earlier are an example, as might be the x_γ^{OBS} distribution. A choice which has been made by ZEUS[1] is to measure $d\sigma/d\bar{\eta}$ for both jets above a given E_T^{jet} cut. Here $\bar{\eta} \equiv (\eta_1 + \eta_2)/2$ is the boost of the dijet system in the lab frame. Rewriting $x_\gamma^{OBS} \approx E_T^{jet} e^{-\bar{\eta}} \cosh \Delta\eta / y E_e$ shows that for small $|\Delta\eta| \equiv |\eta_1 - \eta_2|$) the smallest x values are probed for a given E_T^{jet}. Scanning across $\bar{\eta}$ means scanning across yx_γ and x_p. Low $\bar{\eta}$ means high x_γ and low x_p (typically 0.005). High $\bar{\eta}$ means low x_γ and moderate x_p (typically 0.1). The cross section is measured in two x_γ^{OBS} regions corresponding to direct ($x_\gamma^{OBS} > 0.75$) and resolved ($0.3 < x_\gamma^{OBS} < 0.75$). The data are shown in Fig.4.

3.3. Comments on Jet Cross Sections

The situation now looks a little less clear. There is reasonable agreement between

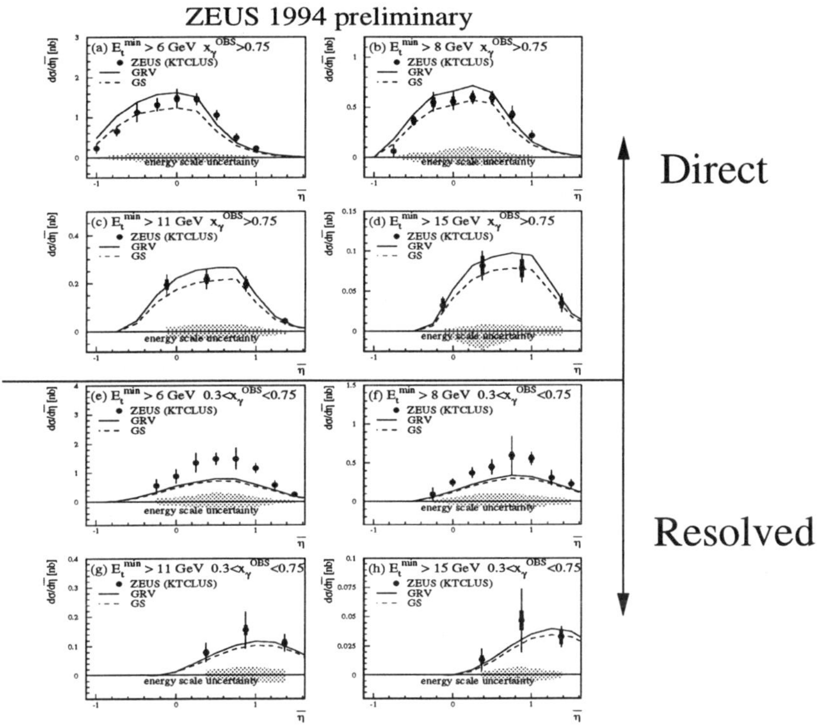

Fig. 4. Dijet cross sections from ZEUS[1]. The uncertainty in the data arising from the energy scale of the calorimeter is correlated between points and is indicated by the shaded band.

data and theory for dijet cross sections at high x_γ^{OBS}, but the calculations are too low at low E_T^{jet} and low x_γ^{OBS}. There is reasonable agreement in backward inclusive jets, but the calculation is too low in the forward direction, particularly for the lowest E_T^{jet} values measured. The next section contains some possible explanations and hints at how these discrepancies might be resolved.

4. What is a jet?

The theory we wish to investigate is QCD. Although MC simulations describe some event properties more successfully by use of phenomenological models, the best calculations available are at next-to-leading order in α_s and do not include effects such as hadronisation. This means that although in nature jets consist of many (typically more than five) hadrons, in the theory they consist of two (at most) partons. These partons are the hardest in the event, and thus can be expected to give a reasonable description of high E_T^{jet} jets. However, the following issues (and maybe others!) must be addressed before strong conclusions can be drawn from comparisons between data and theory. None of them are trivial issues, and the investigation of them should deepen our understanding significantly.

4.1. Jet Algorithms

It is not correct to think of jets as being simply 'smeared' partons. Jets are defined by an algorithm. Are the algorithms the same in experiment and theory? Experiments have used cone algorithms of various flavours or more lately a mode of the so-called k_T cluster algorithm which uses separation in $\eta - \phi$ space as its distance parameter[7]. So far the calculations have used only a cone algorithm.

A major source of ambiguity arises in the seed finding (that is, where do you begin looking for a jet?) and jet merging and/or splitting (that is, at what point does a fairly hard subjet become an extra jet in its own right?). The details of how these are performed can have a large effect on the experiment and a lesser or different one on the theory. No unique treatment is defined in the famous Snowmass convention[8]. A particular example of a problem can be seen by considering two partons or hard particles separated by $\delta r = \sqrt{\delta\phi^2 + \delta\eta^2} = 2$. An experimental cone algorithm running with jet radius $R = 1$ would typically take one of these (or the calorimeter cluster caused by it) and draw a circle of radius one around it. No other particle is inside this radius and so a stable jet is formed. The second particle will form a second jet. However, in a theoretical calculation the partons will be merged (if they have the same E_T) because they *do* both lie with a cone of radius one centred on their midpoint.

The parameter R_{sep} is introduced in the theory to combat this problem[9]. Partons separated by a distance greater than R_{sep} will never be merged by the algorithm. In this way, calculations attempt to mimic the effects of the seed finding stage of experimental jet finders. As a bonus, setting $R_{sep} = R$ makes the results of the cone algorithm identical (for a three parton final state) to the k_T version employed by the experiments.

Therefore one approach is to measure and calculate jet cross sections with various jet algorithms and see how changes in the algorithm affect the comparison between the two[10]. However, the R_{sep} method is in fact badly defined for higher order calculations, and in fact cone algorithms like this are not infrared safe in four-parton final states[11].

4.2. Underlying Events

Jets are not the only things in an event. How does the rest of the event affect them? The term 'underlying event' is often used loosely to refer to the activity in real events caused by the fact that, having provided partons for a hard scattering, the remnants

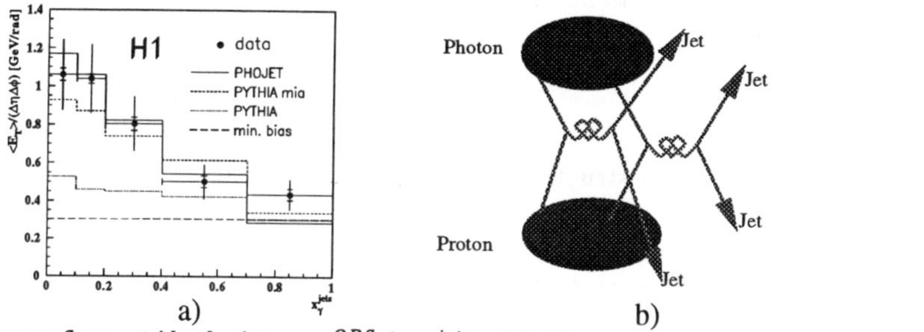

Fig. 5. (a) Energy flow outside the jets vs x_γ^{OBS} ($\equiv x_\gamma^{jets}$). (b) Schematic representation of multiparton interactions.

of the proton and photon do not just go away. The models and language used to describe them vary widely and depending upon taste or convention effects such as initial and final state QCD radiation and soft or hard remnant-remnant interactions may or may not be included. They *do not* however include second photon-proton interactions in a single bunch crossing - the probability for this leading to significant activity in an event is negligible at HERA.

That an underlying event exists in photoproduction at HERA is clearly seen in Fig.5a. Here H1[4] have plotted the mean transverse energy *outside the jet* per unit of $\eta - \phi$ space, as a function of x_γ^{OBS}. First, there is clearly plenty of transverse energy in the event apart from the hard jets, and secondly it is correlated to x_γ^{OBS}. Given the fact that any transverse energy outside the jet must contain some of the photon's momentum, it is plain from the definition of x_γ^{OBS} that such a correlation must exist. However, the amount of transverse energy involved and strength of the correlation are remarkable. The implication is that this underlying event could indeed affect the jets significantly, and that it cannot be treated as independent of the jets. The size of the underlying event is correlated to the hard process, and therefore no technique relying on subtraction of typical 'minimum bias' events will be able to correct for it properly. That such techniques have occasionally been used at hadron-hadron colliders without large ill effects seems to be because all the events are at low x, whereas the correlation becomes important in the range above $x \approx 0.1$.

A possible explanation of this type of effect is offered by multiparton interactions (MI)[12,13]. These are allowed in eikonal models for extra hard (or sometimes soft) scatters in a single γp event, as illustrated in Fig.5b. Such models are available in PYTHIA[14], HERWIG[15] and PHOJET[16] (but not in NLO QCD!). These models improve the description of the data - they increase energy flow around the jet core and in general increase jet cross sections. The price paid is that when MI are allowed, the energy flow outside the jet becomes very sensitive to the parton distributions in the photon and proton, and to $p_T^{\min}$, the cutoff for hard scattering[13]. This is because the average number of partonic interactions in a given γp event goes up with increasing parton density.

4.3. Jet Shapes

Jets have internal structure. Does the theory describe this properly? Measurements of jet shapes provide a useful way of looking at this. The jet shape $\Psi(r)$ is defined as the average fraction of the jet's transverse energy lying within cone of radius r. It is defined such that $\psi(R) = 1$. The rate at which it approaches unity as r approaches R is a measure of how collimated the jet is. ZEUS have measured jet shapes[17] for a sample inclusive jets with $E_T^{jet} > 14$ GeV. An example of the data is shown in Fig.6. In Fig.6a the fraction of E_T^{jet} contained within a sub-cone of radius 0.5 is plotted as a function of η^{jet}. The data show that forward jets are broader; that is, the fraction of E_T^{jet} within the inner cone decreases as η^{jet} increases. Also shown are several curves from PYTHIA. The two continuous lines show the prediction of PYTHIA with (thicker) and without (thinner) MI. Clearly, MI once more improve the description of the data. Also shown separately are the shapes for those PYTHIA jets which are initiated by a gluon or a quark at LO. Gluon jets are broader in the MC, as expected from the fact that their larger colour charge leads to more QCD radiation. The jets in the rear direction look like quark jets, whereas the behaviour is gluon like in the forward direction. The transition between the two is reproduced in the MC, where it arises from the transition from dominantly direct to dominantly resolved processes as η^{jet} increases. In Fig.6b, the same quantity is plotted, but now as a function of E_T^{jet}. Jets get narrower as E_T^{jet} increases. In addition, in the MC the effect of multiparton interactions decreases. Comparisons to QCD calculations are given in Klasen's presentation.

5. Outlook and Conclusion

A theme in this area of physics is the benefit (and difficulty) of making general comparisons. There has been significant progress in our understanding of how to compare data and theory. The choice of jet algorithm has been discussed in this context and an attractive solution to the problems involved seems to be offered by k_T algorithm, which by virtue of being a cluster rather than cone algorithm avoids in a natural manner all seed finding and jet merging ambiguities (and hence the need for $R_{\rm sep}$), but which in the chosen mode preserves the attractive features of cone algorithms in hadronic physics.

Another point to bear in mind is that we need to choose 'theory friendly' cross sections. In particular, demanding two jets above a certain E_T leads to divergences in

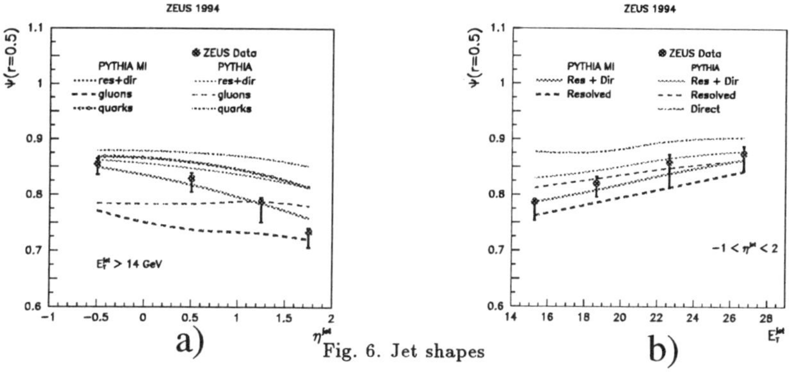

Fig. 6. Jet shapes

NLO QCD which are responsible for discrepancies between theoretical calculations[3,5]. The problem is discussed in Klasen's contribution. Recent preliminary ZEUS data[18] avoid this problem by symmetrising dijet cross sections and cutting on the highest E_T^{jet} whilst applying a different and lower threshold to the second jet, also shown by Klasen.

To make comparisons at lower E_T^{jet} with confidence, a better understanding of the 'underlying event' is required. Multiparton interactions are a hot topic, but have many free parameters and are based upon simplifying assumptions which may not be justified. The eventual answer is surely to measure and constrain them with data. A promising approach is to study multijet events, and the current statistics are now allowing this to begin at HERA[19].

Leaving aside the theory for a moment, comparisons between data from different experiments would be greatly enhanced by the adoption of a standard set of cuts and jet algorithms for a minimal subset of jet cross sections, where this is practical. The comparison of jet physics between photoproduction and DIS is another subject which is gaining momentum, both in jet shapes and cross sections.

Photoproduction at HERA and $\gamma\gamma$ collisions at e^+e^- experiments probe the photon in complementary ways. Global fits to both jet data and F_2^γ data similar to those carried out for the proton will probably place the strongest constraints on the structure of the photon in the end. However, general physics messages can been drawn from the data in other ways. A promising approach here is to study the effective parton distributions in the photon[20]. In resolved photoproduction, the most important matrix elements have the same angular dependence and contribute to the cross section basically according to the colour factors involved. This fact allows the dijet cross section to be written in terms of single effective matrix element and an effective

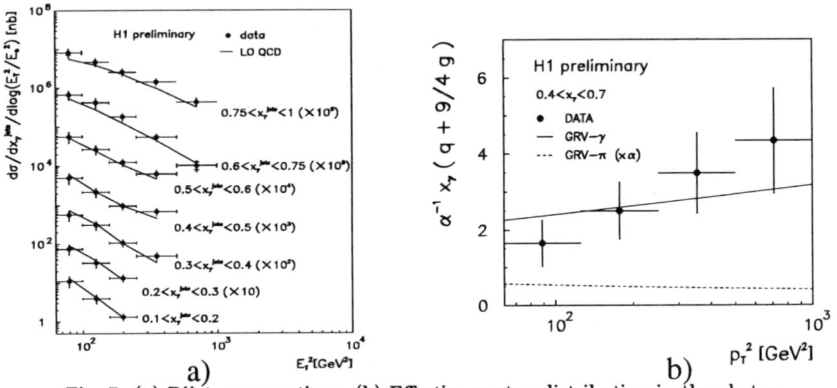

Fig. 7. (a) Dijet cross sections, (b) Effective parton distribution in the photon.

parton distribution:

$$\frac{d^4\sigma}{dy dx_\gamma dx_p d\cos\theta^*} = \frac{1}{32\pi s_{ep}} . \frac{f_{\gamma/e}}{y} . \frac{f^\gamma_{eff}}{x_\gamma} . \frac{f^p_{eff}}{x_p} . |M_{eff}|^2$$

where $f_{\rm eff}(x, p_T^2) = \sum q(x, p_T^2) + \bar{q}(x, p_T^2) + \frac{9}{4} g(x, p_T^2)$. H1[21] have measured the dijet cross section shown in Fig.7a, and extracted $f^\gamma_{\rm eff}(x, p_T^2)$ as shown in Fig.7b. There is a positive scaling violation, as seen in F_2^γ. This is characteristic of the $\gamma \to q\bar{q}$ splitting probability, and is not present in purely hadronic structure functions.

Finally, a nice instance of feedback between the two types of experiments was recently shown in OPAL data[22]. In 1993, ZEUS results showed that the data are better described by MC simulations if the intrinsic k_T in the photon is increased[23]. This is a way of faking the effect of the fact that in the $\gamma \to q\bar{q}$ splitting the quarks can have significant relative p_T. Following this, a similar trick has been used in $e\gamma$ collisions to achieve a significantly better agreement in the spectrum of transverse energy outside the plane defined by the beam and the scattered electron, as well as in the dijet rates for these events.

I hope some flavour of the rapid and continuing progress in field has been conveyed. I am confident that there is a great deal more physics to come from studying jet photoproduction.

6. Acknowledgements

My thanks go to the many people on H1 and ZEUS who provided plots. I would also like to thank the DESY directorate and MPI, as well as the organisers and participants of what was an immensely stimulating and enjoyable meeting.

7. References

1. ZEUS Collab., M. Derrick et al, *Phys. Lett.* **B348** (1995) 665; Paper submitted to ICHEP96 pa02-040.
2. ZEUS Collab., M. Derrick et al, *Phys. Lett.* **B384** (1996) 401.
3. B. W. Harris and J. F. Owens, FSU-HEP-970411, hep-ph/9704324.
4. H1 Collab., S. Aid et al., *Zeit. f. Phys.* **C70** (1996) 17.
5. M. Klasen, G. Kramer DESY-96-246, hep-ph/9611450.
6. ZEUS Collab., M. Derrick et al, *Phys. Lett.* **B342** (1995) 417; Paper submitted to ICHEP96 pa02-041.
7. S. Catani, Yu. L. Dokshitzer, M. H. Seymour and B. R. Webber, *Nucl. Phys.* **B406** (1993) 187; S. D. Ellis, D. E. Soper, *Phys. Rev.* **D48** (1993) 3160.
8. J. E. Huth et al., Proc. of the 1990 DPF Summer Study on High Energy Physics, Snowmass, Colorado, edited by E.L. Berger, World Scientific, Singapore (1992) 134.
9. S. D. Ellis, Z. Kunszt, D.E. Soper, *Phys. Rev. Lett.* **69** (1992) 3615.
10. J. M. Butterworth, L. Feld, M. Klasen and G. Kramer, *hep-ph 9608481* Proceedings of the workshop Future Physics at HERA Editors G.Ingelman, A. DeRoeck, R.Klanner, DESY 1996, p.554.
11. W. B. Kilgore, hep-ph/9705384, to appear in the proceedings of *XXXIInd Rencontres de Moriond.*
12. T. Sjöstrand and M. van Zijl, *Phys. Rev.* **D36** (1987) 2019; G. A. Schuler and T. Sjöstrand, *Phys. Lett.* **B300** (1993) 169; *Nucl. Phys.* **B407** (1993) 539; J. M. Butterworth and J. R. Forshaw, *J. Phys.* **G19** (1993) 1657; R. Engel, *Z. Phys.* **C66** (1995) 203.
13. J. M. Butterworth, J. R. Forshaw and M. H. Seymour, Zeit. f. Phys. **C72** (1996) 637.
14. T. Sjöstrand, *Comp. Phys. Comm.* **82** (1994) 74.
15. G. Marchesini et al, *Comp. Phys. Comm.* **67** (1992) 465.
16. R.Engel, *Zeit. f. Phys.* **C66** (1995) 203.
17. ZEUS Collab., paper submitted to ICHEP96 pa02-043.
18. M. E. Hayes, for the ZEUS Collab., to appear in proc. of Photon '97, World Scientific, hep-ex/9707014.
19. E. Strickland, for the ZEUS Collab., to appear in proc. of Photon '97, World Scientific.
20. B. L. Combridge and C. J. Maxwell, *Nucl. Phys.* **B239** (1984) 429.
21. H1 Collab., paper submitted to ICHEP96 pa02-080.
22. J. A. Lauber, L. Lönnblad and M. H. Seymour, to appear in proc. of Photon '97, World Scientific, hep-ex/9707002.
23. ZEUS Collab., M. Derrick et al, *Phys. Lett.* **B354** (1995) 163.

THEORETICAL OVERVIEW OF JET PHOTOPRODUCTION AT HERA

MICHAEL KLASEN
Deutsches Elektronen-Synchrotron DESY, Theory Group,
Notkestrasse 85, D-22607 Hamburg, Germany
E-mail: michael.klasen@desy.de

ABSTRACT

We review the theoretical framework of jet photoproduction at HERA discussing the conceptual ideas, phenomenological models, and higher order perturbative calculations. Numerically, we study the physically interesting distribution of transverse energy within the observed jet, the real and virtual photon structure, and the proton structure at large x.

1. Theoretical Concepts

In electron-proton scattering at HERA the dominant fraction of the scattering events proceeds through photons with low virtuality Q^2.[a] Experimentally the electron is anti-tagged and remains in the beam-pipe. Theoretically the lepton tensor and phase space can then be factorized with the Weizsäcker-Williams approximation[1], which was improved recently through power-suppressed terms by Frixione et al.[2] The virtuality of the photon is less than 4 GeV^2, and it retains a fraction $y \in [0.2; 0.85]$ of the incident electron energy $E_e = 27.5$ GeV. The proton energy is 820 GeV.

Jet production in photon-proton collisions was first calculated in leading order (LO) of perturbative QCD by Owens[3]. The two-fold nature of direct and resolved processes and their separation on a kinematical basis were pointed out as a means to test the underlying partonic dynamics. It is thus possible to study the distribution of partons in the initial photon and proton and the spin of the exchanged particle.

Since then much effort has been spent on improving the tree-level understanding. Phenomenological models have been implemented into the Monte Carlo event generators PYTHIA[4], HERWIG[5], and PHOJET[6] employing parton showers, fragmentation models, and multiple interactions. Next-to-leading order (NLO) calculations for inclusive single-jet[7] and dijet production[8] provide perturbative correction factors, reduce the scheme and scale dependences, and allow for an implementation of jet definitions. Recently the first NLO analysis of the transition from real to virtual photoproduction of jets has been published[9].

Although several parametrizations exist for the parton distributions in the photon in NLO[10], the gluon is still poorly constrained by the data from $\gamma\gamma$ scattering at e^+e^- colliders and needs complementary measurements from γp. The situation is better for the parton densities in the proton[11], where photoproduction data might improve the

[a]From 1994-1997, HERA operated with positrons instead of electrons. Since we will only be concerned with neutral current exchange, we use the term "electron" for positrons as well.

data from deep inelastic and $p\bar{p}$ scattering at intermediate to large values of x. Except where indicated we use the GS96 and CTEQ4M parametrizations for the photon and proton, respectively.

2. Phenomenological Models

Perturbative QCD describes only the hard partonic scattering process and the scale evolution of the hadronic structure functions. The link between colored partons and real hadrons belongs to the non-perturbative domain and has to be fitted to data or guessed from phenomenological models. These models are implemented in Monte Carlo generators and compared in Table 1.

Table 1. Properties of different Monte Carlo generators.

Monte Carlo Generator	Parton Showers	Fragmen-tation	Multiple Interactions
PYTHIA	Initial+Final	String	Hard (optional)
HERWIG	Initial+Final	Cluster	Soft (optional)
PHOJET	Final	String	Soft and Hard

As a first step, one can attach additional angularly ordered partons to the hard process until the original parton reaches a maximum virtuality of $Q^2_{\max} < E_T^2$. In addition and deviating from the parton model, intrinsic transverse momenta are allowed for the partons in the hadrons up to $k_T < k_T^{\max}$ thereby introducing a second phenomenological parameter.

The second step consists in the fragmentation of the proliferated partons into hadrons. The Lund string model confines the color field between quarks into a color flux tube. The energy increases proportional to the distance between the quarks until the string breaks up and new quark-antiquark pairs are created. Hadrons are formed when the energy is too low for the string to break up according to a fragmentation function with two free parameters. Gluons appear as excitations and produce kinks in the string. Alternatively the cluster model starts with the splitting of gluons into $q\bar{q}$ pairs and the subsequent formation of color neutral clusters. Heavy clusters cascade into light clusters which then transform isotropically into hadrons.

Finally multiple interactions seem to be important at HERA to describe transverse energy flow and cross sections in the direction of the proton remnant at low E_T. They increase the multiplicity and energy flux and model interactions between the photon and the proton remnant. Secondary interactions are softer than the first scattering defined by E_T and can be of partonic or soft nature.

3. Next-to-Leading Order Calculations

The undesirable drawback of phenomenological models is the large number of free parameters that have to be tuned to data. These are not present in next-to-leading order calculations where one calculates virtual corrections with internal particle loops and real corrections with soft and collinear radiation one order higher in the strong coupling constant α_s. The singularities can then be controlled through dimensional regularization and removed consistently through renormalization and factorization procedures. This avoids unphysical cut-off parameters, reduces the scheme and scale dependence, and allows for an implementation of jet definitions.

Real corrections are most easily calculated with the phase space slicing method. After approximation of the invariants and partial fractioning of the $2 \to 3$ matrix elements, we factorized the Born process and integrated the remaining singular kernels analytically up to an invariant mass cut-off $y_{\rm cut}$. Aurenche et al.[8] employed a transverse energy cut-off for the soft/collinear initial state, which does not appear in a related method for the single-jet case, and a cone cut-off for the collinear final state.[b] Harris and Owens integrated the soft and collinear regions separately[8]. The real emission outside the cut-offs is integrated numerically. This removes the dependence on the technical cut-off and introduces the experimental jet definition.

The subtraction method relies on a point-by-point subtraction of singularities in the numerical integration and has only been applied to the direct process by Bödeker[7]. The resolved process could, however, be adapted from an existing program in $p\overline{p}$ scattering[12].

4. Jet Definition Uncertainties

In hadronic collisions cluster algorithms of the JADE type combine particles not only into the hard jets but also into the remnant jets. This is avoided if a cone of size $R = 1$ in azimuth-rapidity space is used to define a jet. The Snowmass[13] accord determines the jet axis from the E_T weighted directions of all particles in the cone. Unfortunately it contains a number of ambiguities which render a matching between theory and experiment difficult[14].

Contrary to fixed cone algorithms, iterative cone algorithms can merge overlapping jets. This is not described by a NLO calculation with just three final state partons. If two partons have a distance between R and $2R$, they can be counted as one or two jets, and one has to avoid double-counting. The phenomenological parameter $R_{\rm sep}$ defining the distance of two partons can be used to model the narrower jets found by iterative cone algorithms.

In a recent study[15], we used this $R_{\rm sep}$ parameter to describe jet shapes $\rho(r)$ as measured by the ZEUS, CDF, and D0 collaborations. The result is compared to preliminary ZEUS data from 1994[16] which was obtained with an iterative cone algorithm and is shown in Figs. 1 and 2. It depends moderately on the rapidity η of the jet (Fig. 1) being broader in the forward direction and narrows slightly with the trans-

[b]See also the contribution by M. Fontannaz in these proceedings.

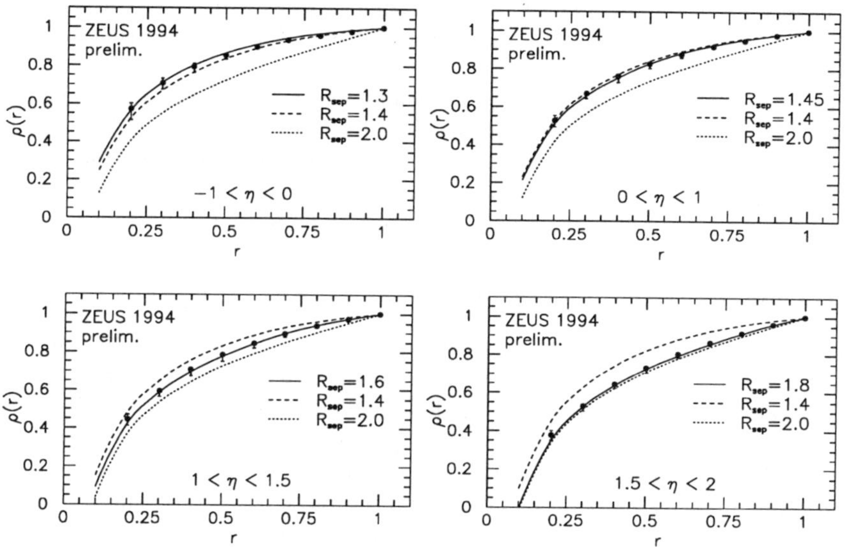

Fig. 1. *Jet shape $\rho(r)$ for single-jet photoproduction integrated over $E_T > 14$ GeV and four different regions of η. We compare our results using the Snowmass convention with $R = 1$ and three different values of $R_{\rm sep}$ to preliminary 1994 data from ZEUS.*

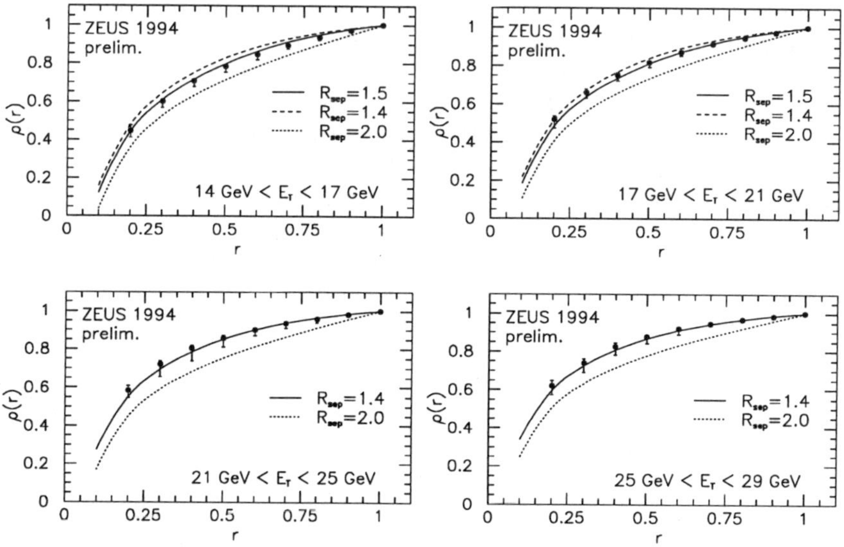

Fig. 2. *Jet shape $\rho(r)$ for single-jet photoproduction integrated over $-1 < \eta < 2$ and four different regions of E_T. We compare our results using the Snowmass convention with $R = 1$ and different values of $R_{\rm sep}$ to preliminary 1994 data from ZEUS.*

verse energy E_T of the jet (Fig. 2). Whereas the curves with $R_{sep} = 2$ correspond to no R_{sep} and predict too broad jets, an average value of $R_{sep} = 1.4$ describes the data rather well.

The uncertainties of the Snowmass cone definition are circumvented in the hadronic implementation of the k_T-cluster algorithm[17] with the same E_T-weighted recombination scheme as for Snowmass, but a different jet condition. One considers the distance of two particles in $\eta - \phi$ space. This corresponds to a unique value of $R_{sep} = R$ in theory and experimentally assigns every hadron to a unique jet.

5. Real and Virtual Photon Structure

The determination of the structure of the photon is clearly one of the most important physics goals in photoproduction. Since the cross section drops rapidly with the transverse energy, most events at HERA have been observed at small E_T so far. They are dominantly produced by resolved photons and give access to the poorly constrained small-x region and the gluon density in the photon.

However at low E_T a separation of hard and soft physics is experimentally and theoretically difficult. Hadronization effects, jet definition uncertainties, energy pedestals around the jets, and multiple interactions between the remnant jets may play an important role. This can be seen in Figs. 3 and 4, where we compare our NLO calculation to preliminary 1994 data from ZEUS, again obtained with the iterative cone algorithm[18]. At $E_T > 14$ GeV, there is a clear excess of data over theory in the forward direction, which decreases continuously for larger E_T. We do not show a comparison with data obtained with a smaller cone size $R = 0.7$ which shows no excess in the forward region. Obviously the hard jets are better separated from the underlying event for smaller cones and there is no need for multiple interaction effects.

A second conclusion from Figs. 3 and 4 is that the jet definition uncertainty is of comparable size to the photon structure function uncertainty. Thus one either has to rely on the fitted value of $R_{sep} = 1.4$ from jet shapes or use the k_T cluster algorithm.

In the backward region, where direct and quark initiated processes dominate, there is fairly good agreement with the data with a slight tendency of GRV to overestimate the measurement. This can be understood from the bigger quark contribution of GRV compared to GS96 at large values of x.

The kinematics of the partonic subprocess can be better constrained in dijet cross sections than in single cross sections, e.g. with the help of the variable $x_\gamma^{OBS} = (E_{T_1}e^{-\eta_1} + E_{T_2}e^{-\eta_2})/(2yE_e)$. As long as one integrates over different bins, this variable is infrared-safe and there is no need to abandon it as put forward by Aurenche et al.[8] With the help of this variable, ZEUS separated direct and resolved regions experimentally and studied the dependence of the cross section on the average rapidity of the observed jets[19]. Unfortunately they constrained the E_T of both jets to the same minimal values. The theoretical predictions are then not infrared safe and

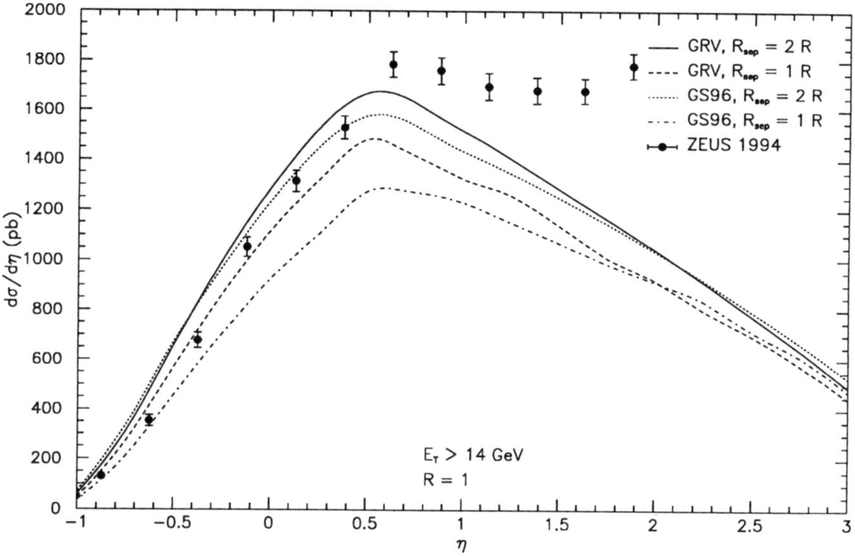

Fig. 3. *Rapidity dependence of the single-jet photoproduction cross section integrated above $E_T > 14$ GeV. We compare our NLO prediction with GRV and GS96 photon parton densities and the two extreme values of $R_{\rm sep} = 1R, 2R$ to preliminary 1994 data from ZEUS.*

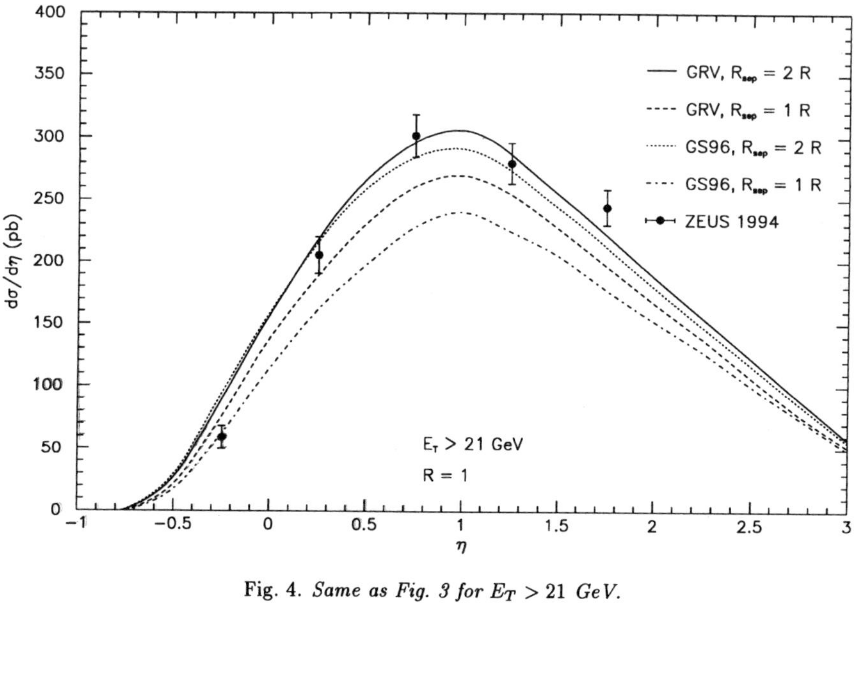

Fig. 4. *Same as Fig. 3 for $E_T > 21$ GeV.*

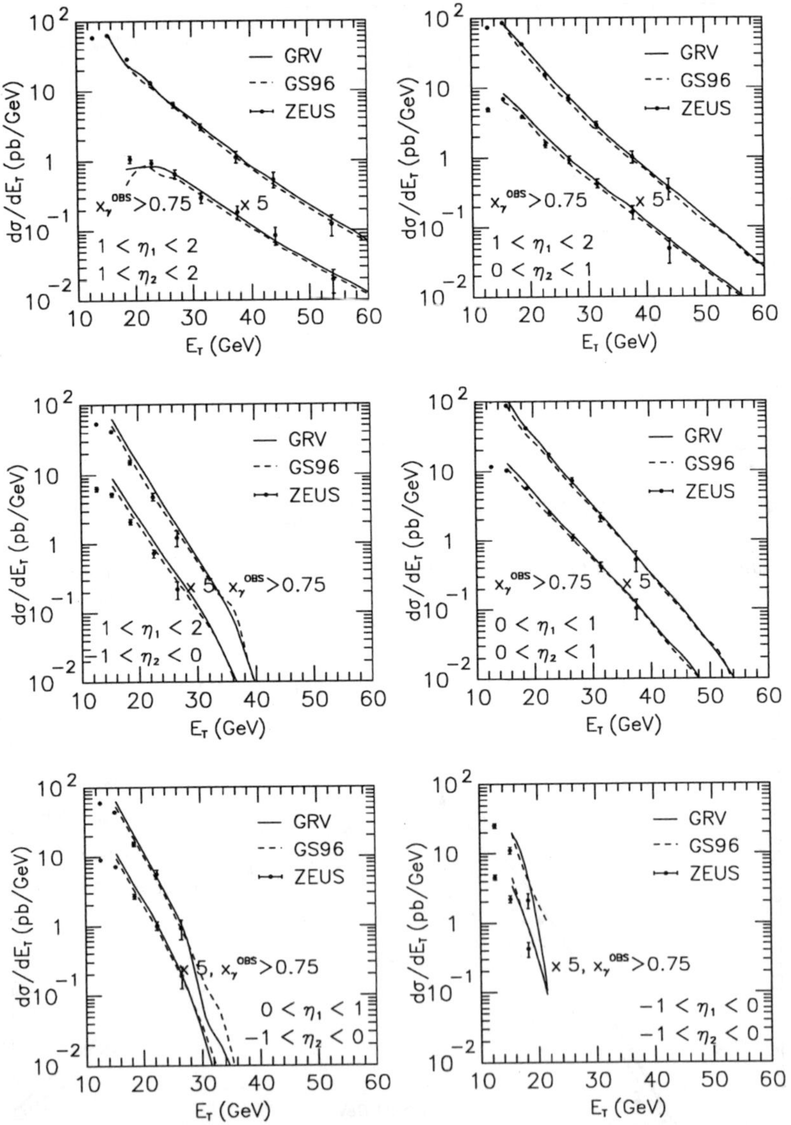

Fig. 5. *E_T dependence of the symmetrized dijet photoproduction cross section integrated over different rapidity bins. We compare our NLO prediction with GRV and GS96 photon parton densities and the full and upper range of $x_\gamma^{\rm OBS}$ to preliminary 1995 data from ZEUS.*

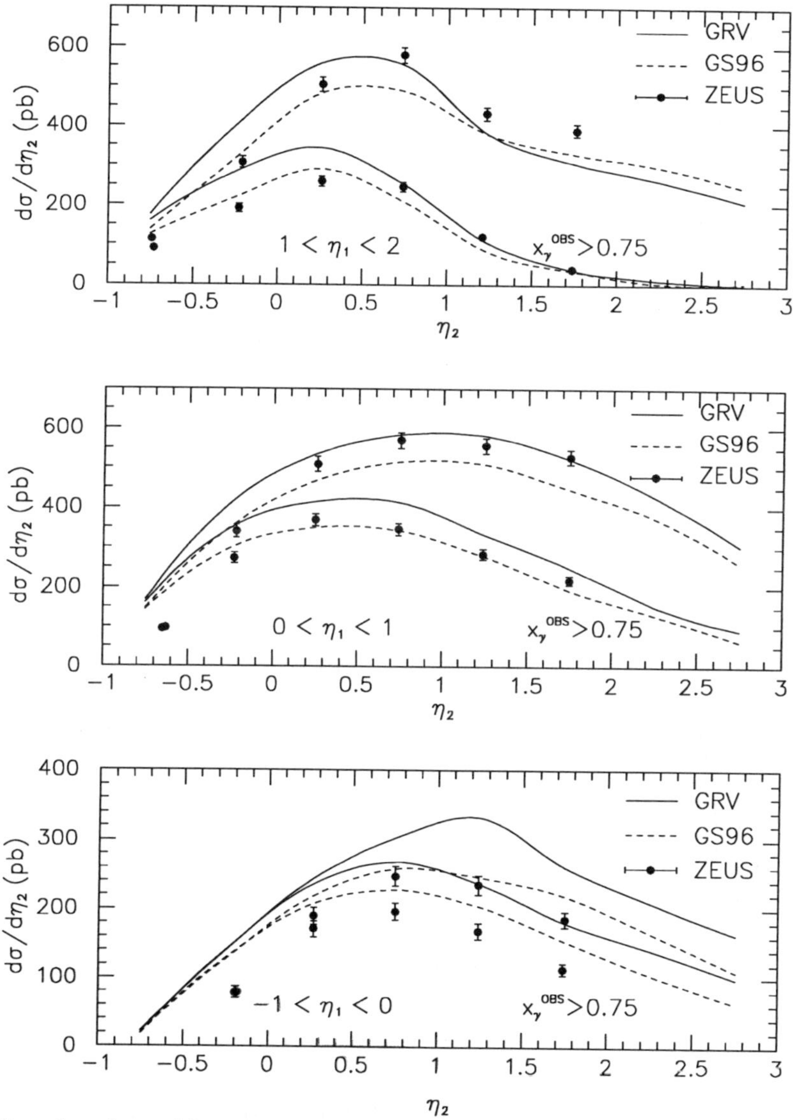

Fig. 6. *η_2 dependence of the symmetrized dijet photoproduction cross section integrated over $E_T > 14$ GeV and $E_{T_2} > 11$ GeV. We compare our NLO prediction with GRV and GS96 photon parton densities and the full and upper range of $x_\gamma^{\rm OBS}$ to preliminary 1995 data from ZEUS.*

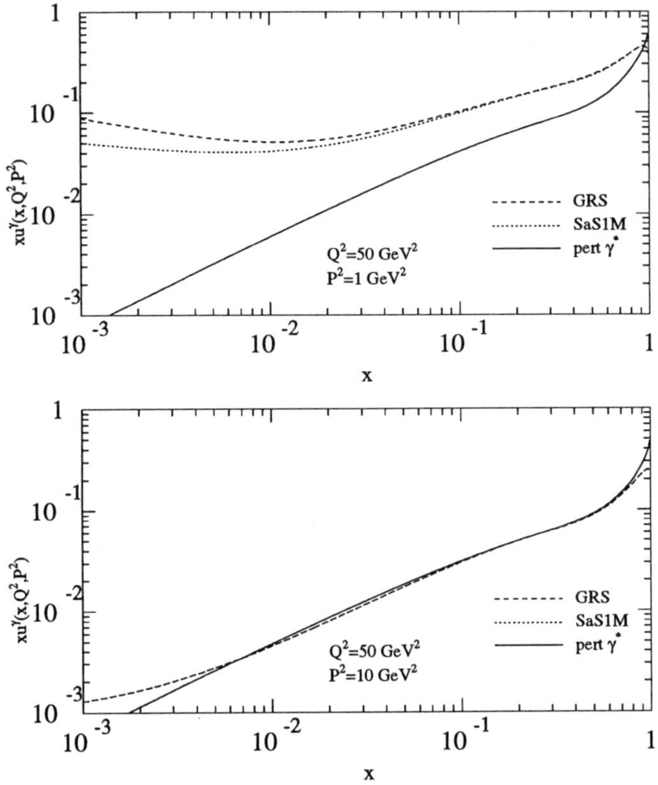

Fig. 7. *Up-quark distributions in the virtual photon as function of x. We compare the GRS and SaS1M parametrizations with the perturbative result for $P^2 = 1$ and $10\ GeV^2$* [22].

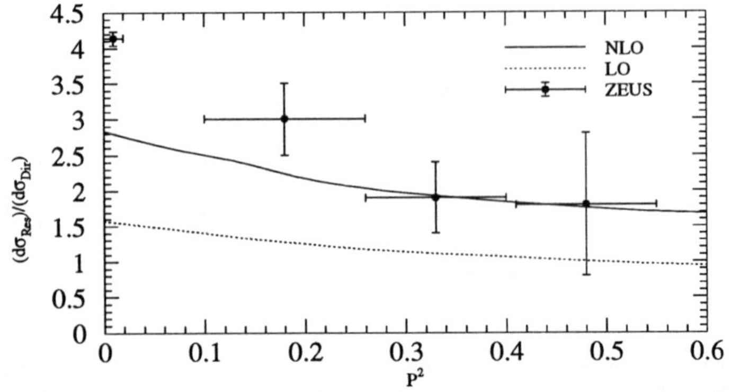

Fig. 8. *Ratio of resolved over direct contributions defined with the x_γ^{OBS} cut at 0.75 for $E_{T_1}, E_{T_2} > 4$ GeV using the SaS1M virtual photon structure with four flavors.*

depend on a phenomenological (Klasen and Kramer) or the technical cut-off (Harris and Owens)[8]. This analysis employed the k_T cluster algorithm with $R = 1$ and exhibited again the excess of data over theory for the resolved dominated regions at low $x_\gamma^{\rm OBS}$ and E_T.

A new preliminary analysis of 1995 ZEUS data has been presented recently[20]. They measured the symmetrized dijet cross section $\mathrm{d}\sigma/\mathrm{d}E_T\mathrm{d}\eta_1\mathrm{d}\eta_2$ with the leading $E_T > 14$ GeV and $E_{T_2} > 11$ GeV in different rapidity bins and the full and upper region of $x_\gamma^{\rm OBS}$. The result is infrared safe and compared to our NLO predictions in Fig. 5 as a function of E_T and in Fig. 6 as a function of η_2. The general agreement is good for both GRV and GS96, even when both jets are in the forward region and the complete range in $x_\gamma^{\rm OBS}$ is covered. This may be due to the fairly large cuts on the E_T of both jets, which suppresses the underlying event. In the backward region, the calculation lies above the data and GS96 is slightly favored as was the case for the single-jet cross sections. In addition the systematic errors not shown here are quite prominent in the backward region. Since the data using the k_T algorithm were not yet available, we simulated the iterative cone algorithm with the optimized value of $R_{\rm sep} = 1.4$ for $R = 1$. This value should be lower in the backward region leading to a reduced theoretical prediction and better agreement with the data.

A new NLO program for virtual photoproduction[9] allows us to study the the transition to photons with virtuality P^2. Apart from using the unintegrated Weizsäcker-Williams approximation, the main difference consists in the analytic integration of the virtual photon initial state singularity. This singularity is then factorized and leads to a scheme- and P^2-dependent finite contribution which reduces to the real expression as $P^2 \to 0$. How this leading-logarithmic singularity compares to the two existing up-quark parametrizations in the virtual photon[21] is shown in Fig. 7. At low $P^2 = 1$ GeV^2 the evolution up to 50 GeV^2 produces many more up-quarks than just the single pair predicted perturbatively. At large $P^2 = 10$ GeV^2, GRS and SaS1M agree with the perturbative box-diagram[22].

ZEUS have also published P^2 dependent data on the ratio of resolved over direct contributions disentangled with the $x_\gamma^{\rm OBS}$ cut at 0.75[23]. As can be seen in Fig. 8 the NLO effects are large at $E_{T_1}, E_{T_2} > 4$ GeV. The lowest P^2 point for real photoproduction has much better statistics than the virtual photoproduction data and lies above the theory. This is to be expected since we are again in the soft region where remnant-remnant interactions are assumed to be important. At larger P^2, theory and data agree well in shape and normalization. We use the SaS1M LO parametrization, since NLO parton densities are not available in parametrized form and GRS does not contain charm.

6. Proton Structure at large x

The excess of high-Q^2 events in DIS at HERA have triggered a lot of speculation

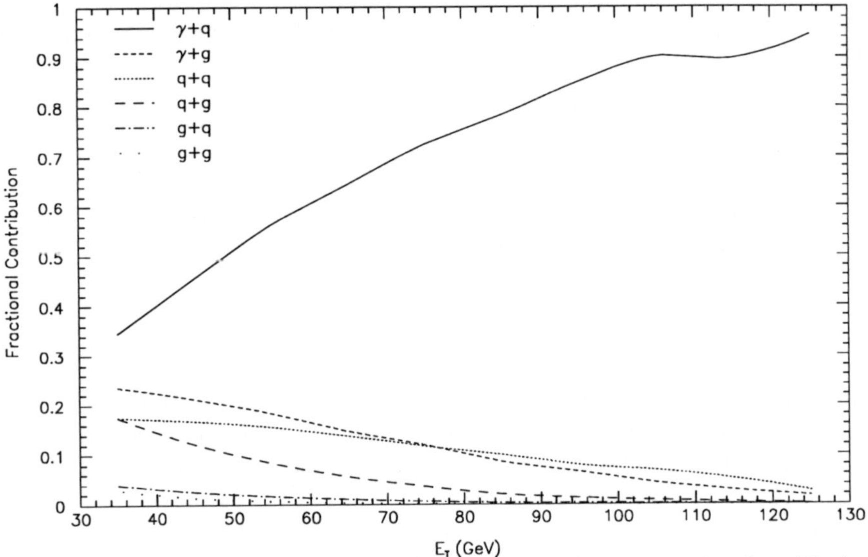

Fig. 9. *Fractional Contribution of direct and resolved partonic subprocesses as function of E_T in the single-jet photoproduction cross section.*

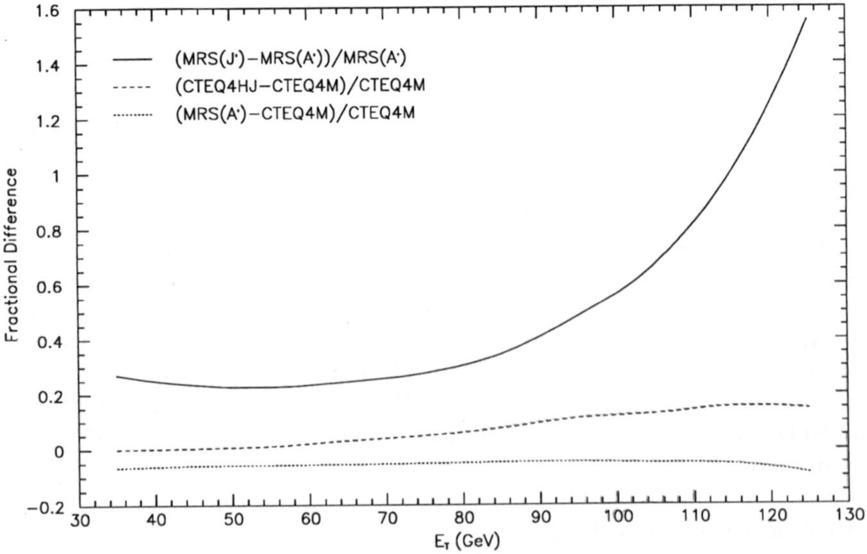

Fig. 10. *Fractional difference of different proton structure functions as function of E_T in single-jet photoproduction compared to standard parametrizations.*

about new physics such as leptoquarks, R-parity violating supersymmetry, or contact interactions. This kinematic regime can and should also be tested in photoproduction at large E_T, e.g. by measurements of $\mathrm{d}\sigma/\mathrm{d}E_T$, $\mathrm{d}\sigma/\mathrm{d}E_\gamma$, or $\mathrm{d}\sigma/\mathrm{d}M_{JJ}$. An unambiguous determination of new physics requires a precise knowledge of the Standard Model background. Due to phase space limitations, NLO corrections become increasingly important at the boundary of phase space at large E_T, whereas the uncertainty from the photon structure function becomes negligible. This can be extracted from Fig. 9, where the fractional contribution of direct photon-quark scattering accounts for more than 90% of the total single-jet cross section above $E_T = 100$ GeV. Therefore a precise knowledge of the quark distribution in the proton is required for $x > 0.4$.

In Fig. 10, we demonstrate the increase in the cross section that can be obtained with the larger quark densities of MRS(J') which were designed to describe the excess of CDF $p\bar{p}$ data at large E_T[24]. A factor of 1.5 seems possible here. Larger gluon distributions as proposed by CTEQ4(HJ)[11] only give a rise of 15%. It should be mentioned that MRS(J') fail to fit the low energy BCDMS data, whereas CTEQ4(HJ) are still consistent. Finally, a study of the scheme dependence might indicate further uncertainties here or in the gluon density.

7. Conclusions

A wealth of new and precise data coming from H1 and ZEUS in photoproduction has triggered increased theoretical interest. Three independent NLO calculations for real dijet production and the first NLO program for virtual photoproduction are now available. The interpretation of jets is improved with smaller cone sizes and the k_T cluster algorithm. This will allow for first stringent tests of the real photon parton density, especially of the gluon. More data and a parametrization of NLO parton densities are needed in virtual photoproduction. The access to large transverse momenta offers interesting studies of the proton structure at large x.

8. Acknowledgements

The author thanks the organizers of the Ringberg workshop for the kind invitation, G. Kramer and B. Pötter for their collaboration, and C. Glasman and Y. Yamazaki for making the preliminary ZEUS data available to him.

9. References

1. C.F.v. Weizsäcker, *Z. Phys.* **88** (1934) 612; E.J. Williams, *Phys. Rev.* **45** (1934) 729.
2. S. Frixione, M.L. Mangano, P. Nason, G. Ridolfi, *Phys. Lett.* **B319** (1993) 339.
3. J.F. Owens, *Phys. Rev.* **D21** (1980) 54.

4. T. Sjöstrand, hep-ph/9508391 (1995).
5. G. Marchesini et al., hep-ph/9607393 (1996).
6. R. Engel, *Z. Phys.* **C66** (1995) 203.
7. H. Baer, J. Ohnemus, J.F. Owens, *Phys. Rev.* **D40** (1989) 2844; L.E. Gordon, J.K. Storrow, *Phys. Lett.* **B291** (1992) 320; P. Aurenche, M. Fontannaz, J.P. Guillet, *Phys. Lett.* **B338** (1994) 98; D. Bödeker, G. Kramer, S.G. Salesch, *Z. Phys.* **C63** (1994) 471; M. Klasen, G. Kramer, S.G. Salesch, *Z. Phys.* **C68** (1995) 113.
8. M. Klasen, G. Kramer, *Z. Phys.* **C72** (1996) 107, hep-ph/9611450 (1996); P. Aurenche, L. Bourhis, M. Fontannaz, J.P. Guillet *Proc. of "Future Physics at HERA"*, Hamburg (1996) 570; B.W. Harris, J.F. Owens, hep-ph/9704324 (1997).
9. M. Klasen, G. Kramer, B. Pötter, hep-ph/9703302 (1997).
10. M. Glück, E. Reya, A. Vogt, *Phys. Rev.* **D45** (1992) 3986; L.E. Gordon, J.K. Storrow, *Phys. Lett.* **B291** (1992) 320, *Nucl. Phys.* **B489** (1997) 405; P. Aurenche, M. Fontannaz, J.P. Guillet, *Z. Phys.* **C64** (1994) 621.
11. A.D. Martin, R.G. Roberts, W.J. Stirling, *Phys. Lett.* **B387** (1996) 419; H.L. Lai et al., CTEQ Collaboration, *Phys. Rev.* **D55** (1997) 1280.
12. S.D. Ellis, Z. Kunszt, D.E. Soper, *Phys. Rev. Lett.* **64** (1990) 2121.
13. J.E. Huth et al., *Proc. of the 1990 DPF Summer Study on High Energy Physics*, Snowmass CO (1992) 134.
14. J. Butterworth, L. Feld, M. Klasen, G. Kramer, *Proc. of "Future Physics at HERA"*, Hamburg (1996) 554.
15. M. Klasen, G. Kramer, hep-ph/9701247 (1997).
16. M. Derrick et al., ZEUS Collaboration, Contribution to the XXVIII. ICHEP, Warsaw (1996) pa 02-043.
17. S.D. Ellis, D.E. Soper, *Phys. Rev.* **D48** (1993) 3160.
18. M. Derrick et al., ZEUS Collaboration, Contribution to the XXVIII. ICHEP, Warsaw (1996) pa 02-041.
19. M. Derrick et al., ZEUS Collaboration, Contribution to the XXVIII. ICHEP, Warsaw (1996).
20. M. Derrick et al., ZEUS Collaboration, Contribution to PHOTON 97, Egmond aan Zee (1997).
21. M. Glück, E. Reya, M. Stratmann, *Phys. Rev.* **D51** (1995) 3220; G.A. Schuler, T. Sjöstrand, *Phys. Lett.* **B376** (1996) 193.
22. B. Pötter, Ph.D. Thesis, Universität Hamburg (1997), DESY 97-138, hep-ph/9707319.
23. M.L. Utley, ZEUS collaboration, *Proc. of "HEP95"*, Brussels (1995) 570; C. Foudas, ZEUS collaboration, *Proc. of "DIS96"*, Rome (1996) 275.
24. E.W.N. Glover, A.D. Martin, R.G. Roberts, W.J. Stirling, *Phys. Lett.* **B381** (1996) 353.

PHOTOPRODUCTION OF HADRONS AT HERA

JAROSLAV CVACH
Institute of Physics, Academy of Sciences of the Czech Republic, Na Slovance 2
Praha, CZ - 180 40, Czech Republic
E-mail: cvach@fzu.cz

On behalf of the H1 and ZEUS collaborations.

ABSTRACT

New experimental results from photoproduction of hadrons at HERA are reviewed.

1. Introduction

The photoproduction of hadrons was in the first years of HERA experimentation the process to be studied most easily because of the large cross section. The recent results tend to investigate the domain of harder physics which can be compared with QCD calculations. In this contribution the latest results are reviewed which were presented[1] on the spring conferences of 1997 and are not covered by more specialized talks of this workshop. Specifically, I shall talk about inclusive particle production, resonances in multi-photon final state and prompt photons.

2. Inclusive particle production

The inclusive production of single hadrons in fixed target experiments and colliding beam experiments has been one of the important testing grounds for the QCD parton model. The HERA collider makes possible such studies in *ep* interactions over a wide range of squared four-momentum transfer, $-Q^2$, from photoproduction $Q^2 \approx 0$ to very high photon virtuality. Recent progress in the determination of parton fragmentation functions enables us to make specific predictions for inclusive particle production in next-to-leading order QCD. The parton distributions of the photon and the proton constitute an important input to the cross section calculations.

2.1. Neutral strange particles

New results for K^0 and Λ inclusive photoproduction were obtained in the H1 experiment[2] for an average γp cms energy W of 187 GeV. K^0_S mesons and Λ baryons are identified through their decay channels $K^0_S \to \pi^+\pi^-$, $\Lambda(\bar{\Lambda}) \to p\pi^-(\bar{p}\pi^+)$, where the pion, the proton and the antiproton are reconstructed in the central jet chamber (CJC). In order to ensure an optimal acceptance of the CJC for neutral strange particles , their pseudorapidity has been restricted to $|\eta| < 1.3$ and their transverse momentum to $0.5 < p_t < 5$ GeV for K^0_S and to $0.6 < p_t < 5$ GeV for Λ. All K^0_S results are multiplied by 2 and therefore correspond to $K^0 + \bar{K}^0$ production and are denoted by K^0. As the Λ cross section is within errors equal to $\bar{\Lambda}$ cross section they are combined and referenced as Λ. Fig. 1 shows a comparison of the measured K^0 and Λ cross sections in p_t and η with the predictions of PYTHIA in combination

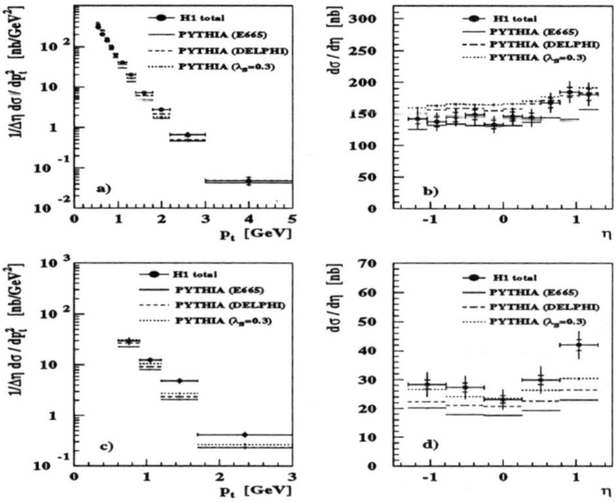

Fig. 1. Measured K^0 (a,b) and Λ (c,d) total ep cross section in photoproduction as a function of p_t and pseudorapidity η compared to PYTHIA in the kinematic range $|\eta| < 1.3$ and $0.5 < p_t < 5$ GeV.

with JETSET using standard parameter settings ($\lambda_s = 0.3$ as well as parameter sets obtained from fits to DELPHI and E665 data).[a] In the case of K^0 mesons the measured cross sections are in reasonable agreement with the Monte Carlo prediction using the DELPHI or E665 settings. However the measured Λ cross section lies significantly above the Monte Carlo prediction for all three settings. The UA5 collaboration has reported[3] a similar discrepancy between Λ production in $\bar{p}p$ interactions at 200 GeV and PYTHIA predictions.

Recent NLO QCD calculations of inclusive spectra in ep scattering use fragmentation functions[4] which have been fitted to e^+e^- data. A comparison of this calculation to data in the region of $p_t > 1.8$ GeV is satisfactory[2,5]. If fragmentation is process independent, the rate of low p_t particles in γp and DIS interactions will be the same. A comparison of K^0 production rates in γp and DIS[6] is shown in Fig 2 (upper

[a] The following JETSET hadronization parameters have been changed with respect to the default settings (default/DELPHI/E665): PARJ(2) $= \lambda_s =$ (0.3/0.23/0.2), PARJ(11) = 0.5/0.365/0.5), PARJ(12) = (0.6/0.41.0.6). PARJ(2) is the strangeness suppression factor, PARJ(11) (PARJ(12)) is the probability that a meson containing u or d (s) quark has spin 1.

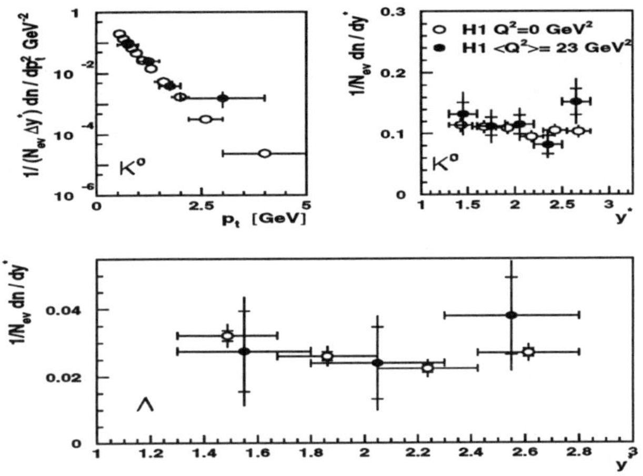

Fig. 2. K^0 rate as a function of p_t and centre of mass rapidity y^* (upper part) and Λ rate as a function of y^* (lower part) for non-diffractive events. γp data are represented by open and DIS data by closed circles. The inner vertical error bars indicate statistical error, the outer error bars the statistical and systematic error added in quadrature.

part) and of Λ production rates in Fig 2 (lower part) for the non-diffractive event samples[b]. The γp and DIS rates are compatible in y^* and at low p_t (y^* is the γp cms rapidity defined with respect to the direction of the exchanged boson). Comparison between K^0 rates of production in $\bar{p}p$ collisions at 200 GeV obtained by the UA5 collaboration[3] and this measurement show good agreement[2]. In contrast, the Λ rates in the photon fragmentation region in photoproduction are significantly lower than in $\bar{p}p$ interactions. This may be attributed to the presence of an initial baryon in $\bar{p}p$ in this hemisphere from the incoming proton or antiproton. A difference between photon and target fragmentation region with respect to baryon production has also been observed in deep inelastic μN scattering[7].

ZEUS examines[8] the features of inclusive K^0 mesons produced with a high-E_t jet. The aim of the study is also to evaluate average longitudinal fragmentation function $D(z)$ for K^0 in jets from hard photoproduction and compare it with the fragmentation functions[4] obtained in high energy quark jets in e^+e^-.

Events with at least one jet of $E_t > 8$ GeV and $|\eta^{jet}| < 0.5$ and K^0_S with $0.5 < p_t <$ 4 GeV and $|\eta| < 1.5$ were selected. The data were corrected to take into account effects due to reconstruction efficiency, event selection efficiency and bin-to-bin migration.

[b] For non-diffractive samples H1 requires the presence of at least 500 MeV of energy in the forward region of the liquid argon calorimeter, $2.03 < \eta < 3.26$, to remove large rapidity gap events.

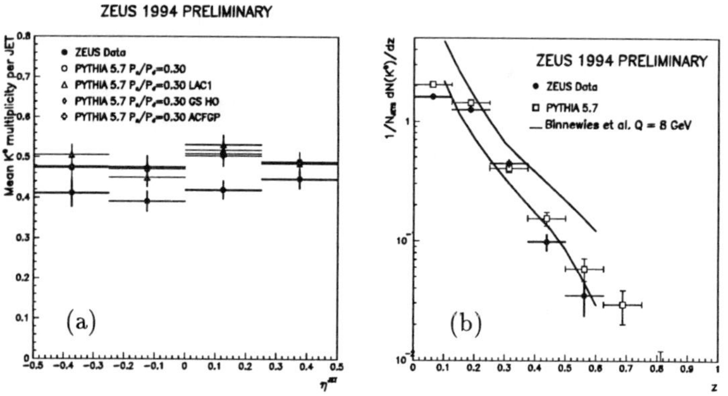

Fig. 3. (a) Corrected numbers of K_S^0 per jet as a function of jet pseudorapidity. (b) Corrected fragmentation function $D(z)$. Both distributions are defined with respect to 8 GeV hadron jets.

Comparison of the mean number of K_S^0 per jet, integrated over $R \leq 1$ (R is the jet radius in cone algorithm in η and ϕ) are shown in Fig. 3a as a function of η^{jet} where the total number of jets was evaluated without the K_S^0 requirement. The shapes of the measured and Monte Carlo distributions are similar but the Monte Carlo is higher than data as observed also in other distributions. This discrepancy which cannot be explained by the size of systematic errors nor variation of MC parameters or photon parton densities is under investigation.

The distributions of particles within a jet may be described by the fragmentation function $D(z)$, where z is the longitudinal fraction of the jet momentum carried by the particle. Fig. 3b shows $D(z)$ for data compared to that from PYTHIA. Again K_S^0 distribution is lower than Monte Carlo by 20%. The curves are upper and lower limits from the phenomenological fit of Binnewies et al.[4] The lower curve corresponds to gluon fragmentation and the upper curve to s and u fragmentations and covers the spread of the fitted $D(z)$ values of fragmentation of u, d, s quarks using the LO and NLO schemes at a QCD scale of 8 GeV. The disagreement between data and phenomenological fit of $D(z)$ can be due to the fact that data are corrected to the hadron level to avoid the strong dependence of the correction to the parton level on the Monte Carlo used. One should keep in mind that e^+e^- data do not provide a sufficient constraint on $g \to K_S^0$ fragmentation and in this place the ep measurement is important.

2.2. Charged particles

Latest results of the inclusive transverse momentum spectra of charged particles in photoproduction events in the laboratory pseudorapidity range $-1.2 < \eta < 1.4$ have been measured up to $p_t = 8$ GeV using the ZEUS detector[9] and in range $|\eta| \leq 1$ and $2 < p_t < 12$ GeV by H1[10] at the average γp c.m. energy of $W = 200$ GeV. The inclusive transverse momentum spectra fall exponentially in the low p_t region. The non-diffractive data show a pronounced high p_t tail departing from the exponential shape. Comparison to hadron–hadron collisions[11] at a similar c.m. energy shows that the photoproduction data are clearly harder and in fact are similar to $\bar{p}p$ at $\sqrt{s} = 1.8$ TeV (see Fig. 4). In Fig. 5, the inclusive charged particle cross section $d\sigma/d\eta$

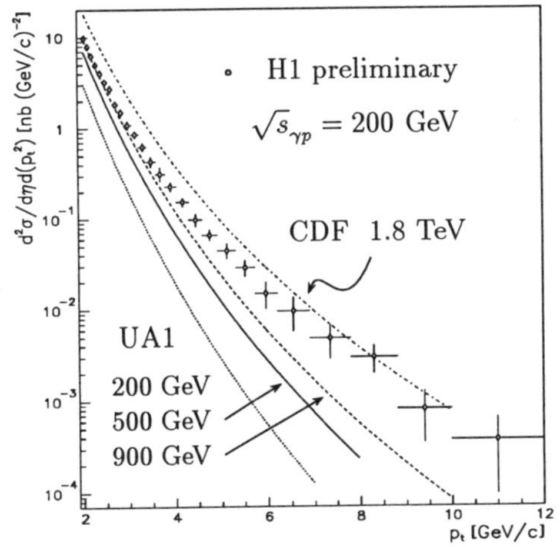

p_t spectrum is harder in γp due to:

- γp is measured in γ fragmentation region whereas $\bar{p}p$ in the central region
- γp behaves like Vp and parton momenta of quarks in the vector meson V are harder than in baryons
- dominance of direct part in γp cross section at high p_t can lead to harder scattering

Fig. 4. Inclusive cross section as a function of p_t for charged particles for $|\eta| < 1$. $\bar{p}p$ data were multiplied by 10^{-4} and fitted to the dependence $(1+\frac{p_t}{p_{0,t}})^{-n}$. The text in the right part of the figure explains possible reasons of the difference between γp and $\bar{p}p$ distributions.

is shown for particles with $p_t > 2$ GeV. A comparison is made with PYTHIA with two different parton distribution functions: GRV-LO(Fig. 5a) and LAC1(Fig. 5b). None of the predictions match the data and they have mutually opposite slopes of the η distribution. From the figure it is clear that the rapidity distribution is sensitive to the photon structure function — the fact which will be pursued in the final analysis of the data with the aim to extract the gluon structure function of the photon. The inclusive distributions of charged particles were compared with NLO QCD calculations. Good agreement is observed in p_t distribution in the whole measured range of $p_t > 2$ GeV as well as in $d\sigma/d\eta$ distribution[5].

3. Scale influence on the energy dependence of photon-proton cross section

The energy dependence of the real photon-proton total cross section is consistently described at high energies by the power law $\sigma_{\gamma p}^{tot} \propto (W^2)^\lambda$, where W is the γp centre of mass energy and the power $\lambda \sim 0.08$. In contrast, the virtual photon-proton cross section is found to rise faster[12,13] with increasing W^2 for Q^2 values larger than a

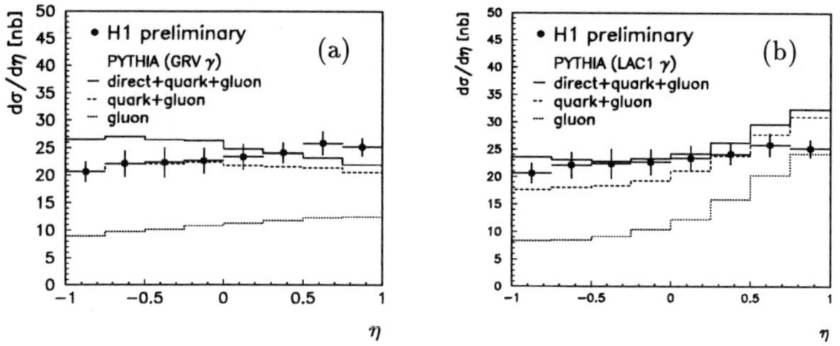

Fig. 5. Inclusive cross section as a function of η for charged particles with $p_t > 2$ GeV compared to PYTHIA with the GRV-LO(a) and LAC1(b) photon parton parametrisations.

few GeV^2. In the HERA regime fits of the form $\sigma^{tot}_{\gamma^* p} \propto (W^2)^\lambda$ lead to values of λ which rise[13,12] from 0.2 to 0.4 in the Q^2 range from 1.5 to 10^3 GeV^2 (see Fig. 6). The change of the slope parameter λ with increasing Q^2 is associated with a transition from the non-perturbative 'soft' to the perturbative 'hard' regime[14]. The perturbative QCD (pQCD) can be applied in the domain where Q^2 is larger than a few GeV^2 and it predicts an increase of λ with Q^2. In the Reggeon Field Theory (RFI) the change of λ can be interpreted as a reduction of screening corrections[15] with increasing Q^2. RFI and pQCD are two complementary approaches which successfully describe physics processes in different regimes - 'soft' and 'hard'.

Whereas in deep inelastic scattering (DIS) the hardness of the interaction (scale) can be characterized by the photon virtuality Q^2, in photoproduction with $Q^2 \approx 0$ the largest p_t of a hadron in the photon fragmentation region can take over the role of the scale. This investigation was the subject of the study disscussed in this chapter the results of which have been published recently[16].

In total about two million photoproduction events were selected in the energy range $150 < W_{\gamma p} < 250$ GeV. The W dependence of the cross section was measured in different bins of the maximal $p_{t,max}$ of the charged particle in an event. Good detector acceptance and high resolution of p_t measurement together with large statistics of photoproduction events, allow a continuous coverage of both soft and hard domains in the pseudorapidity range $1.1 < \eta^* < 3.1$ (η^* is measured with respect to the γ direction in the γp centre of mass system). The energy dependence of the cross section in each $p_{t,max}$ bin was fitted to the form $\sigma(W, p^2_{t,max}) \sim W^{2\lambda(p^2_{t,max})}$. The results of the fit are shown in Fig. 6, where the slope λ is given as function of $4p^2_{t,max}$. A strong increase of λ is observed with increasing $p_{t,max}$ which can be fitted by a straight line. Note that when the scale $4p^2_{t,max}$ is multiplied with a number, this does not change the slope of the straight line. Hence the change of λ with the scale is significant. Comparison with the longitudinal phase space model shows that the phase space plays only a minor role.

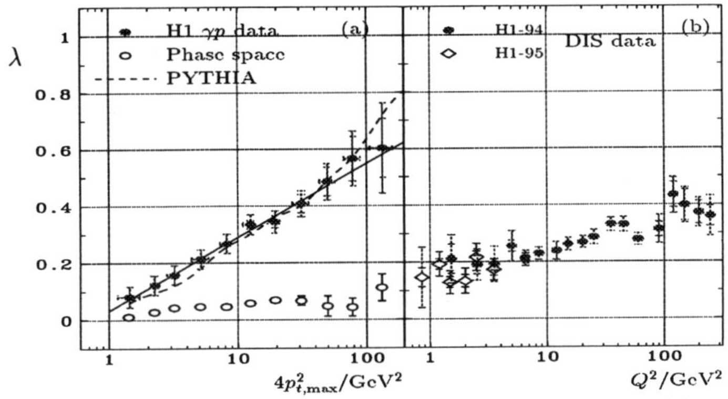

Fig. 6. Slope λ of the W^2 dependence of the photoproduction (a) and DIS (b) cross section as a function of $4p_t^2$ and Q^2 respectively.

On the other hand the $p_{t,max}$ dependence of λ is described by PYTHIA. The model is based on the Monte Carlo version which includes both pQCD for the hard scattering and a Regge inspired soft interaction component. It was checked that the effects of different proton and photon structure function, of parton showers as well as multiple parton-parton scattering are small. Therefore, the change of the slope λ is not associated with the growth of the gluon density in the proton at small x but rather with the leading order matrix element of the hard process in PYTHIA.

4. Resonances in multi-photon final state

As mentioned already in the previous chapter the energy dependence of soft photon-hadron interactions at high energies are well described[17] by the Reggeon Field Theory provided that the exchange of Pomeron trajectory with the quantum numbers close to vacuum, i.e. with the intercept $\alpha(t=0) \approx 1.08$ and C parity $+1$ is introduced. No direct evidence has been found at high energies for an exchange of trajectory with the same intercept but with $C = -1$ called Odderon trajectory. At HERA the existence of an Odderon trajectory could result in the production of pseudoscalar mesons with $C = +1$ in addition to their production via the two photon fusion process. Schäfer et al.[18] suggest to measure the cross section ratio for exclusive production of η's and η''s since the sensitivity to the Odderon contribution differs for η and η'. Assuming the coupling to the meson for photon-Odderon fusion being proportional to the product of the electric charge and the baryon number of the quarks composing the meson, η - being predominantly in the octet state with respect to flavour $SU(3)$ - gets larger contribution compared to η' being dominantly the singlet. The data were taken in 1996 with the H1 detector and amount to an integrated luminosity of about 5 pb^{-1}. The lead/scintillating fibre calorimeter SPACAL[19] has a sampling term of

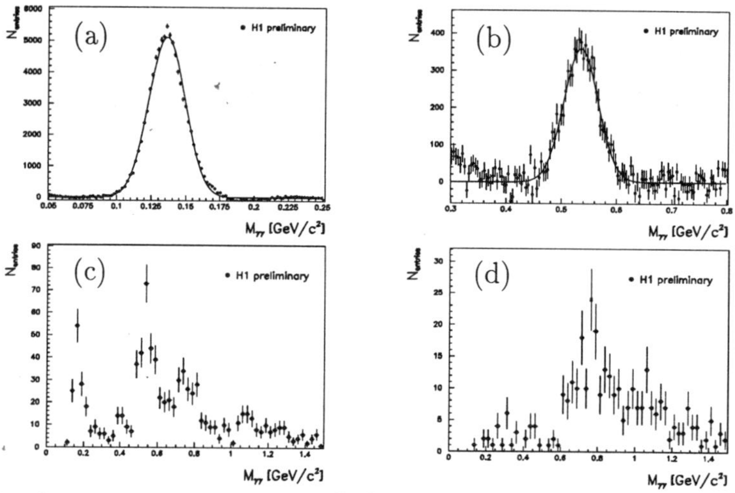

Fig. 7. Invariant mass spectra of 2γ's: (a) π^0 mass region for photon energies $E_\gamma > 1$ GeV after background subtraction (b) η mass region for photon energies $E_\gamma > 1.5$ GeV after background subtraction (c) events with a vertex and $E_\gamma > 0.7$ GeV (d) events without a vertex and $E_\gamma > 0.7$ GeV.

$7\%/\sqrt{E/\mathrm{GeV}}$ and a position resolution of about 5 mm at 1 GeV. Together with the threshold ≈ 100 MeV this allows for a precise measurement of mesons decaying into multi-photon final states, e.g. of π^0 and η mesons in their 2γ decay mode, as shown in Fig 7. The central values of the Gaussian fit amount to 136 MeV for π^0 and 535 MeV for η. The measured width of the invariant mass distribution is $\sigma \approx 13$ MeV for π^0 and $\sigma \approx 31$ MeV for η, demonstrating the good performance of the detector.

The acceptance of the positron calorimeter of the luminosity system restricts the photon energy in the laboratory system to lie between 8 and 20 GeV and to a photon virtuality $Q^2 < 0.01$ GeV^2. The selected events consist of the scattered positron, two photons within the angular acceptance of the SPACAL ($-3.8 < \eta^\gamma < -1.4$, η^γ is photon laboratory rapidity), and an elastically scattered proton which escapes undetected in the main detector.

Selected events with a reconstructed vertex, i.e. with at least one charged track in the central detector, result in the invariant mass spectrum of photon pairs as shown in 7c. A clear signal of π^0 and η mesons are observed, no indication for a signal of the η' meson at the mass 958 MeV is found. By requiring no charged particles in the central detector (events without a reconstructed vertex) is compatible with exclusive meson production. It leads to the invariant mass spectrum of the photon pairs as shown in 7d. No signal of exclusively produced η and η' mesons is observed nor of π^0. The resonance-like structure at an invariant mass of about 800 MeV is interpreted as being due to elastic photoproduction of ω mesons and their subsequent decays into the $\gamma\pi^0$ final state, where only two out of three photons are detected.

The non-observation of η mesons is compatible with the expectation from a pure two photon fusion process, given the small acceptance of the detector. This excludes a large contribution from Odderon induced process with cross section substantially larger than the two photon fusion cross section. In the case of η' the small branching ratio of 2.12% further reduces the expected number of events by an order of magnitude. The analysis continues to quantify the results into the cross section upper bounds for exclusive production of π^0, η and η' mesons.

5. Observation of isolated high E_t photons at HERA

The ZEUS detector identified a class of events showing the characteristics of high E_t isolated photons termed here as "prompt" to distinguish them from those produced via particle decays. This is the first observation of prompt photons at γp centre of mass energies an order of magnitude higher than those previously employed. In the kinematic range covered by HERA detectors, the prompt photons are produced predominantly in two hard subprocesses[20] which in LO are represented as the Compton scattering $\gamma q \to \gamma q$ in the direct channel and by $qg \to \gamma q$ in the resolved channel. NLO QCD adds additional processes where hard photon is radiated from outgoing quarks. Due to the electromagnetic nature of the photon the prompt photon production is suppressed approximately by a factor α with respect to hard processes involving partons only which represent the main source of background. Therefore, a sophisticated data analysis must be done to isolate events with prompt photons from background. A particular virtue of prompt photon processes is that the observed final-state photon emerges from the QCD process directly, without the intermediate hadronisation of final state partons into jets.

The data correspond to an integrated luminosity of 6.36 pb^{-1} collected by the ZEUS detector in 1995 in e^+p collisions with energy $E_e = 27.5$ GeV with protons of energy $E_p = 820$ GeV. Photons were detected in the barrel part of the uranium-scintillator calorimeter in its electromagnetic section (BEMC) which restricts the photon pseudorapidity range to $-0.75 < \eta^\gamma < 1.0$. A standard ZEUS electron finding algorithm was used to identify photon signal with $5 < E_t^\gamma < 10$ GeV. A cone jet finding algorithm was used to identify jets with $E_t^{jet} > 5$ GeV, pseudorapidity $-1.5 < \eta^{jet} < 1.7$ and unit cone radius, using entire ZEUS calorimeter system. Events with an observed DIS electron were removed, restricting incoming photon virtualities to $Q^2 \leq 1$ GeV^2. Isolation cone criteria were applied on the photon candidate to reduce possible background from radiation of electrons or positrons and from energy fluctuation of jets into hard π^0 or η. Two shape-dependent quantities were studied in order to distinguish photon, π^0 and η signals. These were the energy weighted mean width of the BEMC cluster in z direction z_{width} and the energy fraction of the most energetic cell in the cluster F_{max} (Fig. 8a) of the photon candidate. The relative

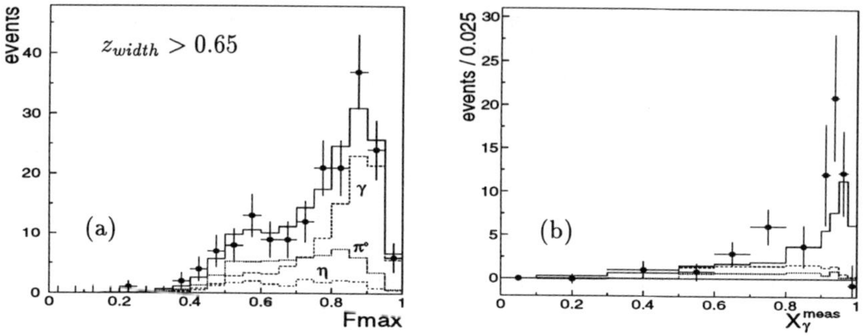

Fig. 8. (a) Distribution of F_{max} for prompt photon candidates and γ, π^0, η fitted Monte Carlo. z_{width} given in BEMC cell units (b) x_γ^{meas} of prompt photon after background subtraction. (•) data points, histogram - Monte Carlo: (···) quark radiation, (___) q rad+resolved, (—) q rad+resolved+direct.

amounts of γ, π^0 and η from single particle Monte Carlo distributions were obtained from fits of z_{width} and F_{max} to data. To get distribution of physical quantities, a background subtraction is applied on the assumption that the data may be expressed as a sum of the photon signal plus neutral meson background. The result for the fraction x_γ of the incoming photon momentum that contributes to the production of the high E_t photon and jet $x_\gamma^{meas} = \sum_{jet+\gamma}(E - p_z)/\sum_{event}(E - p_z)$ is shown in Fig. 8b. The Monte Carlo distributions are normalised to the same integrated luminosity as data. A strong peak near $x_\gamma^{meas} = 1$ corresponds to the Compton process. It is not possible to make conclusions concerning the presence of a resolved photon component. The background distribution is consistent with zero for $x_\gamma^{meas} > 0.9$. The quoted cross section for the process $ep \rightarrow e + \gamma_{prompt} + jet + X$ with $x_\gamma^{meas} \geq 0.8$ is $15.3 \pm 3.8(\text{stat}) \pm 1.8(\text{syst})$ pb in the kinematic range defined above. The systematic error is dominated by 8% contribution due to the uncertainty in the calorimeter calibration. The cross section is in an agreement with the NLO QCD calculation[21].

6. Conclusion

Inclusive particle production provides a complementary view on hard scattering to the commonly more used jet production. Selecting events with a hadron with p_t at least $1 - 2$ GeV, the inclusive spectra are calculable in QCD and are sensitive to structure functions of beam projectiles — photon (real or virtual) and proton. It will be the task of the coming years of HERA hadron physics to use inclusive spectra of hard hadrons to obtain the gluon structure function of the real photon. Events with prompt photons have a sensitivity to the quark content of the photon. With the increase of HERA luminosity they will supplement the inclusive spectra of charged particles in the study of partonic structure of the photon.

It is a pleasure for me to thank my colleagues from ZEUS and H1 collaboration who helped to prepare this talk and provided me with results and figures. I would also like to thank the organizers of the workshop for a very friendly atmosphere and a place for many stimulating discussions during the workshop.

This work is supported by the Grant Agency of Academy of Sciences of the Czech Republic under the grant no. A1010619.

7. References

1. D. Milstead, N. Gogitidze, S. Tapprogge, T. Vaiciulis, to appear in the proceedings of the *International Conference on the Structure and the Interactions of the Photon, Photon97*, World Scientific.
2. H1 Collab., C. Adloff et al., DESY-97-065, to be published in *Z. Phys.* **C**.
3. UA5 Collab., R.E. Ansorge et al., *Nucl. Phys* **B328** (1989) 36; ibid. *Z. Phys.* **C41** (1988) 179.
4. J. Binnewies, B. A. Kniehl, G. Kramer, *Phys. Rev.* **D53** (16996) 3573.
5. B. A. Kniehl, in the contribution to this Proceedings.
6. H1 Collab., S. Aïd et al., *Nucl. Phys.* **B480** (1996) 3.
7. E665 Collab., M. R. Adams et al., *Z. Phys.* **C61** (1994) 539.
8. ZEUS Collab., paper submitted to ICHEP96 pa02-044.
9. ZEUS Collab., M. Derrick et al., *Z. Phys.* **C67** (1995) 227.
10. W. Hoprich, *Diplomarbeit*, I. Phys. Institut, Univ. Heidelberg, 1995; H1 Collab., paper 270 submitted to ICHEP97.
11. CDF Collab., F. Abe et al., *Phys. Rev. Lett.* **61** (1988) 1819; UA1 Collab., C. Albajar et al., *Nucl. Phys.* **B335** (1990) 261.
12. ZEUS Collab., M. Derrick et al., *Z. Phys.* **C69** (1996) 607.
13. H1 Collab., C. Adloff et al., DESY-97-042, to be published in *Nucl. Phys.*
14. A. Levy, TAUP 2349-96, hep-ex/0608009.
15. A. Capella, A. Kaidalov, C. Merino, J. Tran Than Van, *Phys. Lett.* **B337** (1994) 358.
16. H1 Collab., S. Aïd et al., *Phys. Lett.* **B392** (1997) 234.
17. A. Donnachie and P. V. Landshoff, *Phys. Lett.* **B296** (1992) 227.
18. A, Schäfer, L. Mankiewicz and O. Nachtmann, Proceedings of *Future Physics at HERA*, eds. G. Ingelman, A. De Roeck, R. Klanner, DESY 1996, p. 243.
19. H1 SPACAL group, R. D. Appuhn et al., *Nucl. Instrum. Methods* **A374** (1996) 149.
20. P. Aurenche, P. Chiappetta, M. Fontannaz, J. P. Guillet, E. Pilon, *Z. Phys.* **C56** (1992) 589; L. E. Gordon and W. Vogelsang, *Phys. Rev.* **52D** (1995) 58.
21. L. E. Gordon, to appear in the proceedings of the *International Conference on the Structure and the Interactions of the Photon, Photon97*, World Scientific.

THEORETICAL ASPECTS OF HADRON PHOTOPRODUCTION

BERND A. KNIEHL
Max-Planck-Institut für Physik (Werner-Heisenberg-Institut)
Föhringer Ring 6, 80805 Munich, Germany
E-mail: kniehl@vms.mppmu.mpg.de

ABSTRACT

We review recent developments in the theoretical description of inclusive single-hadron production at next-to-leading order in the parton model of quantum chromodynamics. Fragmentation functions are extracted from fits to data of inclusive pion and kaon production in e^+e^- annihilation at different centre-of-mass energies. Exploiting sensitivity to the scaling violation, one can simultaneously fit the asymptotic scale parameter Λ so as to obtain an independent determination of the strong coupling constant α_s. Owing to the factorization theorem, the fragmentation functions only depend on the species of the produced particles, but not on the process by which they are produced. This allows one to make absolute theoretical predictions for inclusive pion and kaon production in other types of experiments such as photon-photon, photon-hadron, or hadron-hadron scattering. Recent data of photoproduction taken by the H1 and ZEUS Collaborations at DESY HERA nicely agree with such next-to-leading-order predictions.

1. Introduction

The Lagrangian of quantum chromodynamics (QCD) contains quarks and gluons as elementary fields. Allowing for these particles to appear as asymptotic states, we can evaluate scattering amplitudes perturbatively, in principle with arbitrary precision. Of course, this picture needs to be complemented by the principle of confinement of colour; experiments detect hadrons rather than quarks and gluons. Nevertheless, this simplified computational procedure is very successful in describing the production of jets of hadrons at high centre-of-mass (c.m.) energies ($\sqrt{s}$) in e^+e^- annihilation and at high transverse momenta (p_T) in scattering processes. Due to parton-hadron duality, clustering partons in the final state according to certain jet definitions yields a useful approximation, although this does not account for any details of hadronization.

On the other hand, experiments are providing us with copious information on the inclusive production of single hadrons, which cannot be interpreted along these lines. In this case, we need a detailed concept for describing how partons turn into hadrons. In the framework of the QCD-improved parton model, this is achieved by introducing fragmentation functions (FF's), $D_a^h(x,\mu^2)$. The value of $D_a^h(x,\mu^2)$ corresponds to the probability for the parton a produced at short distance $1/\mu$ to form a jet that includes the hadron h carrying the fraction x of the longitudinal momentum of a. Unfortunately, it is not yet understood how the FF's can be derived from first principles, in particular for hadrons with masses smaller than or comparable to the asymptotic scale parameter, Λ. However, given their x dependence at some scale μ, the evolution with μ may be computed perturbatively in QCD using the Altarelli-Parisi[1] (AP) equations. This allows us to test QCD quantitatively within one experiment observing single hadrons at different values of $\sqrt{s}$ (in the case of e^+e^- annihilation) or p_T (in the case of scattering). Moreover, the factorization theorem guarantees that the D_a^h functions are independent of the process in which they have been determined, and represent a universal property of h. This enables us to make quantitative predictions for other types of experiments as well. To summarize, having extracted FF's from fits to experimental data, we may test their μ dependence as predicted by the AP equations and their universality as postulated by the factorization theorem.

After the pioneering leading-order (LO) analyses of pion, kaon,[2] and charmed-meson[3] FF's in the late 70's, there had long been no progress on the theoretical side of this field. This may partly be attributed to the advent of general-purpose Monte Carlo (MC) event generators based on LO parton-level matrix elements in the early 80's, which were soon to become very popular in the experimental community. In these computer programs, the fragmentation into hadrons is simulated according to certain phenomenological model assumptions, *e.g.*, the cluster algorithm in the case of HERWIG or the LUND string model in the case of PYTHIA. Although such MC packages often lead to satisfactory descriptions of the data, from the theoretical point of view, their drawback is that this happens at the expense of introducing a number of ad-hoc fine-tuning parameters, which do not originate in the QCD Lagrangian. Furthermore, in the MC approach, it seems impossible to implement the factorization of final-state collinear singularities, which impedes a consistent extension to next-to-leading order (NLO). On the contrary, in the QCD-improved parton model, such singularities are absorbed into the bare (infinite) FF's so as to render them renormalized (finite) in a way quite similar to the procedure for the parton density functions (PDF's) at the incoming legs. Therefore, a meaningful quantitative test of QCD can only be performed in the parton model endowed with FF's at NLO.

Some time ago, NLO FF sets for π^0,[4] $\pi^\pm$, $K^\pm$, and η mesons[5] have been constructed through fits to data of e^+e^- annihilation generated with HERWIG. By contrast, our procedure, and partly that of Ref. 4, has been to fit to genuine experimental e^+e^- data. We have introduced LO and NLO FF sets for charged pions and kaons[6,7] and for neutral kaons.[8] The results presented here are obtained with the up-to-date sets of Refs. 7 and 8, which are based on data from SLAC PEP with $\sqrt{s} = 29$ GeV and CERN LEP1 with $\sqrt{s} = M_Z$. In contrast to Refs. 4 and 5, Refs. 6–8 also provide ready-to-use parameterizations of the μ dependence of the FF's.

This presentation is organized as follows. In Section 2, we shall report the essential features of our fitting procedure for charged pions and kaons[7] and neutral kaons.[8] In Sections 3 and 4, we shall confront recent data of inclusive pion and kaon production collected by the H1[9–11] and ZEUS Collaborations[12] in γp scattering at

DESY HERA and by the UA1 Collaboration[13,14] in $p\bar{p}$ scattering at the CERN $Sp\bar{p}S$ collider, respectively, with NLO predictions based on our FF's. Our conclusions will be summarized in Section 5.

2. Fit to data of e^+e^- annihilation

The most direct way to obtain information on the FF's of hadrons is to analyze their energy spectrum measured in e^+e^- annihilation, where the theoretical predictions are not obscured by additional nonperturbative input, *e.g.*, in the form of PDF's for the incoming particles. At NLO in the parton model with n_f massless quark flavours, the differential cross section of $e^+e^- \to \gamma/Z \to h + X$, normalized to the total hadronic cross section σ_{tot}, is given by

$$\frac{1}{\sigma_{\text{tot}}}\frac{d\sigma}{dx} = \sum_a \int_x^1 \frac{dz}{z} D_a^h\left(\frac{x}{z}, M_h^2\right) \frac{1}{\sigma_{\text{tot}}}\frac{d\sigma_a}{dz}\left(z, \mu^2, M_h^2\right), \tag{1}$$

where $x = 2E_h/\sqrt{s}$ is the fraction of the beam energy carried by h, $a = g, q_1, \ldots, \bar{q}_{n_f}$, μ is the renormalization scale, and M_h is the factorization scale. The parton-level cross sections read

$$\begin{aligned}
\frac{1}{\sigma_{\text{tot}}}\frac{d\sigma_{q_i}}{dx}\left(x, \mu^2, M_h^2\right) &= \frac{e_{q_i}^2}{\sum_{i=1}^{n_f} e_{q_i}^2}\left\{\delta(1-x) + \frac{\alpha_s(\mu^2)}{2\pi}\left[P_{q\to q}^{(0,T)}(x)\ln\frac{s}{M_h^2} + C_q(x)\right]\right\}, \\
\frac{1}{\sigma_{\text{tot}}}\frac{d\sigma_g}{dx}\left(x, \mu^2, M_h^2\right) &= 2\frac{\alpha_s(\mu^2)}{2\pi}\left[P_{q\to g}^{(0,T)}(x)\ln\frac{s}{M_h^2} + C_g(x)\right],
\end{aligned} \tag{2}$$

where e_{q_i} is the effective coupling of q_i to the photon and the Z boson including propagator adjustments and C_a are the NLO corrections.[15] Here, $P_{a\to b}^{(0,T)}$ are the LO terms of the timelike $a \to b$ splitting functions,

$$P_{a\to b}^{(T)}\left(x, \alpha_s(\mu^2)\right) = \frac{\alpha_s(\mu^2)}{2\pi} P_{a\to b}^{(0,T)}(x) + \left(\frac{\alpha_s(\mu^2)}{2\pi}\right)^2 P_{a\to b}^{(1,T)}(x) + \mathcal{O}\left(\alpha_s^3\right), \tag{3}$$

which control the μ evolution of the FF's via the AP equations,

$$\frac{\mu^2 d}{d\mu^2} D_a^h(x, \mu^2) = \sum_b \int_x^1 \frac{dz}{z} P_{a\to b}^{(T)}\left(\frac{x}{z}, \alpha_s(\mu^2)\right) D_b^h(x, \mu^2). \tag{4}$$

Analytic expressions for $P_{a\to b}^{(T)}$ may be found, *e.g.*, in Ref. 16. The integro-differential equation (4) may be solved either via the Mellin transform technique or with brute force as it stands. Good agreement is found between these two methods.[16]

It is natural to choose $\mu = M_h = \sqrt{s}$. This eliminates the terms involving $P_{a\to b}^{(0,T)}$ in Eq. (2). In e^+e^- annihilation, gluon fragmentation occurs only at NLO and beyond. To increase the sensitivity to D_g^h, one may select longitudinal polarization,[17] so that the delta function in Eq. (2) does not contribute, or concentrate on gluon-tagged three-jet events.[17,18] Then, the respective NLO terms in Eq. (2) constitute the Born approximation, and it is desirable to include the next-to-next-to-leading-order corrections[19] in order to reduce the scale and scheme dependences. Unfortunately, these were not yet available when the analyses of Refs. 6–8 were carried out.

The fitting procedure in Refs. 7 and 8 was as follows. For each of the hadron channels $h = \pi^+ + \pi^-, K^+ + K^-, K_S^0 + K_L^0$ and each parton type $a = g, u, d, s, c, b$, we made the ansatz

$$D_a^h(x, \mu_0^2) = N x^\alpha (1-x)^\beta, \tag{5}$$

at the respective starting scale

$$\mu_0 = \begin{cases} \sqrt{2}\ \text{GeV} & \text{if } a = g, u, d, s, \\ M(\eta_c) = 2.98\ \text{GeV} & \text{if } a = c, \\ M(\Upsilon) = 9.46\ \text{GeV} & \text{if } a = b. \end{cases} \tag{6}$$

According to the flavour composition of h, we respectively imposed

$$\begin{aligned} D_u^{\pi^\pm}\left(x, \mu_0^2\right) &= D_d^{\pi^\pm}\left(x, \mu_0^2\right), \\ D_u^{K^\pm}\left(x, \mu_0^2\right) &= D_s^{K^\pm}\left(x, \mu_0^2\right), \\ D_u^{K^0}\left(x, \mu_0^2\right) &= D_s^{K^0}\left(x, \mu_0^2\right). \end{aligned} \tag{7}$$

The analysis of Ref. 7 was based on charged-pion, charged-kaon,[20,21] and unidentified charged-hadron data.[17,18] Apart from charged pions and kaons, mainly protons and antiprotons contribute to the charged-hadron yield. Inspired by Ref. 21, we approximated

$$\frac{d\sigma^{h^\pm}}{dx} = [1 + f(x)] \frac{d\sigma^{\pi^\pm}}{dx} + \frac{d\sigma^{K^\pm}}{dx}, \tag{8}$$

where $f(x) = 0.195 - 1.35\,(x - 0.35)^2$. In their charged-hadron analysis, the ALEPH Collaboration[17] distinguished between *uds*-, *c*-, and *b*-enriched samples. Also exploiting information on identified gluon jets,[17,18] we were thus able to treat $g, u, s, c, b \to \pi^\pm$ and $g, u, d, c, b \to K^\pm$ fragmentation separately. Due to the large gap in $\sqrt{s}$ between PEP[20] and LEP1,[17,18,21] we could simultaneously determine $\Lambda_{\overline{\text{MS}}}^{(5)}$. Thus, we had a total of $2(\pi, K) \times 5(\text{partons}) \times 3(N, \alpha, \beta) + 1(\Lambda_{\overline{\text{MS}}}^{(5)}) = 31$ independent fit parameters. These turned out to be tightly constrained by our LO and NLO fits. In fact, we obtained χ^2/d.o.f. values of $134.4/136 = 0.99$ and $125.3/136 = 0.92$, respectively. We found $\Lambda_{\overline{\text{MS}}}^{(5)} = 108$ MeV (227 MeV) at LO (NLO), which corresponds to $\alpha_s(M_Z^2) = 0.122$ (0.118). This nicely agrees with the latest LEP value, (0.121 ± 0.003).[22]

In our neutral-kaon analysis,[8] we adopted the $\Lambda_{\overline{\text{MS}}}^{(5)}$ values from Ref. 7. Furthermore, appealing to the flavour blindness of the gluon, we assumed that

$$D_g^{K^0}\left(x, \mu_0^2\right) = D_g^{K^\pm}\left(x, \mu_0^2\right), \tag{9}$$

at $\mu_0 = \sqrt{2}$ GeV. Thus, the number of independent fit parameters was 4(partons) $\times$ $3(N, \alpha, \beta) = 12$. Our combined fit to the MARK II[23] and ALEPH[24] neutral-kaon samples yielded $\chi^2/\text{d.o.f.} = 9.9/20$ at LO and 8.6/20 at NLO.

3. Comparison with data of photoproduction in *ep* scattering

According to present HERA conditions, $E_e = 27.5$ GeV positrons collide with $E_p = 820$ GeV protons in the laboratory frame, so that $\sqrt{s} = 300$ GeV is available in the c.m. frame. It has become customary to take the rapidity of hadrons travelling in the proton direction to be positive. The rapidities measured in the ep laboratory and c.m. frames are related through

$$y_{\text{c.m.}} = y_{\text{lab}} - \frac{1}{2} \ln \frac{E_p}{E_e}. \tag{10}$$

In photoproduction, the electron or positron beam acts like a source of quasi-real photons, with low virtualities $-Q^2$, so that HERA is effectively operated as a γp collider. The appropriate events may be discriminated from deep-inelastic-scattering events by electron tagging or anti-tagging. The photon flux is well approximated by the Weizsäcker-Williams[25] formula,

$$F_\gamma^e \left(x, Q^2_{\text{max}}\right) = \frac{\alpha}{2\pi} \left[\frac{1 + (1-x)^2}{x} \ln \frac{Q^2_{\text{max}}}{Q^2_{\text{min}}} + 2m_e^2 x \left(\frac{1}{Q^2_{\text{max}}} - \frac{1}{Q^2_{\text{min}}} \right) \right], \tag{11}$$

where $x = E_\gamma/E_e$, $Q^2_{\text{min}} = m_e^2 x^2/(1-x)$, and $Q^2_{\text{max}} = 0.01\ \text{GeV}^2$ ($0.02\ \text{GeV}^2$) for tagged events in the case of H1 (ZEUS). The cross section of $ep \to h + X$ emerges from the one of $\gamma p \to h + X$ by convolution with F_γ^e. By kinematics, $x_{\text{min}} \le x \le 1$, where $x_{\text{min}} = p_T \exp(-y_{\text{c.m.}})/[\sqrt{s} - p_T \exp(y_{\text{c.m.}})]$. In conformity with the H1 and ZEUS event-selection criteria, we impose $0.3 < x < 0.7$ and $0.318 < x < 0.431$, respectively.

It is well known that $\gamma p \to h + X$ proceeds via two distinct mechanisms. The photon can interact either directly with the partons originating from the proton (direct photoproduction) or via its quark and gluon content (resolved photoproduction). Both contributions are formally of the same order in the perturbative expansion. Leaving aside the proton PDF's, F_b^p, and the FF's, D_c^h, which represent common factors, the LO cross sections are of $\mathcal{O}(\alpha\alpha_s)$ in both cases. In the case of the resolved mechanism, this may be understood by observing that the $ab \to cd$ cross sections, which are of $\mathcal{O}(\alpha_s^2)$, get dressed by photon PDF's, F_a^γ, whose leading terms are of the form $\alpha \ln(M_\gamma^2/\Lambda^2) \propto \alpha/\alpha_s$, with M_γ being the corresponding factorization scale. Here, a, b, c, d denote quarks and gluons. In fact, the two mechanisms also compete with each other numerically. Resolved photoproduction dominates at small p_T and positive y_{lab}, while direct photoproduction wins out at large p_T and negative y_{lab}.

LO calculations suffer from significant theoretical uncertainties connected with the freedom in the choice of the renormalization scale, μ, in α_s and the factorization scales, M_γ, M_p, and M_h, in F_a^γ, F_b^p, and D_c^h, respectively. In order to obtain reliable predictions, it is indispensable to proceed to NLO. Let us first consider resolved photoproduction, which is more involved. Starting out from the well-known LO cross section of $\gamma p \to h + X$, one needs to include the NLO corrections, $K_{ab\to c}$, to the parton-level cross sections, $d\sigma_{ab\to c}/dt$, to substitute the two-loop formula for α_s, and to endow F_a^γ, F_b^p, and D_c^h with NLO evolution. This leads to

$$\frac{d^3\sigma}{dy\,d^2p_T} = \frac{1}{\pi}\sum_{a,b,c}\int dx_\gamma dx_p \frac{dx_h}{x_h^2} F_a^\gamma\left(x_\gamma, M_\gamma^2\right) F_b^p\left(x_p, M_p^2\right) D_c^h\left(x_h, M_h^2\right)\left[\frac{d\sigma_{ab\to c}}{dt}(s,t,\mu^2)\right.$$
$$\left.\times\,\delta\left(1+\frac{t+u}{s}\right) + \frac{\alpha_s(\mu^2)}{2\pi}K_{ab\to c}\left(s,t,u,\mu^2,M_\gamma^2,M_p^2,M_h^2\right)\theta\left(1+\frac{t+u}{s}\right)\right], \qquad (12)$$

where $a,b,c = g, q_1, \ldots, \bar{q}_{n_f}$, $s = (p_a + p_b)^2$, $t = (p_a - p_c)^2$, and $u = (p_b - p_c)^2$. The parton momenta are related to the photon, proton, and hadron momenta by $p_a = x_\gamma p_\gamma$, $p_b = x_p p_p$, and $p_c = p_h/x_h$. The $K_{ab\to c}$ functions may be found in Ref. 26 for $M_\gamma = M_p$. This restriction was relaxed in Ref. 27.

The NLO cross section of direct photoproduction emerges from Eq. (12) by substituting $F_a^\gamma(x_\gamma, M_\gamma^2) = \delta(1 - x_\gamma)$, replacing $d\sigma_{ab\to c}/dt$ and $K_{ab\to c}$ with $d\sigma_{\gamma b\to c}/dt$ and $K_{\gamma b\to c}$, respectively, and omitting the sum over a. The $K_{\gamma b\to c}$ functions were first derived in Ref. 28 setting $M_\gamma = M_p = M_h$ and taking the spin-average for incoming photons and gluons to be 1/2. In Ref. 27, the scales were disentangled and the spin-average convention was converted to the $\overline{\text{MS}}$ scheme, *i.e.*, to be $1/(n-2)$, with n being the dimensionality of space-time. Analytic expressions for the $K_{\gamma b\to c}$ functions are listed in Ref. 29.

The theoretical predictions presented here are calculated at NLO in the $\overline{\text{MS}}$ scheme with $n_f = 5$ quark flavours and $\Lambda_{\overline{\text{MS}}}^{(5)} = 202$ MeV. Unless otherwise stated, we use the CTEQ4M[30] proton PDF's, the GRV[31] photon PDF's, and the BKK FF's for charged hadrons[7] and neutral kaons.[8] We set $\mu = M_\gamma = M_p = M_h = \xi p_T$, with $\xi = 1/2, 1, 2$, in order to estimate the theoretical uncertainty.

In Fig. 1, we compare the ZEUS[12] and preliminary H1[9] data on $ep \to h^\pm + X$ via photoproduction with the corresponding NLO predictions. We find good agreement as for both normalization and shape. In the upper p_T range, our central prediction slightly overshoots the ZEUS data (see Fig. 1a), while it tends to be a tiny bit below the centres of the H1 data points, but well within their errors bars (see Fig. 1b). The study of the y_{lab} spectrum in Fig. 1c nicely illustrates the interplay of direct and resolved photoproduction. The direct-photon (resolved-photon) contribution peaks at negative (positive) y_{lab}. At NLO, both contributions strongly depend on the factorization scheme and scale associated with the incoming photon leg, while their sum represents a meaningful physical observable. In the $\overline{\text{MS}}$ scheme with $\xi = 1$, the resolved-photon contribution clearly dominates for a minimum-p_T cut as low as 2 GeV. This offers the opportunity to probe the photon PDF's. In particular, F_g^γ is only feebly constrained experimentally. In the forward direction, at $y_{\text{lab}} > 1$, it makes

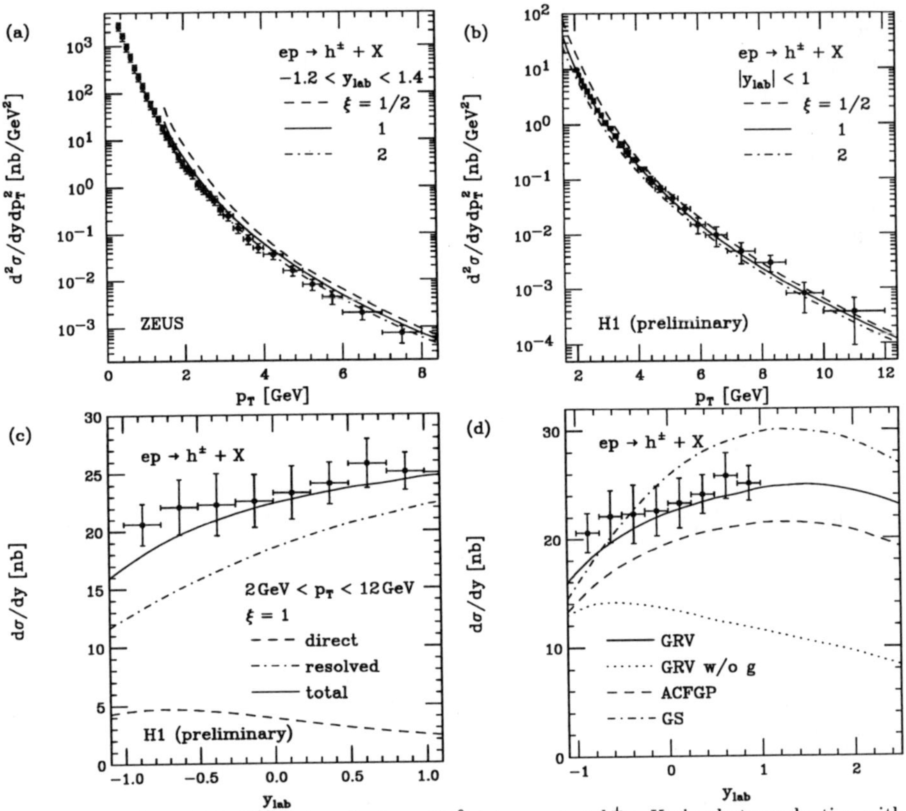

Fig. 1. Comparison of ZEUS[12] and preliminary H1[9] data on $ep \to h^\pm + X$ via photoproduction with NLO predictions.

up more than 50% of the cross section (see Fig. 1d). Thus, a dedicated experimental study in the forward direction could help to pin down F_g^γ and to distinguish between the various available photon PDF sets. Figure 1d also shows the predictions for the ACFGP[32] set with massless charm quark and for the updated GS[33] set. Unfortunately, the minimum-p_T cut in the data is still too low to allow for a sufficiently precise theoretical description within the QCD-improved parton model.

The H1[10,11] data on $ep \to K^0/\overline{K}^0 + X$ via photoproduction are in good agreement with the corresponding NLO predictions (see Fig. 2). In Fig. 2b, the vertical error bars include the statistical errors quoted in Table 4.11 of Ref. 11 and an overall systematic uncertainty of[11] 10% added in quadrature. Strictly speaking, the minimum-p_T cut of 2 GeV is too low for the parton-model prediction to be usefully precise. Nevertheless, Fig. 2b nicely exposes the potential of the $y_{\rm lab}$ distribution to probe F_g^γ and $D_g^{K^0}$. In view of the general circumstance that the gluon FF's are less tightly constrained by

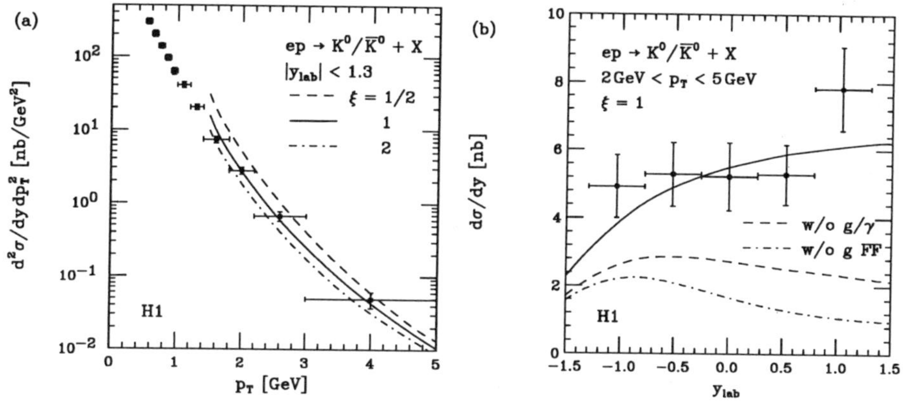

Fig. 2. Comparison of H1[10,11] data on $ep \to K^0/\overline{K}^0 + X$ via photoproduction with NLO predictions.

the e^+e^- data than the quark FF's and the particular assumption underlying Eq. (9), an independent experimental determination of $D_g^{K^0}$ would be especially desirable.

4. Comparison with data of $p\bar{p}$ scattering

The high-statistics data on $p\bar{p} \to h^\pm + X$ and $p\bar{p} \to K_S^0 + X$ recently published by the UA1 Collaboration[13,14] offer yet another opportunity to test the universality of the FF's predicted by the factorization theorem in a nontrivial way (see Fig. 3). At low p_T, the bulks of the cross sections are due to the gluon FF's, which makes $p\bar{p}$ scattering complementary to e^+e^- annihilation.[8] The charged-hadron data are spread over a wide range in p_T, way up to $p_T = 25$ GeV, and thus carry intrinsic information on the scaling violation of fragmentation. This is nicely illustrated by the dotted line in Fig. 3a, which emerges from the upper solid line by suspending the AP evolution of the FF's and evaluating them at the fixed scale $\mu_0 = \sqrt{2}$ GeV instead. This leads to a significant increase at high p_T, by more than a factor of 5 at $p_T = 25$ GeV. The experimental data clearly favour the scaling violation encoded in the solid line.

5. Conclusions

The comparative study of inclusive single-hadron production in e^+e^-, ep, and $p\bar{p}$ collisions allows for a meaningful quantitative test of the QCD-improved parton model and, in particular, of the scaling violation and the universality of fragmentation. Furthermore, photoproduction experiments at HERA provide useful information on the interplay of the direct- and resolved-photon mechanisms. Extensions of these measurements to higher values of p_T and $y_{\rm lab}$ would render an independent determination of α_s possible and considerably improve our knowledge of F_g^γ.

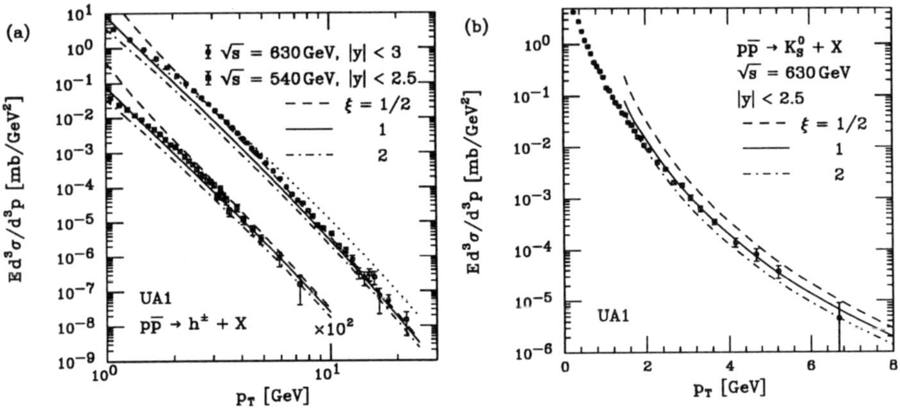

Fig. 3. Comparison of UA1 data on[13] $p\bar{p} \to h^{\pm} + X$ and[14] $p\bar{p} \to K_S^0 + X$ with NLO predictions.

ACKNOWLEDGMENTS

I would like to thank Janko Binnewies and Gustav Kramer for their collaboration on the work presented here.

1. V.N. Gribov and L.N. Lipatov, *Yad. Fiz.* **15** (1972) 781 [*Sov. J. Nucl. Phys.* **15** (1972) 438]; G. Altarelli and G. Parisi, *Nucl. Phys.* **B126** (1977) 298; Yu.L. Dokshitser, *Zh. Eksp. Teor. Fiz.* **73** (1977) 1216 [*Sov. Phys. JETP* **46** (1977) 641].
2. V. Barger, T. Gottschalk, and R.J.N. Phillips, *Phys. Lett.* **70B** (1977) 51; R. Baier, J. Engels, and B. Petersson, *Z. Phys.* **C2** (1979) 265; M. Anselmino, P. Kroll, and E. Leader, *Z. Phys.* **C18** (1983) 307.
3. L.M. Sehgal and P.M. Zerwas, *Nucl. Phys.* **B108** (1976) 483; C. Peterson, D. Schlatter, I. Schmitt, and P.M. Zerwas, *Phys. Rev.* **D27** (1983) 105.
4. P. Chiappetta, M. Greco, J.-Ph. Guillet, S. Rolli, and M. Werlen, *Nucl. Phys.* **B412** (1994) 3.
5. M. Greco and S. Rolli, *Z. Phys.* **C60** (1993) 169; *Phys. Rev.* **D52** (1995) 3853; M. Greco, S. Rolli, and A. Vicini, *Z. Phys.* **C65** (1995) 277.
6. J. Binnewies, B.A. Kniehl, and G. Kramer, *Z. Phys.* **C65** (1995) 471.
7. J. Binnewies, B.A. Kniehl, and G. Kramer, *Phys. Rev.* **D52** (1995) 4947.
8. J. Binnewies, B.A. Kniehl, and G. Kramer, *Phys. Rev.* **D53** (1996) 3573.
9. I. Abt *et al.* (H1 Collaboration), *Phys. Lett.* **B328** (1994) 176; M. Erdmann, private communication.
10. C. Adloff *et al.* (H1 Collaboration), Report Nos. DESY 97–095 and hep–ex/9705018 (May 1997), *Z. Phys.* **C** (in press).
11. K. Johannsen, Ph.D. thesis, University of Hamburg, Internal Report No. DESY FH1–96–01 (June 1996).

12. M. Derrick *et al.* (ZEUS Collaboration), *Z. Phys.* **C67** (1995) 227.
13. G. Arnison *et al.* (UA1 Collaboration), *Phys. Lett.* **118B** (1982) 167; G. Bocquet *et al.* (UA1 Collaboration), *Phys. Lett.* **B366** (1996) 434.
14. G. Bocquet *et al.* (UA1 Collaboration), *Phys. Lett.* **B366** (1996) 441.
15. G. Altarelli, R.K. Ellis, G. Martinelli, and S.-Y. Pi, *Nucl. Phys.* **B160** (1979) 301; R. Baier and K. Fey, *Z. Phys.* **C2** (1979) 339.
16. J. Binnewies, B.A. Kniehl, and G. Kramer, Report Nos. DESY 97–012, MPI/PhT/97–009, and hep–ph/9702406 (February 1997), *Z. Phys.* **C** (in press).
17. G.D. Cowan, in *Proceedings of the XXVII International Conference on High Energy Physics*, 20–27 July 1994, Glasgow, Scottland, UK, edited by P.J. Bussey and I.G. Knowles (IOP, Bristol, 1995), p. 883; D. Buskulic *et al.* (ALEPH Collaboration), *Phys. Lett.* **B357** (1995) 487; **B364** (1995) 247 (E); C. Padilla Aranda, Ph.D. thesis, University of Barcelona (September 1995).
18. P.D. Acton *et al.* (OPAL Collaboration), *Z. Phys.* **C58** (1993) 387.
19. P.J. Rijken and W.L. van Neerven, *Phys. Lett.* **B386** (1996) 422; *Nucl. Phys.* **B487** (1997) 233; J. Binnewies, Report Nos. DESY 97–128 and hep–ph/9707269 (July 1997).
20. H. Aihara *et al.* (TPC/Two Gamma Collaboration), Report No. LBL–23737 (March 1988); *Phys. Rev. Lett.* **61** (1988) 1263.
21. R. Akers *et al.* (OPAL Collaboration), *Z. Phys.* **C63** (1994) 181; D. Buskulic *et al.* (ALEPH Collaboration), *Z. Phys.* **C66** (1995) 355.
22. R. Clare *et al.* (LEP Electroweak Working Group), Report No. LEPEWWG/97–02 (August 1997).
23. H. Schellman *et al.* (MARK II Collaboration), *Phys. Rev.* **D31** (1985) R3013.
24. D. Buskulic *et al.* (ALEPH Collaboration), *Z. Phys.* **C64** (1994) 361.
25. E.J. Williams, *Proc. Roy. Soc. London* **A139** (1933) 163; C.F. v. Weizsäcker, *Z. Phys.* **88** (1934) 612.
26. F. Aversa, P. Chiappetta, M. Greco, and J.Ph. Guillet, *Phys. Lett.* **B210** (1988) 225; **B211** (1988) 465; *Nucl. Phys.* **B327** (1989) 105.
27. B.A. Kniehl and G. Kramer, *Z. Phys.* **C62** (1994) 53.
28. P. Aurenche, R. Baier, A. Douiri, M. Fontannaz, and D. Schiff, *Nucl. Phys.* **B286** (1987) 553.
29. L.E. Gordon, *Phys. Rev.* **D50** (1994) 6753.
30. H.L. Lai, J. Huston, S. Kuhlmann, F. Olness, J. Owens, D. Soper, W.K. Tung, and H. Weerts, *Phys. Rev.* **D55** (1997) 1280.
31. M. Glück, E. Reya, and A. Vogt, *Phys. Rev.* **D46** (1992) 1973.
32. P. Aurenche, P. Chiappetta, M. Fontannaz, J.P. Guillet, and E. Pilon, *Z. Phys.* **C56** (1992) 589.
33. L.E. Gordon and J.K. Storrow, *Nucl. Phys.* **B489** (1997) 405.

Chapter 7

Heavy-flavour Production

OPEN HEAVY FLAVOR PRODUCTION IN DEEPLY INELASTIC *ep* SCATTERING AT HERA*

KARIN DAUM†

Universität Wuppertal, Rechenzentrum, Gaußstraße 20
D-42097 Wuppertal, Germany
E-mail: daum@mail.desy.de

(On behalf of the H1 and ZEUS collaborations)

ABSTRACT

Recent results on inclusive D^0 and $D^{*\pm}$ production in deeply inelastic *ep* scattering at $\sqrt{s} = 301$ GeV are summarized. The data have been collected by the H1 and ZEUS experiments at HERA. The total and the differential cross sections are discussed in the framework of LO and NLO QCD predictions. The data exhibit clear evidence for boson gluon fusion dominating open heavy flavor production in the kinematic range currently explored at HERA. The measurements of $F_2^{c\bar{c}}(x, Q^2)$ at small Bjorken x are presented. The prospects for future analyses of open charm and beauty production including detector upgrades and anticipated high luminosities are investigated.

1. Introduction

The study of heavy flavor production in deeply inelastic lepton nucleon scattering *(DIS)* is of importance for the understanding of the parton densities in the nucleon [1,2,3,4]. In accordance with early measurements by the EMC collaboration [5], neutral current charm quark production is expected to be dominated by the *boson gluon fusion (BGF)* process $\gamma(Z^0)g \to c\bar{c}$ at large Bjorken x. Recent results on inclusive production of D^0 and $D^{*\pm}$ mesons from H1 [6] and ZEUS [7] at HERA enable the charm production mechanism at small x to be investigated.

The measurement of the charm contribution $F_2^{c\bar{c}}(x, Q^2)$ to the proton structure function F_2 is supposed to provide information on this mechanism. Present parton density calculations include heavy flavors in the proton using different assumptions. Charm is either considered as *flavor excitation (FE)* ,i.e. $\gamma(Z^0)q \to qX$, in the massless quark evolution starting with a charm density of zero at a scale Q_0^2 [8,9] or it is exclusively treated as BGF taking into account the charmed quark mass [10]. In terms of phenomenological inclusive analyses [9,10,11] BGF is predicted to dominate near threshold, while FE should lead to a good description well above threshold. Various schemes to match these two regions have been proposed[4,12]. Both approaches lead to similar predictions for $F_2^{c\bar{c}}(x, Q^2)$ but differences in exclusive heavy quark spectra are expected.

If charmed hadron production is dominated by BGF, it provides direct information

*To appear in Proc, of "New Trends in HERA Physics", Ringberg, May 1997
†presently at DESY, D-22603 Hamburg, Germany

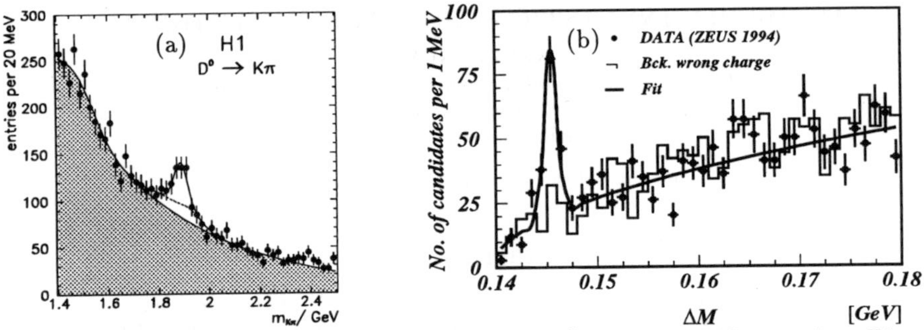

Fig. 1. *(a) The $K^-\pi^+$ mass distribution observed in the D^0 analysis of DIS events from H1, and (b) the Δm distribution obtained in the D^{*+} analysis of DIS events from ZEUS.*

on the gluonic content of the proton, $x_g g(x_g, Q^2)$, complementary to what is obtained from scaling violation analyses of F_2. Since the reconstruction of hadrons with heavy flavor content unequivocally tags the parent quark, heavy quark spectra become directly accessible. Compared to two jet studies the analysis of charmed hadrons allows (a) to investigate the BGF process down to the phase space limit $\hat{s} = 4m_c^2$ which extends the range for the determination of the gluon density in the proton significantly towards smaller x_g, and (b) to tag the gluon in the proton almost free of background from FE of light quarks in the proton, since heavy flavor production from gluon splitting is suppressed by heavy quark masses.

2. Results

Published results on charm production in deeply inelastic ep scattering exist from both experiments H1 and ZEUS based on an integrated luminosity of roughly 3 pb^{-1} collected with each experiment at HERA in 1994. In addition preliminary results from ZEUS are available based on an integrated luminosity of 6.4 pb^{-1} from the HERA 1995 running. The H1 collaboration [6] has performed the tagging of heavy quark events by reconstructing $D^0(1864)$[a]and $D^{*+}(2010)$ mesons, while the ZEUS collaboration [7] gives results for the inclusive $D^{*+}(2010)$ analysis.

The D^0 is identified via its decay mode $D^0 \to K^-\pi^+$ and the D^{*+} through the decay chain $D^{*+} \to D^0\pi^+_{slow} \to K^-\pi^+\pi^+_{slow}$. For the latter use is made of the low Q-value in the decay of $D^{*+} \to D^0\pi^+_{slow}$ which leads to a better resolution for the mass difference $\Delta m = m(D^0\pi^+_{slow}) - m(D^0)$ than for the D^{*+} mass itself.

Figure 1 shows the $m_{K\pi}$ distribution obtained in the inclusive D^0 analysis of H1 and the Δm distribution as observed in the inclusive D^{*+} of ZEUS. Evidently the number of observed events containing heavy quarks is small. Only of the order of 100 to 200 charmed mesons are identified in any of the different analyses.

[a]Charge conjugate states are always included.

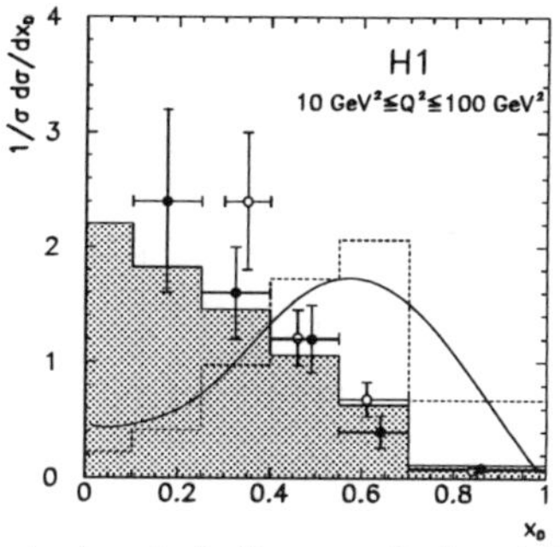

Fig. 2. *Normalized x_D distribution in deep inelastic ep scattering at $\langle W \rangle \approx 125$ GeV for $|\eta_D| < 1.5$. The open/closed points represent the D^0/D^* data of H1. See text for further explanations.*

2.1. Mechanism of charm production in DIS

Due to the hard fragmentation of D mesons its momentum is strongly correlated with the parent charm quark momentum. Therefore the distribution of the scaled D momentum in the γ^*p system $x_D = 2|\vec{P}_D^*|/W$ depends on the charm quark momentum spectrum and thereby on the mechanism of charm production. In BGF a $c\bar{c}$ pair is produced recoiling against the proton remnant in the γ^*p system, whilst in FE a single charmed quark is emitted opposite to the proton remnant, confining the second charmed quark necessary for quantum number conservation. Consequently the mean $\langle x_D \rangle$ of D mesons in the current and central fragmentation region is expected to be roughly a factor of 2 smaller for BGF compared to FE.

Figure 2 shows the distributions $1/\sigma\ \mathrm{d}\sigma/\mathrm{d}x_D$ obtained from the D^0 and D^{*+} analysis of H1 in comparison to the expectation for the BGF and FE process. The LO BGF prediction from AROMA [13] (shaded histogram) agrees well with the data. The figure also includes the FE expectations for charmed mesons either by using the LEPTO 6.1 generator [14], from which only FE events are selected (dashed histogram), or by extrapolating the results from charm production in charged current $\overset{(-)}{\nu}$ N scattering [16] to HERA energies (full line). Large differences in the shape are observed for the data and these FE expectations. From a fit to the data of BGF and FE in arbitrary proportions it is concluded that more than 95% of neutral current charm production in DIS is due to the BGF process at $\langle Q^2 \rangle \approx 25$ GeV. This observation seems to contradict recent inclusive calculations of charm production in DIS [4], from which it was concluded that for the kinematic range of the current analyses at HERA charmed quarks may already be considered as massless partons in the proton.

2.2. Differential inclusive D^* cross sections

Figure 3 shows the differential $e^+p \to D^{*\pm}X$ cross sections as a function of a)

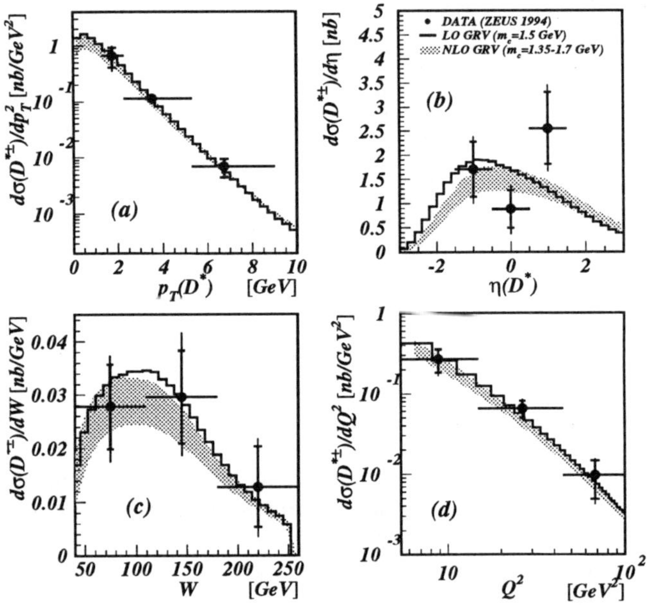

Fig. 3. *Differential* $e^+p \to D^{*\pm}X$ *cross sections for* $5 < Q^2 < 100$ *GeV*^{2}, $y < 0.7$ *in the kinematic region* $1.3 < p_T(D^*) <$ *GeV* $|\eta(D^*)| < 1.5$ *as a function of a)* $p_T(D^*)$*, b)* $\eta(D^*)$*, c)* W *and d)* Q^2 *from ZEUS. The inner/outer error bars correspond to the statistical/total errors respectively. The band and histogram are explained in the text.*

$p_T(D^*)$, b) $\eta(D^*)$, c) W and d) Q^2 measured by ZEUS. The shaded band represents the result of the analytic NLO calculation[3] using the GRV-HO 94 [10] gluon density of the proton which assumes charm being exclusively produced by BGF and treats charm quarks as massive particles. Both the renormalization scale and the factorization scale are set to $\mu = \sqrt{Q^2 + 4m_c^2}$. The largest uncertainties in the calculations arise from the ignorance of m_c. The upper (lower) limit corresponds to a charm quark mass $m_c = 1.35(1.7)$ GeV. This calculation reproduces well the shapes of the differential distributions. The figure also includes the analytic LO calculation as histograms. The LO calculation describes the shapes of the distributions equally well.

2.3. Integrated charm cross sections

All differential distributions discussed in the previous sections are found to be well described by the BGF process.Therefore, it is used to extrapolate the inclusive D cross sections, $\sigma(ep \to eDX)$, from the visible kinematic region to the full phase space in order to determine the charm production cross section, $\sigma(ep \to ec\bar{c}X)$, in DIS.

Table 1. Measured and predicted integrated charm cross sections $\sigma(ep \to ec\bar{c}X)$ for different bins in Q^2 in the region $y < 0.7$ (ZEUS) and $0.01 < y < 0.7$ (H1).

	m_c [GeV]	$5 < Q^2 < 10$ GeV^2	$10 < Q^2 < 100$ GeV^2
ZEUS $D^{*\pm}$		$13.5 \pm 5.2 \pm 1.8^{+1.6}_{-1.2}$ nb	$12.5 \pm 3.1 \pm 1.8^{+1.5}_{-1.1}$ nb
LO MC	1.5	12.5 nb	14.2 nb
LO GRV	1.5	11.0 nb	12.4 nb
NLO GRV	1.5	9.4 nb	11.1 nb
H1 $D^{*\pm}$			$15.1 \pm 1.8^{+2.4}_{-2.0} \pm 1.2$ nb
H1 D^0			$19.5 \pm 2.6^{+2.7}_{-2.5}{}^{+1.6}_{-1.2}$ nb
NLO MRSH	1.3		11.3 nb
NLO MRSH	1.7		8.1 nb
H1 F_2 fit	1.5		13.6 nb

Table 1 summarizes the experimental results together with the predictions of the NLO calculations [2,3] for different parton densities and different values of m_c. The extrapolation in phase space is calculated in NLO using the GRV-H0 gluon density in case of the ZEUS data while the acceptance for the H1 data is calculated using the MRSH gluon density. In both cases $m_c = 1.5$ GeV and $\mu = \sqrt{Q^2 + 4m_c^2}$ are chosen. The charm fragmentation is described by the Peterson function with $\epsilon_c = 0.035 \pm 0.009$ (ZEUS) and the symmetric Lund function with the standard setting of JETSET 7.4 (H1). The errors correspond to the statistical error, the experimental systematic uncertainty and the model dependent uncertainty respectively. The latter accounts for the possible variation of the relevant parameters of the model used for extrapolation, i.e. $xg(x, Q^2)$, m_c, μ and the fragmentation function. In the region of overlap the cross section inferred from the $D^{*\pm}$ analyses are in reasonable agreement. The cross section from the D^0 analysis is higher but the few statistics do not allow any conclusion to be drawn.

In table 1 the data is also compared with predictions from analytic NLO calculations for different gluon densities and different values of m_c. Also given are LO analytic calculations and the result from the LO Monte Carlo. In addition H1 has performed a NLO calculation using the gluon density as derived from the analysis of scaling violations of F_2 [17]. Within reasonable variations of the most relevant parameter, m_c, the NLO calculations tend to be below the experimental results except for the gluon density obtained from the NLO F_2 fit. The LO analytic calculation and the LO MC yield cross sections generally in better agreement with the data.

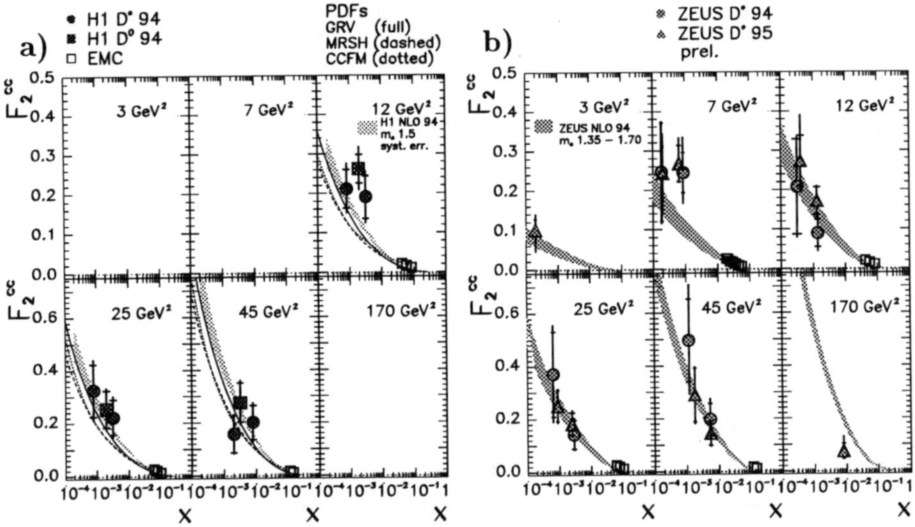

Fig. 4. *The charm contribution* $F_2^{c\bar{c}}$ *to the proton structure function as derived (a) from the* D^{*+} *(dots) and* D^0 *analysis (squares) by the H1 collaboration and (b) from the* D^{*+} *by the ZEUS collaboration from the 1994 data (dots) and the 1995 data (triangles). The inner/outer error bars refer to the statistical/total errors. The EMC data are also shown (open boxes). The shaded band represent the predictions from the NLO fits to the H1 and ZEUS* F_2 *measurements respectively. The ZEUS 1995 data and the prediction of the NLO fit to the* F_2 *measurement are preliminary. In (a) predictions from NLO calculations based on GRV-HO (full line), MRSH (dashed line) and CCFM (dotted line) gluon densities using a charmed quark mass of* $m_c = 1.5$ *GeV are included*

2.4. Charm contribution $F_2^{c\bar{c}}$ *to the proton structure*

The charm contribution $F_2^{c\bar{c}}$ to the structure function is obtained by using the expression for the one photon exchange cross section for charm production

$$\frac{d^2\sigma^{c\bar{c}}}{dx dQ^2} = \frac{2\pi\alpha^2}{Q^4 x}\left[1 + (1-y)^2\right]\ F_2^{c\bar{c}}(x, Q^2)\ . \tag{1}$$

The contribution from Z exchange is expected to be small in the kinematic range explored presently. The $F_L^{c\bar{c}}$ contribution has been estimated to be negligible [18,19]. Therefore both $F_3^{c\bar{c}}$ and $F_L^{c\bar{c}}$ are neglected.

The $F_2^{c\bar{c}}$ measurements from H1 and ZEUS are displayed in figure 4 together with the results from the EMC collaboration at large x. Within the limited statistics the results on $F_2^{c\bar{c}}$ from H1 and ZEUS are in good agreement. The measurements at HERA extend the range of the $F_2^{c\bar{c}}$ measurements by two orders of magnitude towards smaller x values. The comparison of the HERA and EMC data reveals a steep rise of $F_2^{c\bar{c}}$ with decreasing x. H1 has derived a ratio of $F_2^{c\bar{c}}/F_2$ of

$$\left\langle F_2^{c\bar{c}}/F_2 \right\rangle = 0.237 \pm 0.021 \pm 0.041\ , \tag{2}$$

which is an increase of about one order of magnitude of the overall charm contribution compared to the EMC result. This is consistent with the measured rise of the gluon distribution towards low x and the dominance of the BGF process observed here. Within the accuracy of the charm data this ratio seems to be constant in the kinematic range explored in the H1 analysis.

The data are also compared to the NLO calculations of $F_2^{c\bar{c}}$ based on the gluon densities extracted from the NLO fits to the inclusive F_2. In figure 4a the prediction from the H1 F_2 data, using a charmed quark mass of $m_c = 1.5$ GeV, is shown by an error band which includes the propagation of the statistical and the uncorrelated systematic errors on the total F_2 data through the fitting procedure. The uncertainty in the prediction of $F_2^{c\bar{c}}$ due to variations of m_c is indicated by the error band in figure 4b for the ZEUS F_2 data. The upper and lower limits correspond to $m_c = 1.35$ GeV and $m_c = 1.7$ GeV respectively. No error propagation in the fit is performed. In the region of overlap in Q^2 the H1 result of this fit favors a slightly larger gluon density at small x compared to the ZEUS measurement.

Figure 4a also includes the predictions from the NLO calculation based on GRV-HO, MRSH and CCFM [20] gluon density calculations. Although the procedures for treating heavy flavors and for determination of the the gluonic content in the proton are quite different for these three parton density calculations, the resulting predictions for $F_2^{c\bar{c}}$ are very similar. The accuracy of the data in terms of statistical and systematic uncertainty is by far not sufficient to be able to get more insight on the gluon in the proton from purely inclusive charm measurements.

3. Future

Although the charm production cross section is found to be about 25% of the total DIS cross section at HERA, the number of identified charm events is fairly small at present. For the results presented here heavy flavor tagging is performed by the reconstruction of charmed hadrons. From the experimental point of view one advantage of this method is the small contribution to the systematic uncertainties due to hadronization effects. Furthermore, as a consequence of the hard fragmentation of charm quarks, the reconstructed charmed hadron gives direct access to the parent charm quark. This allows the study of the production dynamic of heavy quarks almost free of background from other processes. The drawback of this tagging method is the small tagging efficiency ($\epsilon_{tag} \approx \mathcal{O}(10^{-3})$) due to the small probability that a heavy quark fragments into a specific heavy flavored hadron which subsequently decays into the desired decay mode, e.g. for the chain $c \to D^{*\pm}X \to D^0\pi_s^{\pm}X \to (K^{\mp}\pi^{\pm})\pi_s^{\pm}X$ this probability is only $6.7 \cdot 10^{-3}$.

In order to increase ϵ_{tag} of heavy flavor production a Silicon Vertex Detector is installed in H1 this year (1997) and will be installed in ZEUS together with the luminosity upgrade of the HERA machine in 1999/2000. Such a device enables the re-

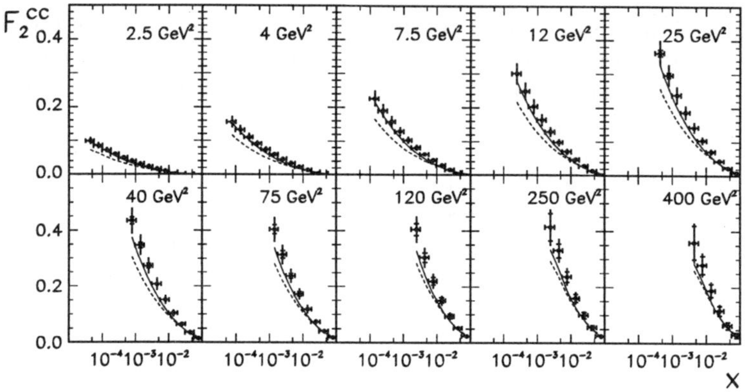

Fig. 5. *The charm contribution $F_2^{c\bar{c}}$ as expected for an integrated luminosity of $\mathcal{L}_{int} \approx 200\ pb^{-1}$ using fully reconstructed charmed hadrons together with the Silicon Vertex Detector.*

construction of the decay vertices of heavy flavored hadrons and will thereby suppress the combinatorial background in the mass reconstruction of heavy flavored hadrons. The reduction of background will give access to a large variety of decay modes with good signal to background ratios. Considering the analysis of all accessible decay modes together ϵ_{tag} will be increased by about one order of magnitude.

Alternatively heavy flavor events may be tagged by secondary vertices (SVTX) having high significance without explicit reconstruction of heavy flavored hadrons. This method will lead to tagging efficiencies comparable or even superior to the efficiency expected from the mass reconstruction method. Due to the absence of a mass constraint the SVTX method is expected to have larger background from faked secondary vertices. A much better understanding of fragmentation, detector and reconstruction performance is needed in this case to achieve experimental systematic uncertainties comparable to those of the mass reconstruction method. Furthermore it will become more difficult to access exclusive quantities of the production dynamics since the strong correlation between the heavy flavored hadron and the parent quark gets lost or will be weakened in partly reconstructed heavy flavored hadrons.

Figure 5 shows the expected accuracy of $F_2^{c\bar{c}}$ measurements for an integrated luminosity of $\mathcal{L}_{int} \approx 200$ pb^{-1} using fully reconstructed charmed hadrons together with the Silicon Vertex Detector. The data points show the expectation based on the gluon density extracted from the H1 NLO fit to the total F_2 for $m_c = 1.5$ GeV and $\mu^2 = Q^2 + 4m_c^2$. The inner/outer error bars refer to the statistical/total errors. Also shown are the NLO predictions for the GRV-HO (full line) and MRSH (dashed line) parton densities and the same charm mass and scales. At low Q^2 the measurement will be limited by the experimental uncertainty while at large Q^2 the precision will be limited by statistics. The ultimate precision of 10 % will allow these parameterizations of the gluon density to be distinguished.

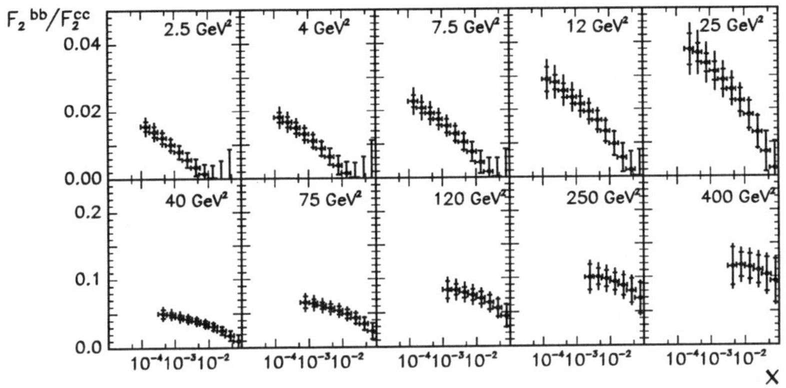

Fig. 6. *The ratio* $F_2^{b\bar{b}}/F_2^{c\bar{c}}$ *as expected for an integrated luminosity of* $\mathcal{L}_{int} \approx 500$ pb^{-1} *using fully reconstructed charmed hadrons together with the Silicon Vertex Detector.*

Figure 6 shows the expected accuracy of the ratio $F_2^{b\bar{b}}/F_2^{c\bar{c}}$ for an integrated luminosity of $\mathcal{L}_{int} \approx 500$ pb^{-1} using fully reconstructed charmed hadrons together with the Silicon Vertex Detector [19]. Even for a luminosity of $\mathcal{L}_{int} \approx 500$ pb^{-1} the measurement of this ratio will be limited by statistics in the full x and Q^2 plane. For a given Q^2 or x the ratio is expected to rise with decreasing x or increasing Q^2, respectively. A mean value of $\left\langle F_2^{b\bar{b}}/F_2^{c\bar{c}} \right\rangle \approx 0.02$ is predicted.

4. Conclusions

First results on charm production in deeply inelastic *ep* scattering at HERA have been presented. The investigation of differential D cross sections has shown that the production dynamics of charmed mesons in the current and central fragmentation region may be described by the boson gluon fusion process. The contribution from flavor excitation of charm in this region is found to be negligible at $\langle Q^2 \rangle \approx 25$ GeV ^{2}. This observation sets constraints on the kinematic region where charm may be treated as massless parton in the proton.

The measurement of $F_2^{c\bar{c}}(x, Q^2)$ at small x reveals a strong rise of $F_2^{c\bar{c}}(x, Q^2)$ from the fixed target to the HERA regime. Agreement is observed between the measured $F_2^{c\bar{c}}$ and the results from the NLO fits to the inclusive F_2. Averaged over the kinematic range currently explored a ratio $\left\langle F_2^{c\bar{c}}/F_2 \right\rangle = 0.237$ is observed which is one order of magnitude larger than at large x.

An estimation on the accuracy for the exploration of charm and beauty contribution $F_2^{c\bar{c}}$ and $F_2^{b\bar{b}}$ to the structure of the proton for an integrated luminosity of more than 200 pb^{-1} has been summarized. For $Q^2 \leq 100$ GeV^2 the charm analysis will be systematically limited. The heavy flavor data will allow a direct measurement of the gluon density in the proton to be performed. A precision on $xg(x, Q^2)$ of 10 % may be achieved from charm data, which is competitive with other methods. Despite of the smallness of the ratio $F_2^{b\bar{b}}/F_2^{c\bar{c}}$ high luminosity running will allow detailed studies of beauty production dynamics.

5. References

1. A. Ali *et al.,* in *Proc. of the HERA Workshop*, Hamburg, Vol. 1,393 (1988), A. Ali and D. Wyler, in *Proc. of the Workshop "Physics at HERA"*, Hamburg, Vol. 2, 667 (1991) (and references therein).
2. E. Laenen *et al.*, *Nucl. Phys.***B392**162, 229, S. Riemersma *et al.*, *Phys. Lett.* **B347** (1996) 143.
3. B.W. Harris and J. Smith, *Nucl. Phys.***B452** (1995) 109, FSU-HEP-970527, [hep-ph/9706334].
4. M.A.G. Aivazis *et al.*, *Phys. Rev.* **D50** (1994) 3085, 3102.
5. EMC Coll., J.J. Aubert *et al., Nucl. Phys.***B231** (1983) 31.
6. H1 Coll. C. Adloff *et al.*, *Z. Phys.* **C72** (1996) 539.
7. ZEUS Coll. J. Breitweg*et al.*, *DESY preprint, DESY 97-089.*
8. A.D. Martin, R.G. Roberts, and W.J. Stirling: *Phys.Lett.* **B306** (1993), 145, *Phys. Rev.* **D50** (1994) 6743, *Phys. Rev.* **D50** (1994) 6734.
9. A.D. Martin, *this proceedings* (and references therein).
10. M. Glück, E. Reya, and A. Vogt: *Z. Phys.* **C 53** (1992) 127, M. *Z. Phys.* **C 67** (1995) 433, A. Vogt, *this proceedings* (and references therein).
11. J. Smith, *this proceedings* (and references therein).
12. A.D. Martin *et al.*,DPT-96-102 and hep-ph/9612449, M. Buza *et al.*, DESY 96-258 and hep-ph/9612398.
13. G. Ingelman, J. Rathsman, and G.A. Schuler:*DESY preprint*, DESY 96-058, and HEP-PH/9605285.
14. G. Ingelman: *"LEPTO version 6.1 - The Lund Monte Carlo for Deep Inelastic Lepton-Nucleon Scattering"*, TSL/ISV-92-0065.
15. T. Sjöstrand: *"PYTHIA 5.7 and JETSET 7.4 Physics and Manual"*, CERN-TH.7112/93.
16. H. Abramowic *et al.*, CDHS Coll.: *Z. Phys.* **C15** (1982) 19, N. Ushida *et al.*, E531 Coll.: *Phys. Lett.* **B206** (1988) 380.
17. S. Aid *et al.*, H1 Coll.:*Nucl. Phys.* **B470** (1996) 3.
18. E. Leanen *et al.*,*Nucl. Phys.* **B392** (1993) 162, S. Riemersma *et al.*, *Proc. of the Workshop "Future Physics at HERA"*, DESY, Hamburg Vol. 1 (1995/96) 393.
19. K. Daum *et al.*, *Proc. of the Workshop "Future Physics at HERA"*, DESY, Hamburg Vol. 1 (1995/96) 89.
20. J. Kwieciński, A.D. Martin and P.L. Sutton, *Phys. Rev.* **D53** (1996) 6094, *Z. Phys.* **C71** (1996) 585, G.P. Salam, *this proceedings* (and references therein).

THEORETICAL ASPECTS OF HEAVY-FLAVOUR ELECTROPRODUCTION

J. SMITH *

Deutsches Electronen-Synchrotron DESY, Theory Group,
Notkestrasse 85, D-22607 Hamburg, Germany
E-mail: jacksmit@mail.desy.de

ABSTRACT

We discuss three theoretical schemes to describe charm quark electroproduction.

1. Introduction

The study of charm electroproduction has become an important issue in the extraction of parton densities in the proton. The reason is that the charm content $F_{2,c}(x,Q^2,m^2)$, where $m = m_{\rm charm}$, has grown from around one percent of the contribution to the structure function $F_2(x,Q^2)$ in the x and Q^2 region of the EMC experiment[1] to around twenty-five percent in the x and Q^2 region of the H1[2] and ZEUS[3] experiments at HERA. Therefore the analysis of parton densities in the proton can no longer be done without considering c-quark production.

Let us begin with a brief review of some of the technical points in the calculations of c-quark electroproduction in QCD. We consider $F_{2,c}(x,Q^2,m^2)$ which is usually much larger than $F_{L,c}(x,Q^2,m^2)$. We work in a three flavour number scheme (TFNS) where the u, d and s are light mass quarks ($n_f = 3$). The c $\bar{\rm c}$ pair is produced from the gluon by the Bethe-Heitler process (photon-gluon fusion) in leading order (LO). In next-to-leading order (NLO) both the Bethe-Heitler and the Compton processes involve all light mass quarks (and antiquarks) and the gluon. A NLO calculation organizes the contributions from the various Feynman diagrams according to the following formula

$$F_{2,c}(3,x,Q^2,m^2) = x\int_x^{z_{\rm max}} \frac{dz}{z}\Big\{\frac{1}{3}\sum_{k=1}^{3} e_k^2\Big[\Sigma\Big(3,\frac{x}{z},\mu^2\Big)L_{2,q}^{\rm S}\Big(3,z,\frac{Q^2}{m^2},\frac{m^2}{\mu^2}\Big) \tag{1}$$
$$+G\Big(3,\frac{x}{z},\mu^2\Big)L_{2,g}^{\rm S}\Big(3,z,\frac{Q^2}{m^2},\frac{m^2}{\mu^2}\Big) + \Delta\Big(3,\frac{x}{z},\mu^2\Big)L_{2,q}^{\rm NS}\Big(3,z,\frac{Q^2}{m^2},\frac{m^2}{\mu^2}\Big)\Big]\Big\}$$
$$+xe_c^2\int_x^{z_{\rm max}}\frac{dz}{z}\Big[\Sigma\Big(3,\frac{x}{z},\mu^2\Big)H_{2,q}^{\rm S}\Big(3,z,\frac{Q^2}{m^2},\frac{m^2}{\mu^2}\Big) + G\Big(3,\frac{x}{z},\mu^2\Big)H_{2,g}^{\rm S}\Big(3,z,\frac{Q^2}{m^2},\frac{m^2}{\mu^2}\Big)\Big],$$

where e_c is the charge of the c-quark. The variable z is the partonic longitudinal momentum fraction and $z_{\rm max} = Q^2/(Q^2+4m^2)$. The function Δ is the non-singlet (with respect to the flavour SU(3) group) combination of light-mass parton densities, at the scale μ. The function Σ is the singlet combination of these densities while

---

*On leave from The Institute for Theoretical Physics, SUNY at Stony Brook, NY 11794-3840, USA

G is the density for the gluon. Further $L_{2,k}$ and $H_{2,k}(k = q, g)$ are the singlet and non-singlet heavy-quark coefficient functions at scale μ. Since they contain factors of α_s we have explicitly indicated that they depend on $n_f = 3$. The $L_{2,k}$ describe the reactions where the virtual photon couples to the light quarks (u, d, and s) and the $H_{2,k}$ the reactions where it couples to the c-quark.

The NLO contributions to Eq.(1) were originally calculated in[4] yielding single-particle inclusive transverse momentum and rapidity distributions of the c-quark[5]. The functions $L_{2,k}$ and $H_{2,k}$ were only available in the form of two-dimensional integrals. To speed up the computation of Eq.(1) we made two-dimensional grids of values for $L_{2,k}$ and $H_{2,k}$ together with an interpolation routine in[6]. The NLO calculation was repeated in a completcly exclusive fashion in[7] so that one can plot all the distributions containing the c-quark, the $\bar{c}$ antiquark and the additional parton. This program incorporates a fragmentation function so that one can calculate distributions for $D^{*\pm}$ mesons. These programs have been used by many authors[8,9,10,11] to discuss the sensitivity of the NLO results to changes in parton densities, renormalization/factorization scales, and the mass of the c-quark. Finally the asymptotic formulae for $L_{2,k}$ and $H_{2,k}$ in the limit $Q^2 \gg m^2$ were calculated in[12]. The latter formulae enable one to discuss the transition from the calculated c-quark rate in NLO perturbation theory in the TFNS with $n_f = 3$ to a four flavour number scheme (FFNS) with $n_f = 4$ and a massless c-quark[13,14].

Now we outline the scheme used for these NLO calculations. The renormalization of the virtual graphs with loops containing light quarks was done in the $\overline{\mathrm{MS}}$-scheme while loops with heavy quarks were subtracted at zero external momentum. This scheme was originally proposed in[15] so that the light quark sector remains the same as in a purely massless pQCD calculation. Hence there is a matching condition at the scale $\mu = m$ for both α_s and for the parton densities. To explain this consider QCD with only massless quarks. Then the number of flavours enters via the factors n_f in the β-function and in the gluon splitting function P_{gg}. We solve the equation

$$\frac{\partial}{\partial \ln \mu^2}\Big(\frac{\alpha_s}{\pi}\Big) = -\beta_0\Big(\frac{\alpha_s}{\pi}\Big)^2 - \beta_1\Big(\frac{\alpha_s}{\pi}\Big)^3 - \dots \tag{2}$$

where

$$\beta_0 = \frac{1}{4}\Big(\frac{11}{3}C_A - \frac{4}{3}T_f n_f\Big) \quad , \quad \beta_1 = \frac{1}{16}\Big(\frac{34}{3}C_A^2 - 4C_F T_f n_f - \frac{20}{3}C_A T_f n_f\Big) , \tag{3}$$

by introducing the $\Lambda_{\overline{\mathrm{MS}}}(n_f)$. The solution of Eq.(2) is

$$\frac{\alpha_s(n_f,\mu)}{\pi} = \frac{1}{\beta_0 \ln(\mu^2/\Lambda^2_{\overline{\mathrm{MS}}})} - \frac{\beta_1}{\beta_0^3}\frac{\ln(\ln(\mu^2/\Lambda^2_{\overline{\mathrm{MS}}}))}{\ln^2(\mu^2/\Lambda^2_{\overline{\mathrm{MS}}})} + .. \tag{4}$$

A matching condition is therefore necessary to go from a TFNS with $n_f = 3$ to a FFNS with $n_f = 4$. We have to change the value of $\Lambda_{\overline{\mathrm{MS}}}$ as we cross the scale $\mu = m$ to keep the condition that $\alpha_s(4, \mu = m) = \alpha_s(3, \mu = m)$. However heavy particles cannot be ignored in QCD because they exist in vacuum fluctuations. It is well-known[16] that effects from very heavy particles can be absorbed into an equivalent field theory at a smaller scale by an appropriate renormalization of the coupling constants and fields (decoupling of heavy flavours). QCD is an unbroken gauge theory which does not have heavy particle decoupling when renormalized in the $\overline{\mathrm{MS}}$-scheme as all quarks have $m = 0$. However the top quark cannot be considered massless and virtual top-quark loops yield explicit terms in $\ln \mu^2/m^2$ (here $m = m_{\mathrm{top}}$) so there has to be a matching condition between say a five flavour theory and a six flavour theory. This two-loop matching condition on α_s was first worked out in[17] and corrected in[18]. It reads

$$\frac{\alpha_s(6,\mu)}{\pi} = \frac{\alpha_s(5,\mu)}{\pi} + \left(\frac{\alpha_s(5,\mu)}{\pi}\right)^2 \frac{1}{3} T_f \ln\left(\frac{\mu^2}{m^2}\right) + \left(\frac{\alpha_s(5,\mu)}{\pi}\right)^3 \Big[\frac{1}{9} T_f^2 \ln^2\left(\frac{\mu^2}{m^2}\right)$$
$$+ \frac{1}{12}\left(5C_A T_f - 3C_F T_f\right) \ln\left(\frac{\mu^2}{m^2}\right) + \frac{13}{48} T_f C_F - \frac{2}{9} T_f C_A\Big] + \ldots . \quad (5)$$

Note that $\alpha_s(6, \mu = m) = \alpha_s(5, \mu = m)$ in order α_s but not in order α_s^2.

The parton densities must satisfy similar matching conditions to Eq. (5). In a NLO calculation the collinear singularities are regularized by n-dimensional regularization. We set the mass factorization scale equal to the renormalization scale μ for simplicity. Only the gluon-gluon splitting function

$$P_{gg}(n_f, z) = 2C_A\Big[z\Big(\frac{1}{1-z}\Big)_+ + \frac{1-z}{z} + z(1-z)\Big] + \frac{1}{6}(11C_A - 4T_f n_f)\delta(1-z)\,, \quad (6)$$

depends on the number of light quarks n_f. The c-quark is considered heavy with mass $m = m_{\mathrm{charm}}$ and we calculate the $L_{2,k}$ and $H_{2,k}$ in Eq. (1) in the TFNS. There is no charm density at the scale $\mu = m$ but it does exist for $\mu > m$. This means that the light flavour densities $f_k(x, \mu)$, $k = u, d, s, \bar{u}, \bar{d}, \bar{s}$ satisfy

$$f_k(4, x, \mu = m) = f_k(3, x, \mu = m)\,. \quad (7)$$

One can examine the DGLAP equations[19] for the parton densities when one chooses a scale with $\mu > m$. By taking the difference of the solutions with $n_f = 4$ and $n_f = 3$ and using Eq. (6) one finds that only the gluon density changes in order α_s, namely

$$f_g(4, x, \mu^2) = f_g(3, x, m^2)\Big[1 - \frac{1}{3\pi} T_f \alpha_s(3, \mu) \ln \frac{\mu^2}{m^2}\,,\Big] \quad (8)$$

and there is a c-quark density, proportional to $\alpha_s(3, \mu)\ln(\mu^2/m^2)$, which grows at the expense of the gluon density. The changes in the light flavour densities begin in order α_s^2. Hence at scales $\mu \gg m$ we can switch to a FFNS which contains a (massless) c-quark density and a different momentum sum rule. Eq.(8) is a matching condition on the gluon density between the TFNS and the FFNS at the scale $\mu = m$.

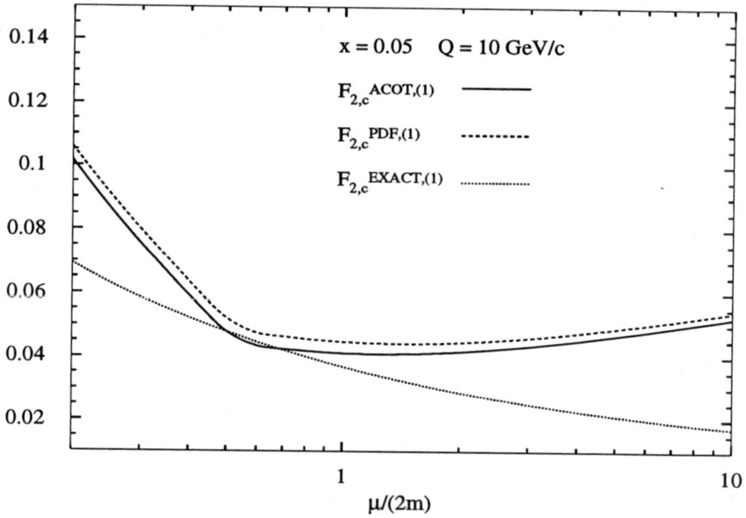

Figure 1: Scale dependence of $F_{2,c}(x = 0.05, Q = 10\,\mathrm{GeV}/c)$.

A way to implement a smooth matching of $F_{2,c}(x, Q^2, m^2)$ from the TFNS in Eq. (1) to a charm density description in the FFNS was proposed in[20] and we call it the ACOT scheme. Their scale choice, which we will use in the rest of this article, is

$$\begin{aligned} \mu^2 &= m^2 + kQ^2(1 - m^2/Q^2)^n \quad \text{for} \quad Q^2 > m^2\,, \\ &= m^2 \quad \text{for} \quad Q^2 \le m^2\,, \end{aligned} \tag{9}$$

with $k = 0.5$, $n = 2$ and $m = 1.5\,(\mathrm{GeV}/c^2)$. They proposed that $F_{2,c}^{\mathrm{ACOT},(1)}(x, Q^2, m^2)$ should consist of three terms. The first one is the TFNS LO process in Eq.(1), which has the correct kinematics near threshold. It contains the convolution of the gluon density in the proton with the LO hard scattering cross section. The third term is a FFNS charm density, i.e., the coefficient function in the proton structure function $F_2(x, Q^2)$ is simply a δ-function, and this should be the best description at $\mu^2 \gg m^2$. Finally the second term is the product of two convolutions. Namely that of the gluon density in the proton with the massless c-density in the gluon and with the LO hard scattering term (a δ-function). This second term is taken with a negative sign so that it cancels the FFNS c-density term at small scales $\mu^2 \approx m^2$ while it tends to cancel the TFNS term when $\mu^2 \gg m^2$. The sum of these three terms is more stable under a variation of the scale μ than each term separately. In Fig.1 we show the scale dependence of $F_{2,c}(x, Q^2, m^2)$ for the TFNS LO term from Eq.(1) (labelled EXACT,(1)), the ACOT description (labelled ACOT,(1) and the FFNS charm density (labelled PDF,(1)). We choose the same CTEQ parton densities[21] as in[20]. The monotonic decrease in the EXACT result, which is due to the decrease in α_s, is modified in the other two approaches. In Fig.2 we plot the three results for

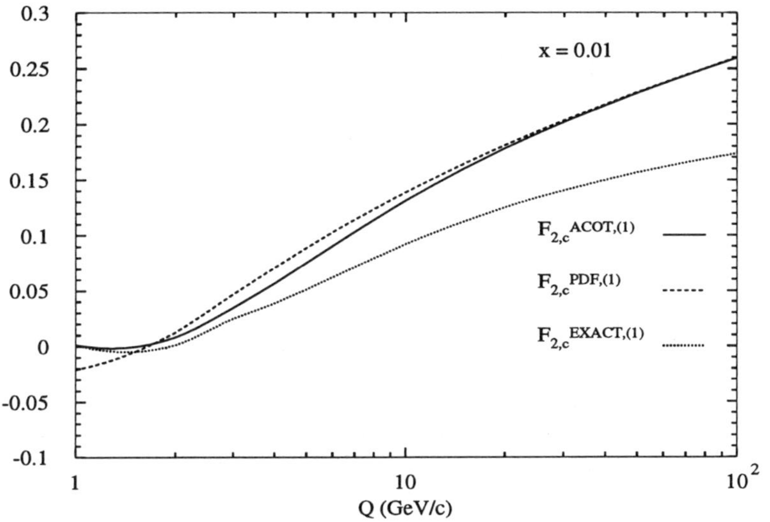

Figure 2: $F_{2,c}(x = 0.01)$ as a function of Q.

$F_{2,c}(x = 0.01, Q^2)$ versus $Q = \sqrt{Q^2}$. One sees that there is a large difference between the EXACT result (TFNS) and the PDF charm density result (FFNS) especially as Q increases. However the charm density result cannot be used at small Q where it is negative. As expected the ACOT result interpolates between the EXACT result at small Q and the charm density PDF result at large Q. The ACOT formula has been incorporated into recent parton densities (the set CTEQ4HQ[22]). Also there are now three light flavour (CTEQ4F3) and four light flavour (CTEQ4F4) global fits to parton densities. The former is basically equivalent to the parton density set in GRV94[23]. An examination of the CTEQ4HQ set shows that the gluon is diminished from previous sets[21] and that all the light quark densities have changed, not just the one for charm. This is due to the interplay between the gluon and the light flavour densities in making a global fit to all the experimental data.

As the ACOT description contains algebraic terms in m, it is not obvious how to generalize it to higher orders in α_s. (One proposal has been made in [24] which we cannot discuss here for lack of space). One would like to retain as much as possible a pQCD description based on massless β and γ functions in the renormalization group equation, and work in the $\overline{\text{MS}}$ scheme. In[13] we have worked out the two-loop matching conditions on the flavour densities by the following procedure. Analytic formulae for the functions $H_{2,k}$ in Eq.(1) are not known. In the limit that $Q^2 \gg m^2$ they reduce to expressions like $a(z)\ln^2(Q^2/m^2)+b(z)\ln(Q^2/m^2)+c(z)$ (where we call them $H_{2,k}^{\text{ASYMP}}$). If we can find $a(z), b(z)$ and $c(z)$ then we can mass factorize the logarithms into terms involving $\ln(Q^2/\mu^2)$ and $\ln(\mu^2/m^2)$. If we then add the result to the two-loop light-flavour corrections[25] to the proton structure function $F_2(3, x, Q^2, \mu^2)$ then the terms

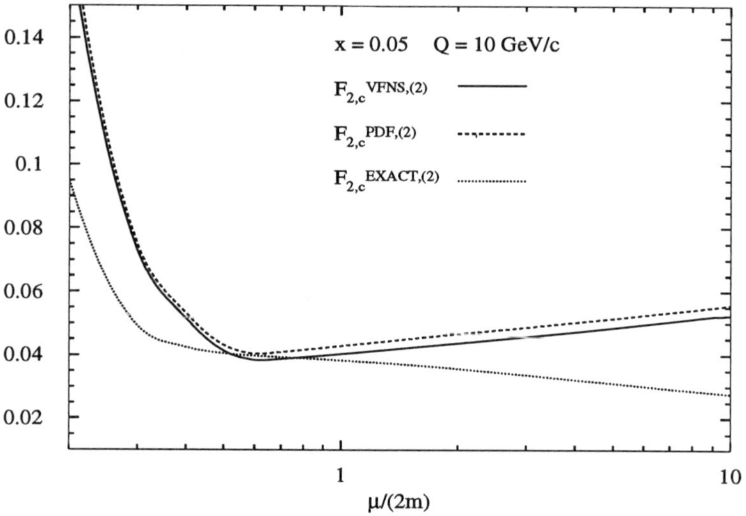

Figure 3: Scale dependence of $F_{2,c}(x = 0.05, Q = 10\,\mathrm{GeV}/c)$.

in $\ln(Q^2/\mu^2)$ should recombine to yield the two-loop corrections to $F_2(4, x, Q^2, \mu^2)$. Note that $F_{2,c}$ must be totally inclusive so we have to add the light quark vertex corrections where the gluon propagator contains a charm quark loop, which yields terms in $\delta(1-z)$. When the correct pieces have been identified we are left over with terms in $\ln^2(\mu^2/m^2)$ and $\ln(\mu^2/m^2)$ which cannot be absorbed into the known matching conditions on α_s. We can then define two-loop matching conditions for the parton densities to incorporate these pieces, analogous to the two loop relations for α_s in Eq. (5). We actually derived $L_{2,k}^{\mathrm{ASYMP}}$, $H_{2,k}^{\mathrm{ASYMP}}$ and the relations between the parton densities by calculating all two-loop matrix elements of the operators in the OPE, where one loop contains a massless quark and the other loop a massive quark.

The result is that above the scale $\mu = m$ we have new FFNS parton densities ($n_f = 4$) which are *calculable* from a TFNS set ($n_f = 3$). All densities are changed. The light quark densities pick up additional terms in order α_s^2, which are numerically rather small, but are required for the FFNS parton densities to satisfy the momentum sum rule. The FFNS charm density is given in terms of the previous TFNS densities and involves terms in order α_s and α_s^2, i.e.,

$$\begin{aligned}
f_{c+\bar{c}}^{\mathrm{FFNS}}(4, x, \mu^2) &\equiv f_4(4, x, \mu^2) + f_{\bar{4}}(4, x, \mu^2) \\
&= \Big(\frac{\alpha_s(\mu^2)}{4\pi}\Big)^2 \int_x^1 \frac{dz}{z} \Sigma\Big(3, \frac{x}{z}, \mu^2\Big) \tilde{A}_{cq}^{\mathrm{PS},(2)}\Big(z, \frac{\mu^2}{m_c^2}\Big) \\
&+ \int_x^1 \frac{dz}{z} G\Big(3, \frac{x}{z}, \mu^2\Big) \Big[\Big(\frac{\alpha_s(\mu^2)}{4\pi}\Big) \tilde{A}_{cg}^{\mathrm{S},(1)}\Big(z, \frac{\mu^2}{m_c^2}\Big) + \Big(\frac{\alpha_s(\mu^2)}{4\pi}\Big)^2 \tilde{A}_{cg}^{\mathrm{S},(2)}\Big(z, \frac{\mu^2}{m_c^2}\Big)\Big], \quad (10)
\end{aligned}$$

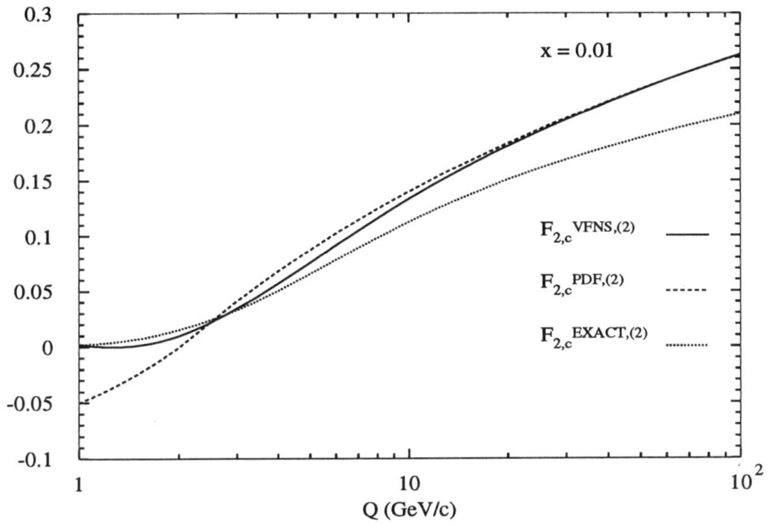

Figure 4: $F_{2,c}(x = 0.01)$ as functions of Q.

where

$$\tilde{A}_{cq}^{\mathrm{PS},(2)}\left(\frac{m^2}{\mu^2}\right) = C_F T_f \Bigg\{ \left[-8(1+z)\ln z - \frac{16}{3z} - 4 + 4z + \frac{16}{3}z^2 \right] \ln^2 \frac{m^2}{\mu^2}$$
$$+ \left[8(1+z)\ln^2 z - \left(8 + 40z + \frac{64}{3}z^2 \right) \ln z - \frac{160}{9z} + 16 - 48z + \frac{448}{9}z^2 \right] \ln \frac{m^2}{\mu^2}$$
$$+(1+z)\left[32\mathrm{S}_{1,2}(1-z) + 16\ln z \mathrm{Li}_2(1-z) - 16\zeta(2)\ln z - \frac{4}{3}\ln^3 z \right] + \left(\frac{32}{3z} + 8 \right.$$
$$\left. -8z - \frac{32}{3}z^2 \right) \mathrm{Li}_2(1-z) + \left(-\frac{32}{3z} - 8 + 8z + \frac{32}{3}z^2 \right) \zeta(2) + \left(2 + 10z + \frac{16}{3}z^2 \right) \ln^2 z$$
$$- \left(\frac{56}{3} + \frac{88}{3}z + \frac{448}{9}z^2 \right) \ln z - \frac{448}{27z} - \frac{4}{3} - \frac{124}{3}z + \frac{1600}{27}z^2 \Bigg\}. \tag{11}$$

The other contributions to Eq. (10) and the relations between the other parton densities are available in[13]. Eq. (11) shows that the charm density has a discontinuity at the scale $\mu = m$ when one considers order α_s^2 terms, just as in Eq. (5).

The above formula is for a FFNS renormalized in the $\overline{\mathrm{MS}}$ prescription. Using the FFNS one can now calculate $F_{2,c}^{\mathrm{PDF},(2)}(4, x, Q^2, m^2)$ in the region where $Q^2 \gg m^2$ and include the higher order QCD[25]. We expect this description to be valid in the region $Q^2 \gg m^2$ where the TFNS formula in Eq. (1) is dominated by the mass factorization logarithms. We have checked numerically that the asymptotic result for the NLO formulae, which we call $F_{2,c}^{\mathrm{ASYMP},(2)}(3, x, Q^2, m^2)$ is a good approximation to the exact NLO result in Eq.(1), which we call $F_{2,c}^{\mathrm{EXACT},(2)}(3, x, Q^2, m^2)$ when $Q^2 \approx 10\,m^2$. Hence

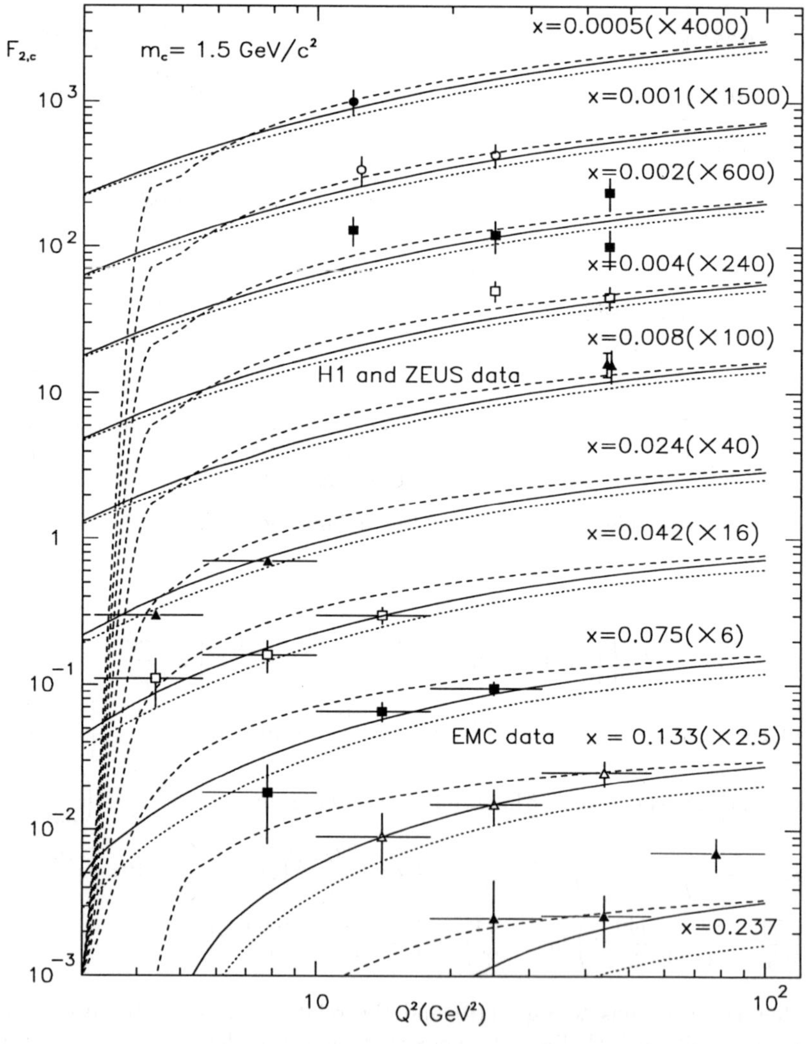

Figure 5: The functions $F_{2,c}^{\mathrm{PDF},(2)}(4, x, Q^2)$ dashed line, $F_{2,c}^{\mathrm{VFNS},(2)}(x, Q^2, m^2)$ solid line and $F_{2,c}^{\mathrm{EXACT},(2)}(3, x, Q^2, m^2)$ dotted line, plotted as a functions of Q^2 for fixed x. The data points are from the EMC, H1 and ZEUS experiments.

above $Q^2 \approx 10\,m^2$ we have the option of retaining the TFNS description for charm or to switching to the FFNS which sums large logarithms in $\ln(Q^2/\mu^2)$.

It now remains to find a formula which allows us to interpolate between the TFNS and the FFNS and is valid in all orders in α_s. In[13] and[14] we have proposed the variable flavour number scheme (VFNS) formula

$$F_{2,c}^{\mathrm{VFNS}}(x,Q^2,m^2) = F_{2,c}^{\mathrm{EXACT}}(3,x,Q^2,m^2) - F_{2,c}^{\mathrm{ASYMP}}(3,x,Q^2,m^2) + F_{2,c}^{\mathrm{PDF}}(4,x,Q^2)\ . \quad (12)$$

This formula is designed like the ACOT one. We have all the pieces to test it in order α_s^2. (Full details are available in[14]). The TFNS ASYMP term cancels the TFNS EXACT term at large μ leaving the FFNS charm density PDF term. However neither the ASYMP term nor the charm density PDF term have the correct kinematics (they also have different n_f and electric charges) to cancel each other exactly at small μ (near threshold). Using as input the same CTEQ densities[21] as in[20], we have calculated (1) $F_{2,c}^{\mathrm{EXACT},(2)}$, the TFNS formulae in Eq.(1) (2) $F_{2,c}^{\mathrm{VFNS},(2)}$, the VFNS formula in Eq.(12) and (3) $F_{2,c}^{\mathrm{PDF},(2)}$, the massless FFNS PDF result. In (3) we have evaluated the two convolutions to go from a TFNS to a FFNS and then to add the higher order corrections[25]. In Figs. 3 and 4 we show the analogous results to those in Figs. 1 and 2. One sees that the scale sensitivity is mostly reduced for the TFNS EXACT result. The PDF result is worse at small Q, which is a consequence of the fact that some of the higher order contributions in QCD are negative. The VFNS result Eq. (12) interpolates nicely between the other two. Finally in Fig. 5 we compare the three descriptions for $F_{2,c}$ with the data from the EMC, H1 and ZEUS experiments. Here we use the GRV92[26] parton densities. One sees that the VFNS interpolation is quite smooth as compared to that in[24]. The massless FFNS results fall dramatically at small Q, due to the choice of μ in Eq. (9). All three curves are within the present experimental errors of the H1 and ZEUS experiments. Therefore we will have to wait for more precise data before we can make a firm statement about which theoretical approach is the best.

2. Acknowledgements

The work reported here was carried out in collaboration with M. Buza, Y. Matiounine and W.L. van Neerven. The author thanks the Alexander von Humbolt Stiftung for an award to allow him to spend his Sabbatical leave in DESY.

1. J.J. Aubert *et al* (EMC Collab.), *Nucl. Phys.* **B452**, (1995) 109; see also B.W. Harris, J. Smith and R. Vogt, *Nucl. Phys.* **B461** (1996) 181.
2. C. Adloff *et al* (H1 Collab.), *Z. Phys.* **C72** (9196) 593.
3. M. Derrick *et al* (ZEUS Collab.), *Phys. Lett.* **B349** (1995) 225; M. Derrick *et al* (ZEUS Collab), XXVII Int. Conf. on HEP '96, Warsaw (1996); J. Breitweg *et al* (ZEUS Collab.) DESY-97-089.

4. E. Laenen, S. Riemersma, J. Smith and W.L. van Neerven, *Nucl. Phys.* **B392** (1993) 162.
5. E. Laenen, S. Riemersma, J. Smith and W.L. van Neerven, *Nucl. Phys.* **B392** (1993) 229.
6. S. Riemersma, J. Smith and W.L. van Neerven, *Phys. Lett.* **B347** (1995) 143.
7. B.W. Harris and J. Smith, *Nucl. Phys.* **B452** (1995) 109; B.W. Harris in *Proceedings of the American Physical Society, Division of Particles and Fields 1996 Meeting* in press; B.W. Harris and J. Smith, hep-ph/9706334.
8. M. Glück, E. Reya and M. Stratmann, *Nucl. Phys.* **B222** (1994) 37.
9. F.I. Olness and S. Riemersma, *Phys. Rev.* **D51** (1995) 4746.
10. A. Vogt, DESY 96-012, hep-ph/9101352.
11. K. Daum, S. Riemersma, B.W. Harris, E. Laenen and J. Smith in *Proceedings of the Workshop 1995/1996 on Future Physics at HERA, Hamburg, 1996*, edited by G. Ingelman et al. (Deutsches Electronen-Synchrotron, Hamburg, 1996), p. 89; E. Laenen, M. Buza, B.W. Harris, Y. Matiounine, R. Migneron, S. Riemersma, J. Smith and W.L. van Neerven ibid p. 393.
12. M. Buza, Y. Matiounine, J. Smith, R. Migneron and W.L. van Neerven, *Nucl. Phys.* **B472** (1996) 611.
13. M. Buza, Y. Matiounine, J. Smith, and W.L. van Neerven, DESY 96-278, to be published in *Z. für Physik* **C**.(1997).
14. M. Buza, Y. Matiounine, J. Smith, and W.L. van Neerven, DESY 97-124.
15. J.C. Collins, F. Wilczek and A. Zee, *Phys. Rev.* **D18** (1978) 42; J.C. Collins and W.K. Tung, *Nucl. Phys.* **B278** (1986) 934.
16. T. Appelquist and J. Carrazone, *Phys. Rev.* **D11** (1975) 2856.
17. W. Bernreuther and W. Wetzel, *Nucl. Phys.* **B197** (1982) 228; W. Bernreuther, *Annals of Physics* **151** (1983) 127.
18. S.A. Larin, T. van Ritbergen and J.A.M. Vermaseren, *Nucl. Phys.* **B438** (1995) 278; see also K.G. Chetyrkin, B.A. Kniehl and M. Steinhauser, hep-ph/9706430.
19. V.N. Gribov and L.N. Lipatov, *Sov. J. Nucl. Phys.* **15** (1972) 438, 675; G. Altarelli and G. Parisi, *Nucl. Phys.* **B126** (1977) 298; Yu. Dokshitser, *Sov. Phys. JETP* **46** (1977) 641.
20. M.A.G. Aivazis, J.C. Collins, F.I. Olness and W.K. Tung, *Phys. Rev.* **D50** (1994) 3102.
21. H.L. Lai et al. (CTEQ-collaboration), *Phys. Rev.* **D51** (1995) 4763.
22. H.L. Lai and W.K. Tung, *Z. Phys.* **C74** (1997) 463.
23. M. Glück, E. Reya and A. Vogt, *Z. Phys.* **C67** (1995) 433.
24. A.D. Martin, R.G. Roberts, M.G. Ryskin and W.J. Stirling, DTP/96/102.
25. E.B. Zijlstra and W.L. van Neerven, *Nucl. Phys.* **B383** (1992) 525.
26. M. Glück, E. Reya and A. Vogt, *Z. Phys.* **C53** (1992) 127.

INCLUSIVE D^* PHOTOPRODUCTION IN ep COLLISIONS AT HERA

C. COLDEWEY
Particle Physics Department, Weizmann Institute of Science
Rehovot 76100, Israel
E-mail: coldewey@mail.desy.de

On behalf of the H1 and ZEUS collaborations.

ABSTRACT

Results on inclusive D^* photoproduction at HERA are presented using 1994 and 1995 data samples of the H1 and ZEUS experiments. The total cross section $\sigma(\gamma p \to c\bar{c}X)$ has been measured in three different, but partly overlapping, kinematical regions with γp center-of-mass energies of $\langle W_{\gamma p}\rangle \approx 230\,\text{GeV}$, $\langle W_{\gamma p}\rangle \approx 200\,\text{GeV}$ and $\langle W_{\gamma p}\rangle \approx 115\,\text{GeV}$. With an integrated luminosity of about $6\,\text{pb}^{-1}$, differential D^* cross sections as a function of $p_\perp^{D^*}$, η^{D^*} and W, are given. The data are compared with two different next-to-leading order QCD calculations, which are based on a massive and a massless charm scheme. First results from a measurement of x_γ^{OBS} using a D^* dijet sample are given and compared with leading order Monte Carlo predictions.

1. Introduction

Charm quark production at HERA is dominated by photoproduction, where a quasi-real photon with negative four momentum squared, Q^2, close to zero is emitted by the incoming positron and interacts with the proton. At the high energy range of HERA, with a γp center-of-mass energy of $W \approx 200\,\text{GeV}$, charm photoproduction can be used to probe perturbative QCD calculations with a hard scale given by the charm mass m_c or the transverse mass $m_\perp = (m_c^2 + p_{\perp,c}^2)^{\frac{1}{2}}$ in kinematic regions with high transverse momenta of the produced partons.

1.1. Inclusive charm production

In leading order (LO) QCD, D^* photoproduction can be divided into two types of processes: the direct processes, where the photon behaves as a point-like particle, which couples directly to the quarks within the proton, and the resolved processes, where the photon is a source of partons, one of which scatters off a parton from the proton. Thus resolved photon-proton collisions are similar to the parton-parton interactions seen at hadron colliders. Charm quarks present in the parton distribution of the photon, as well as of the proton, lead to processes like $cg \to cg$, which are called charm flavour excitation. Diagrams of the boson-gluon fusion and two examples of resolved processes are shown in Fig.1.

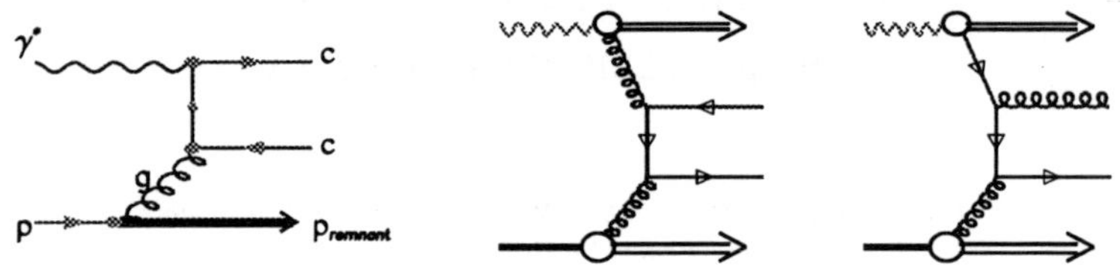

Fig. 1. Examples of leading order Feynman graphs contributing to charm photoproduction in e^+p collisions. Shown are the direct boson-gluon fusion and two resolved processes, gluon-gluon fusion and charm flavour excitation with charm from the photon.

In next-to-leading order (NLO) QCD only the sum of direct and resolved processes is unambiguously defined. Two types of NLO calculations using different approaches are available for comparison with measurements of charm photoproduction at HERA. In the massless charm approach[1,2] charm is treated as a massless parton and it is used as an active flavour within the structure function of the proton and the photon. In contrast, the massive charm approach[3] assumes light quarks to be the only active flavours and charm quarks appear only in the final state. Predictions of the total charm photoproduction cross section, can be calculated only in the massive charm scheme, where the charm quark mass $m_c >> \Lambda_{QCD}$ can be used as the scale for the perturbative calculation.

1.2. Charm fragmentation

Traditionally, inclusive charm production is studied from charmed hadron events, mostly by reconstruction of $D^{*\pm}$ mesons. The fragmentation of heavy quarks into heavy hadrons occurs at large distances where non-perturbative effects become important. Fragmentation is of both theoretical and experimental interest and different fragmentation schemes[1,2] are applied in the NLO calculations available for comparison with H1 and ZEUS data. In the massive charm approach, the fragmentation of charm quarks into D^* mesons was performed using the Peterson formula[4]:

$$f(z) \propto \left[z \left(1 - \frac{1}{z} - \frac{\epsilon_c}{(1-z)} \right)^2 \right]^{-1}$$

with $\epsilon_c = 0.06$ and using the branching ratio $B(c \to D^{*+}) = 0.260 \pm 0.021$ as measured by the OPAL collaboration[5]. Here z is the fraction of the charm quark momentum taken by the D^*. The value for the free parameter ϵ_c is taken from Chrin's analysis[6] of charm-meson production in the PETRA/PEP energy range. For the massless charm scheme predictions[1], as shown in Fig.5, the fragmentation process is described by the Peterson fragmentation function with evolution to higher order scales via Altarelli-Parisi equations[7]. The Peterson function is used only as input at the starting scale $\mu_0 = 2m_c$. In this calculation the fragmentation parameter was also taken to be $\epsilon_c = 0.06$. In a recent publication[8] within the same NLO fragmentation scheme the epsilon parameter was found to be $\epsilon_c = 0.0674$ by fitting LEP1 data on inclusive $D^{*\pm}$ production in e^+e^- annihilation. However, using a different input function of the form $f(z) \propto x^\alpha(1-x)^\beta$, as conventionally used for light-meson production, leads also to excellent agreement compared with results obtained by use of the Peterson fragmentation function.

2. Inclusive $D*$ production

Inclusive photoproduction of $D^{*\pm}(2010)$ mesons has been studied via the decay chain $D^{*+} \to D^0\pi_s^+$, where the D^0 decays via $K^-\pi^+$ and $K^-\pi^+\pi^-\pi^+$.[a] The small mass difference $\Delta M = M(D^*) - M(D^0) = 145.4\,\text{MeV}$ yields a soft pion from the D^* decay (π_s) and prominent signals just above the ΔM threshold, where phase space is highly suppressed. Pairs of tracks with opposite charge considered in turn to be a kaon or pion are selected in the mass range aound the D^0 and a track (π_s) with opposite charge to that of the kaon is added. Both experiments use a restricted kinematical region where tracking efficiency is high and systematic effects are well understood. The ZEUS region is $p_\perp(D^*) > 3.0\,\text{GeV}$, $-1.5 < \eta(D^*) < 1.0$ and $115 < W < 280\,\text{GeV}$. The pseudorapitity η is defined as $-\ln(tan(\frac{\theta}{2}))$, where the polar angle θ is measured with respect to the proton beam direction. The ZEUS analysis is restricted to photoproduction by removing Deep Inelastic Scattering D^* candidates, yielding events with $Q^2 \leq 4\,\text{GeV}^2$. H1 measured three data samples with $p_\perp(D^*) > 2.5\,\text{GeV}$ in the rapidity region $-1.5 < \hat{y}(D^*) < 1.0$:

44 m-tagged: $Q^2 \leq 0.006\,\text{GeV}^2$ and $67 < W < 150\,\text{GeV}$,

33 m-tagged: $Q^2 \leq 0.01\,\text{GeV}^2$ and $159 < W < 242\,\text{GeV}$,

untagged: $Q^2 \leq 4\,\text{GeV}^2$ and $95 < W < 268\,\text{GeV}$.

The events of the 33 m- and 44 m-tagged samples are required to have a positron signal in one of the positron tagger, located 33 m and 44 m from the interaction point in the positron beam direction, respectively. The ZEUS ΔM distribution for $1.80 \leq M(K\pi) \leq 1.92\,\text{GeV}$ and the $M(D^0)$ distribution for $0.143 \leq \Delta M \leq 0.148\,\text{GeV}$ of both decay modes, $D^0 \to K\pi$ and $D^0 \to K\pi\pi\pi$, are shown in Fig.2. In the $D^0 \to K\pi$ channel a clean D^* sample of ≈ 320 events is obtained after subtracting

[a] In this analysis the charge conjugated processes are also included.

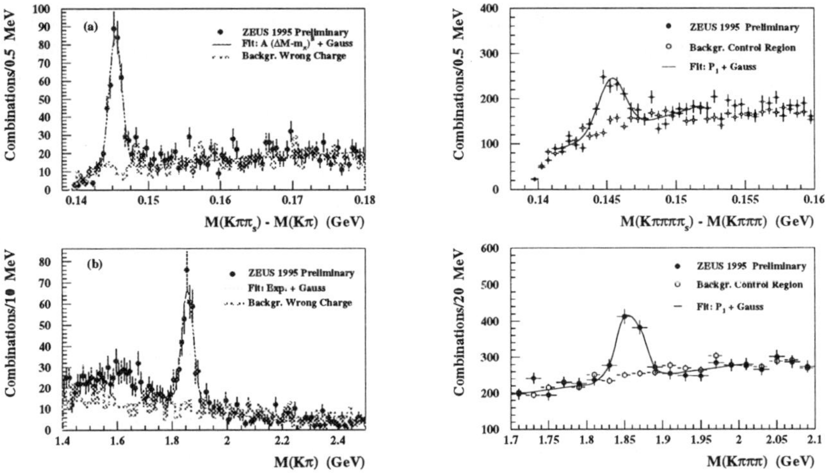

Fig. 2. ΔM and $M(D^0)$ distributions for the $D^0 \to K\pi$ and $D^0 \to K\pi\pi\pi$ decay channels of the ZEUS 1995 photoproduction data. The backgrounds are estimated by wrong charge combinations for the $D^0 \to K\pi$ mode and by control region in the $D^0 \to K\pi\pi\pi$ channel. The full lines are fits to the sum of a Gaussian and a functional form for the combinatorial background.

the background, estimated with pairs of same charge $K\pi$ tracks (wrong charge). A similar signal of about 400 events is seen in the $D^0 \to K\pi\pi\pi$ mode, where a control region has been used for background subtraction. In the ZEUS kinematic region the integrated cross section obtained from 1995 data in the $D^0 \to K\pi$ channel is $\sigma(ep \to D^*X) = 10.1 \pm 0.8^{+0.7}_{-0.4}$ nb (prel.), in agreement with a previous ZEUS measurement using 1994 data[9].

2.1. Total charm cross section $\sigma(\gamma p \to c\bar{c}X)$

Extrapolating to the full kinematic region and converting the ep to γp cross sections using the Weizsäcker-Williams Approximation[10], H1[11] obtains the total charm cross section $\sigma(\gamma p \to c\bar{c}X)$ higher by one magnitude compared with low energy experiments, in agreement with the ZEUS 1993 data analysis[12]. The total charm cross section measured by the 33 m-tagged and untagged 1994 data samples, the 44 m-tagged 1995 sample, as well as the ZEUS 1993 result are shown in Fig.3. The W dependence is in agreement with a NLO massive charm scheme calculation[3], within the large systematic uncertainties due to the extrapolation and due to the sensitivity of the NLO prediction on the charm mass, the choice of renormalization scale and the

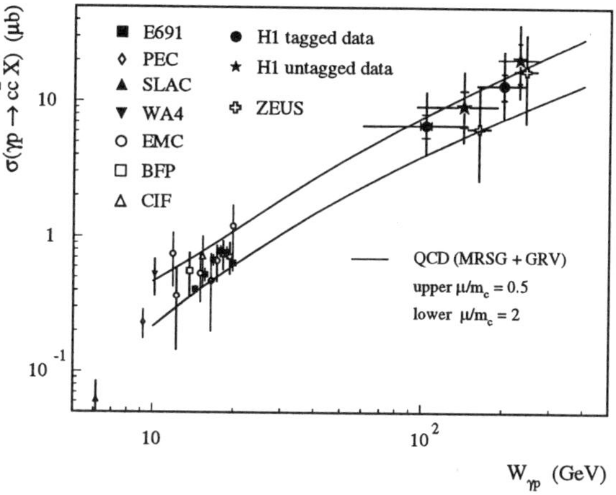

Fig. 3. Total charm photoproduction cross section as a function of W. Inner error bars for H1 are statistical and systematic errors added in quadrature. Outer bars include uncertainties due to the choice of parton density parametrizations. The data point at $W \approx 115\,\text{GeV}$ is obtained from the 44 m-tagged sample. The other H1 tagged data points correspond to the 33 m-tagged sample. Low energy points are from fixed target measurements. Solid lines are NLO predictions with MRSG/GRV-G HO for the proton/photon density with renormalization scale between half and twice the charm mass.

parton density parametrizations of the photon and the proton.

2.2. Differential D^* cross sections

In Fig.4 differential γp cross sections in $p_\perp(D^*)$ and $\hat{y}(D^*)$ are shown for the $D^0 \to K\pi$ channel of the 33 m-tagged and untagged H1 samples. Preliminary differential e^+p cross sections as a function of $p_\perp^{D^*}$, η^{D^*} and W using the ZEUS 1995 data are presented in Fig.5.

The H1 data are compared with calculations[3] using the massive charm scheme. The different curves are predictions where the charm mass or the parton density functions of photon and proton are changed. It is shown that such changes cause large effects in the low $p_\perp(D^*)$ region, while the sensitivity to these parameters is strongly reduced in the restricted kinematic range of the measurement.

Both the H1 and the ZEUS data are above the predicted cross sections of the calculation using a massive charm scheme, while the massless charm approach is in agreement with the data. The shapes of the differential cross sections $d\sigma/dp_\perp^{D^*}$, $d\sigma/d\eta^{D^*}$, $d\sigma/d\hat{y}^{D^*}$ and $d\sigma/dW$ are reasonably reproduced by both models. The rapidity (pseudorapidity) distribution of both H1 and ZEUS exhibit some excess in the forward direction.

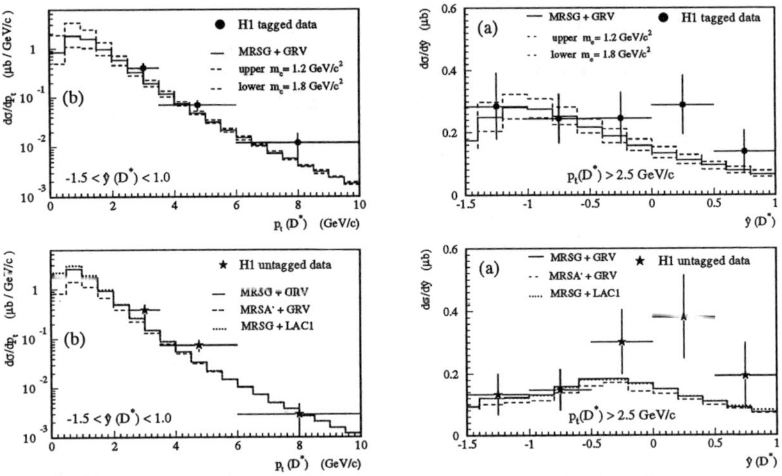

Fig. 4. Differential cross sections in $p_{\perp}^{D^*}$ and $\hat{y}$ for the tagged (full circles) and untagged (solid stars) sample. In the upper plots, the full line shows the NLO QCD prediction using a massive charm mass approach. For the proton the MRSG parton density parametrization and a charm mass of 1.5 GeV are used. The upper (lower) dashed histogram indicates the effect of changing the charm quark mass to 1.2 (1.8) GeV. In the lower plots, the histograms show predictions for various parton density parametrizations of the photon and proton.

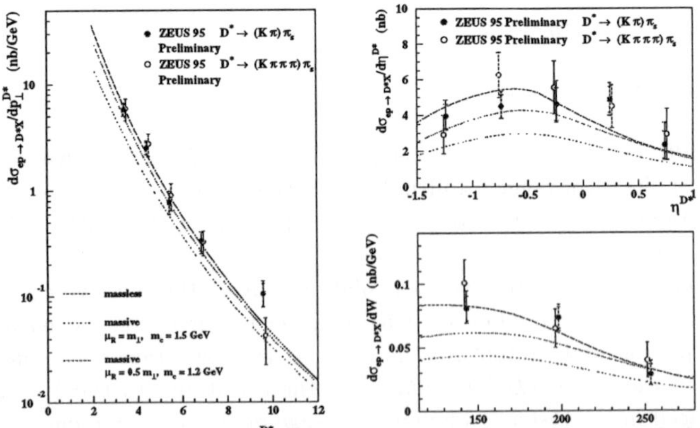

Fig. 5. Differential D^* cross sections as a function of $p_{\perp}^{D^*}$, η^{D^*} and W are shown. For the $D^0 \to K\pi$ decay mode the inner part of the vertical error bars gives the statistical error, while the outer one shows the statistical and systematic errors added in quadrature. For the $D^0 \to K\pi\pi\pi$ channel only statistical errors are shown. The predictions of NLO perturbative QCD calculation from a massive charm approach[3] is given by the dot-dashed curve, using MRSG and GRV-HO as parton density parametrizations for the proton and photon respectively, fragmentation parameter $\epsilon_c = 0.06$, renormalization scale $\mu_R = m_{\perp}$ and $m_c = 1.5\,\text{GeV}$. The dotted curve is from the same calculation, but for $\mu_R = 0.5 m_{\perp}$ and $m_c = 1.2\,\text{GeV}$. The full curve comes from the massless charm approach calculation[1], using the same parameters as for the dot-dashed curve, but with CTEQ4M taken for the parton density parametrization of the proton.

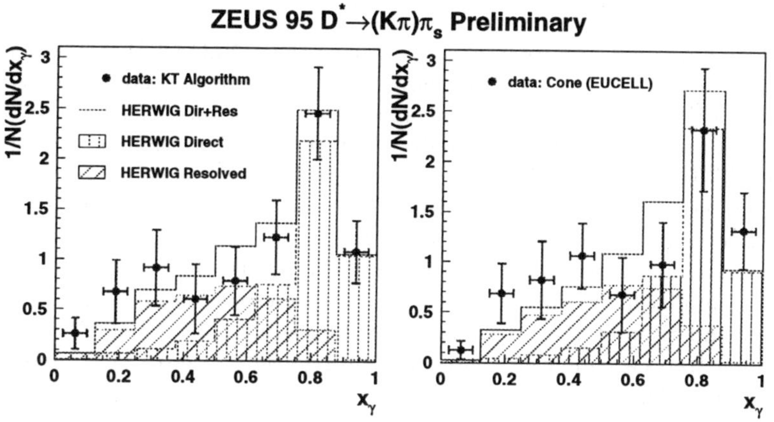

Fig. 6. Distributions of x_γ^{OBS} are shown for events with a D^* in the kinematic region $p_\perp^{D^*} > 3\,\mathrm{GeV}$, $-1.5 < \eta^{D^*} < 1.0$, $115 < W < 280\,\mathrm{GeV}$ and at least two jets with $E_T^{jet} > 4\,\mathrm{GeV}$ and $\eta^{jet} < 2.4$. The distributions are not corrected for detector effects. For the left plot, a KT algorithm has been used, while the right plot shows results with the EUCELL cone algorithm using a cone radius of $R = 1.0$. The ZEUS 1995 data (full circles) are compared with direct, resolved and a combined sample of resolved and direct HERWIG events. The combined sample is mixed according to the cross sections given by HERWIG for resolved and direct contributions.

3. Charm Dijet Study

To obtain additional information on the charm production mechanism a jet analysis of charm events has been performed. Using the EUCELL cone algorithm[13] or the KT cluster algorithm[14] and requiring at least two jets with $E_T^{jet} > 4\,\mathrm{GeV}$ and $\eta^{jet} < 2.4$ the fraction of the photon energy contributing to the production of the two highest E_T^{jet} jets can be estimated by

$$x_\gamma^{OBS} = \frac{E_T^{jet1} e^{-\eta^{jet1}} + E_T^{jet2} e^{-\eta^{jet2}}}{2E_e y}$$

where yE_e is the initial photon energy. Direct processes as defined at leading order have a high x_γ^{OBS} since all the photon's energy participates in the production of the hard jets, while resolved processes as defined at leading order have low x_γ^{OBS} since part of the photon's energy goes into the photon remnant. A comparison of the x_γ^{OBS} distribution (not corrected for detector effects), see Fig.6, with leading order HERWIG Monte Carlo predictions demonstrates significant contribution of resolved processes ($x_\gamma^{OBS} < 0.75$) to D^* production in the measured kinematic range. In the MC sample the resolved component is dominated by charm production via the excitation process ($c^\gamma g^p \to c^\gamma g^p$) with charm coming from the photon.

4. Summary

Charm photoproduction is an effective tool to check NLO QCD calculations. Inclusive charm and D* cross sections have been measured and are compared with NLO QCD calculations using different approaches. The total charm cross section $\sigma(\gamma p \rightarrow c\bar{c}X)$ obtained from photoproduced D^* mesons rises strongly with W, reasonably well described by a NLO QCD calculation using a massive charm scheme. Within the restricted kinematical regions of H1 and ZEUS, where NLO predictions are stable to variations of the parameters of the theory, differential D^* cross sections are reproduced in shape and magnitude by NLO QCD calculations in the massless charm approach. Predictions from the massive charm approach are below the data. A comparison of x_γ^{OBS} distributions with leading order Monte Carlo simulations demonstrates a significant contribution of resolved processes to D^* production, in the measured kinematic range.

5. Acknowledgements

It is a pleasure to thank all the colleagues from the H1 and the ZEUS collaboration who provided material. I would also like to thank the participants and the organisers of the Ringberg Workshop for such an enjoyable and stimulating atmosphere.

6. References

1. B.A. Kniehl, G. Kramer and M. Spira, *DESY* **96–210**, hep-ph/9610267.
2. M. Cacciari et al., *DESY* **96–146**, hep-ph/9608213.
   M. Cacciari and M. Greco, *DESY* **97–029**, hep-ph/9702389.
3. S. Frixione, P. Nason and G. Ridolfi, *Nucl. Phys.* **B454** (1995) 3.
4. C. Peterson et al., *Phys. Rev.* **D27** (1983) 105.
5. OPAL Collaboration, R. Akers et al., *Z. Phys.* **C67** (1995) 27.
6. J. Chrin, *Z. Phys.* **C36** (1987) 163.
7. G. Altarelli, G. Parisi, *Nucl. Phys.* **B126** (1977) 298.
8. J. Binnewies, B.A. Kniehl and G. Kramer, *DESY* **97–012**, hep-ph/9702408.
9. ZEUS Collaboration, J. Breitweg et al., *Phys. Lett.* **B401** (1997) 192.
10. C.F. Weizsäcker, *Z. Phys.* **88** (1934) 612.
    E.J. Williams, *Phys. Rev.* **45** (1934) 729.
11. H1 Collaboration, S. Aid et al., *Nucl. Phys.* **B472** (1996) 32.
12. ZEUS Collaboration, M. Derrick et al., *Phys. Lett.* **B349** (1995) 225.
13. ZEUS Collaboration, M. Derrick et al., *Phys. Lett.* **B348** (1995) 665.
14. S. Catani et al., *Nucl. Phys.* **B406** (1993) 187.

ON HEAVY QUARKS PHOTOPRODUCTION AND $c \to D^*$ FRAGMENTATION FUNCTIONS

MATTEO CACCIARI
Deutsches Elektronen-Synchrotron DESY
D-22603, Hamburg, Germany
E-mail: Matteo.Cacciari@desy.de

ABSTRACT

The state of the art of the theoretical calculations for heavy quarks photoproduction is reviewed. The full fixed order next-to-leading order massive calculation and the resummation of large $\log(p_T/m)$ terms for differential cross sections are described. The implementation of a non-perturbative fragmentation function describing the $c \to D^*$ meson transition is also discussed.

1. Introduction

Heavy quarks production processes provide a powerful insight into our understanding of Quantum Chromodinamics. The large mass of the heavy quark can make the perturbative calculations reliable, even for total cross sections, by cutting off infrared singularities and by setting a large scale at which the strong coupling can be evaluated and found – possibly – small enough. On the experimental side, the possibility to tag heavy flavoured hadrons by means of microvertex detectors can on the other hand provide accurate measurements.

All these potentialities must of course be matched by accurate enough theoretical evaluations of the production cross section. In this talk I shall describe the state of the art of such calculations for heavy quarks photoproduction. I shall first review the next-to-leading order (NLO) QCD evaluations recently presented by Frixione, Mangano, Nason and Ridolfi. These calculations, available for total cross sections, one-particle and two-particles distributions, are now a consolidated result and provide a benchmark for future developments.

Large logarithms appear in the NLO fixed order calculations and potentially make it less reliable in some regimes: $\log(S/m^2)$ and $\log(p_T^2/m^2)$ become large when the center of mass energy $\sqrt{S}$ or the transverse momentum p_T of the observed quark is much larger than its mass. I shall describe the resummation of $\log(p_T^2/m^2)$ terms, leaving the high energy resummation to Marcello Ciafaloni's talk[1].

The perturbative fragmentation function technique used in the resummation of the large $\log(p_T^2/m^2)$ terms has a non-perturbative extension which can be used to describe the transition from c quarks to D^* mesons. I shall therefore also discuss the determination of these non-perturbative fragmentation functions and their inclusion into the heavy quarks photoproduction calculation, showing a comparison with data from HERA.

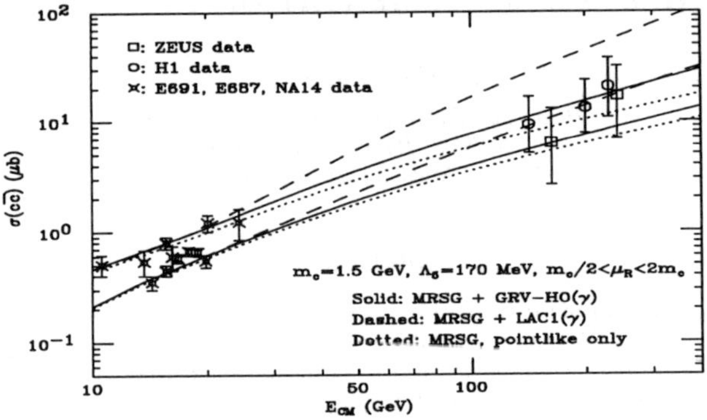

Fig. 1. Total cross section for $c\bar{c}$ photoproduction[6].

2. Fixed Order NLO Calculation

Heavy quarks photoproduction at leading order in the strong coupling α_s looks a very simple process: only the tree level diagram $\gamma g \to Q\bar{Q}$ contributes at the partonic level, and the final answer for the total cross section is simple and well behaved, being finite everywhere.

At a deeper thinking, however, problems seem to arise. For instance, one may ask himself why not to include initial state heavy quarks, coming from the hadron and to be scattered by the photon, like $\gamma Q \to Qg$. To include consistently such a diagram is not an easy task, especially if one wants to keep the quark massive. Taking it massless, on the other hand, would not only be a bad approximation but would also produce a divergent total cross section.

A way out of this problem was provided by Collins, Soper and Sterman[2], who argued that the following factorization formula holds for heavy quarks hadroproduction total cross sections:

$$\sigma(\sqrt{S}, m) = \sum_{ij} \int f_{i/H_1} f_{j/H_2} \hat{\sigma}(ij \to Q\bar{Q}; \sqrt{S}, m). \tag{1}$$

The sum on the partons runs only on i and j being gluons or light quarks, and the heavy quarks are only generated at the perturbative level by gluon splitting. There is therefore no need to try to accommodate them in the colliding hadrons and the relevant kinematics can be kept exact. Eq. (1) provides the basis for an exact perturbative calculation of heavy quarks production to NLO. For what concerns photoproduction, such a calculation has been first performed by P. Nason and K. Ellis, and subsequently confirmed by J. Smith and W.L. van Neerven [3].

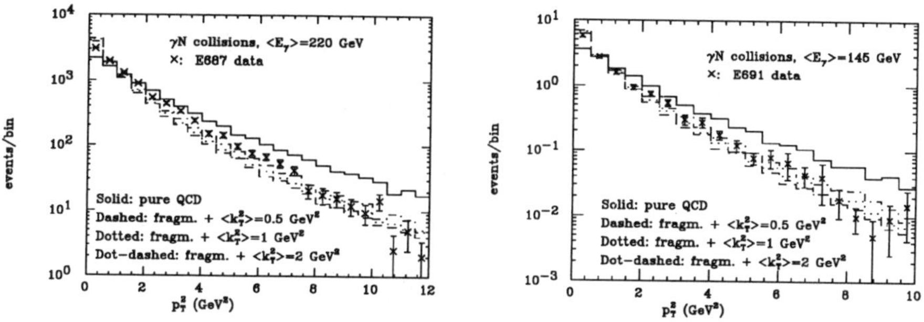

Fig. 2. Differential p_T distributions for charm production in fixed target experiments[6].

When going to order $\alpha\alpha_s^2$ in photon-hadron collision, however, a new feature appears. The photon can now couple directly to massless quarks, for instance in processes like $\gamma q \to Q\bar{Q}q$, and in a given region of phase space a collinear singularity will appear. It can be consistently factored out, but this requires the introduction of *photon* parton distribution functions (PDF) which, pretty much like the hadron ones, will describe the probability that before the interaction the photon splits into hadronic components (light quarks or gluons, in this case). Such a behaviour is sometimes called *resolved photon* (as opposed to *direct*). A full NLO calculation for heavy quark photoproduction will therefore also require a NLO calculation for hadroproduction[4], where one of the PDF's will be the photon's one. A factorization scale μ_γ, related to the subtraction of the singularity at the photon vertex, will link the two pieces and its dependence on the result will only cancel when both are taken into account.

Frixione, Mangano, Nason and Ridolfi[5] (FMNR) have recently presented Montecarlo integrators for these two calculations, thereby allowing detailed comparisons with experimental data. A very extensive collection of such comparisons is presented in a recent review[6], from which we select some plots to be shown here.

A comparison of total cross section experimental results and theoretical predictions for $c\bar{c}$ photoproduction is shown in fig. 1. Although large uncertainties are present, the comparison suggests agreement between theory and experiment. The new HERA data, at large center of mass energy, can be seen to appear larger than the pointlike (= direct) photon prediction only. This suggests the need for a resolved photon component, but by no means can determine it precisely.

One-particle transverse momentum (p_T) distributions are shown in fig. 2. The pure QCD predictions can be seen to be significantly harder than the data. However, when corrected with two non-perturbative contributions they can be matched to the data. These non-perturbative addictions are meant to represent a primordial transverse momentum k_T of the colliding partons, other than the one already taken into account by the QCD radiative corrections, and the effect of the fragmentation of

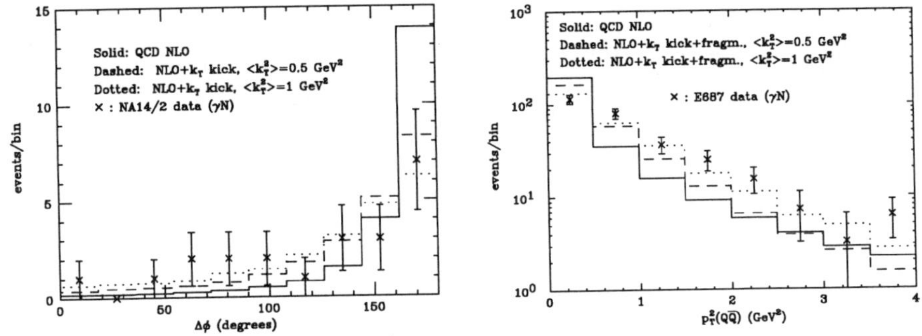

Fig. 3. Two particles correlations in fixed target experiments[6].

the produced heavy quark into the observed heavy flavoured hadrons, here described by the so-called Peterson fragmentation function with $\epsilon = 0.06$.

Comparisons between data and theory for two-particle correlations, like the azimuthal difference $\Delta\phi$ or the relative transverse momentum $p_T(Q\overline{Q})$ of the produced heavy quark pair, are shown in fig. 3. Distributions like these are trivial in leading order QCD, since the Q and the $\overline{Q}$ are produced back-to-back. Hence, $\Delta\phi = \pi$ and $p_T(Q\overline{Q}) = 0$. NLO corrections (as well as non-perturbative contributions) can broaden these distributions, and one could think of being able to perform a direct measurement of $O(\alpha_s^3)$ effects. The plots do however show that non perturbative contributions play a key role in allowing a good description of the data. One can, however, still check that the same inputs allow for a good description of both one- and two-particles distributions, as seems to be the case here.

The overall result of these comparisons can therefore be summarized as follows. Total cross sections seem to be well reproduced by the calculation both at fixed target and HERA regimes, but the huge uncertainties present both on the experimental and the theoretical side do not allow the study of finer details like, for instance, the determination of the resolved component at HERA. For what concerns transverse momentum distributions at fixed target, they can be reproduced after allowing for heavy quark fragmentation effects and for a primordial transverse momentum of the incoming partons of the order of 1 GeV. These same non-perturbative corrections also allow for a description of two-particles correlations, thereby pointing towards a consistent picture.

3. Large Transverse Momentum Resummation

Like any perturbative expansion, the NLO calculation for heavy quarks photoproduction is only reliable and accurate as long as the coefficients of the coupling constant remain small. Large terms of the kind $\log(p_T^2/m^2)$ do however appear in the

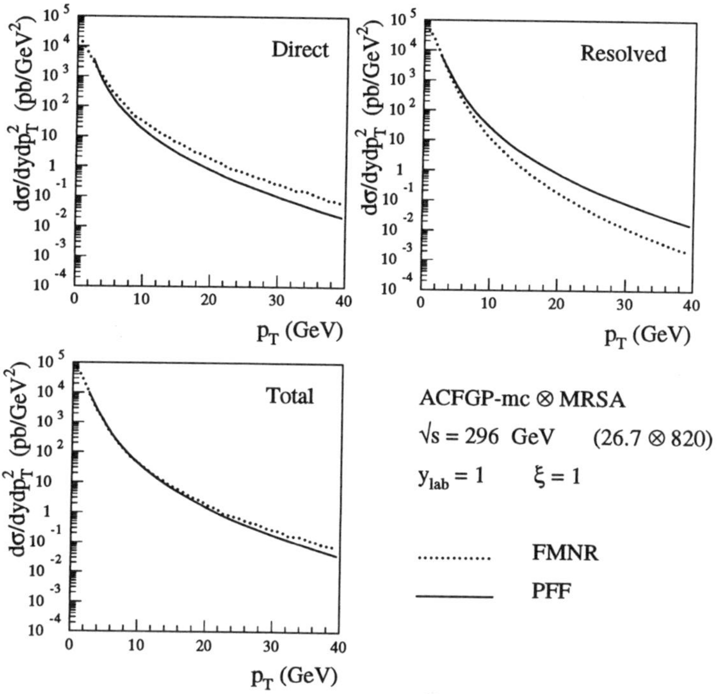

Fig. 4. Comparison between fixed order (FMNR) and resummed (PFF) calculation for charm photoproduction p_T distribution[7]. It is worth noticing how the two calculations describe differently the (unphysical) resolved and direct components, but agree on their sum (a physical observable).

cross section, and for growing p_T they will eventually became large enough to spoil the convergence of the series. Such terms need therefore to be resummed to all orders to allow for a sensible phenomenological prediction. Such a resummation has been performed along the following lines[7].

One observes that in the large-p_T limit ($p_T \gg m$) the only important mass terms are those appearing in the logs, all the others being power suppressed. This means that an alternative description of heavy quark production can be achieved by using *massless* quarks and providing at the same time perturbative distribution and fragmentation functions also for the heavy quark, describing the logarithmic mass dependence. The factorization formula becomes

$$d\sigma(p_T) = \sum_{ijk} \int F_{i/H_1}(\mu,[m]) F_{j/H_2}(\mu,[m]) d\hat{\sigma}(ij \to k; p_T, \mu) D_k^Q(\mu, m), \qquad (2)$$

with parton indices i,j and k also running on Q, taken massless in $\hat{\sigma}$, now an $\overline{\mathrm{MS}}$ subtracted cross section for light partons production. The dependence on m of the parton distribution functions $F_{i/H}$, shown among square brackets in eq. (2), is only there when i or j happens to be the heavy quark Q.

The key point is that the large mass of the heavy quark allows for the evaluation in perturbative QCD (pQCD) of its distribution and fragmentation functions. Initial state conditions for $F_{Q/H}(\mu_0 = m)$[8] and $D_k^Q(\mu_0 \simeq m)$[9] can be calculated in pQCD at NLO level in the $\overline{\rm MS}$ scheme:

$$F_{Q/H}(x, \mu_0 = m) = 0 \quad (3)$$

$$D_Q^Q(x, \mu_0) = \delta(1-x) + \frac{\alpha_s(\mu_0) C_F}{2\pi} \left[\frac{1+x^2}{1-x} \left(\log \frac{\mu_0^2}{m^2} - 2\log(1-x) - 1 \right) \right]_+ \quad (4)$$

$$D_g^Q(x, \mu_0) = \frac{\alpha_s(\mu_0) T_F}{2\pi} (x^2 + (1-x)^2) \log \frac{\mu_0^2}{m^2} \quad (5)$$

$$D_{q,\bar{q},\bar{Q}}^Q(x, \mu_0) = 0 \quad (6)$$

The massive logs will hence appear only through these function, which can then be evolved with the Altarelli-Parisi equations up to the large scale set by $\mu \simeq p_T$. This evolution will resum to all orders the large logarithms previously mentioned.

It is important to mention that due to the neglecting of power suppressed mass terms this approach becomes unreliable when $p_T \simeq m$. In this region only a case by case comparison with the full NLO massive calculation – here reliable and to be taken as a benchmark – can tell how accurate the resummed result is.

Phenomenological analyses show that the effect of the resummation becomes sizeable only at very large p_T, say greater than 20 GeV for charm photoproduction. Fig. 4 shows the effect of such a resummation for a fixed photon energy in HERA-like kinematics. The resummed calculation can be seen to match the fixed order one at $p_T \sim m$, where resummation effects are not expected to be important, and to behave more softly in the large p_T region. This particular theoretical refinement should therefore not be phenomenologically overly relevant for present-day HERA physics, data being only available up to $p_T \simeq 12$ GeV.

4. On the Inclusion of $c \to D^*$ Fragmentation Effects

When comparing theory with data, one always faces the problem of describing as closely as possible what the experiments do observe. With heavy quarks production the problem lies in the experiments actually seeing the decay products of heavy flavoured hadrons rather than the heavy quark itself. This is due to the heavy quark strong and non-perturbative binding into a hadron prior to decay. This binding involves the exchange and radiation of low-momentum (order Λ_{QCD}) gluons, and typically degrades the momentum of the hadron with respect to the one of the original quark. Such a degradation can be described with the help of a non-perturbative fragmentation function (FF) which, lacking the theoretical tools to calculate, can be extract by fitting experimental data.

An often employed parametrization for such a function is the so called Peterson[10] one, which reads

$$D_{np}(z;\epsilon) \sim \frac{1}{z\left[1 - 1/z - \epsilon/(1-z)\right]^2}. \tag{7}$$

The value of ϵ is predicted to scale like Λ^2_{QCD}/m^2. For charm to D^* fragmentation a global analysis[11] based on leading order Montecarlo simulations gives the value $\epsilon \simeq 0.06$. This value has so far usually been taken as the reference one, and used for instance together with the NLO fixed order calculation by FMNR in the plots shown in Section 2.

One should however carefully consider how ϵ has been extracted from e^+e^- experimental data. Experiments usually report the energy or momentum fraction (x_E or x_p) of the observed hadron with respect to the beam energy. On the other hand the fraction which appears as the argument of the non-perturbative FF is rather to be taken with respect to the fragmenting quark momentum, usually denoted by z (see for instance [11] for a discussion on this point). These two fractions are not coincident, due to hard radiation processes which lower the momentum of the quark before it fragments into the hadron. In order to deconvolute these effects one usually runs a Montecarlo simulation of the collision process at hand, including both the perturbative parton showers and the subsequent hadronization of the partons into the observed hadrons. The latter can be parametrized in the Montecarlo by the Peterson fragmentation function, and the value of ϵ which best describes the data can be extracted. Clearly this procedure leads to a resulting value for ϵ which depends on the details of the description of the perturbative part. Indeed, the showering softens the momentum distribution of the heavy quark, producing an effect qualitatively similar to that of the non-perturbative FF. On the quantitative level, the amount of softening (and hence the value of ϵ) required by the non-perturbative FF to describe the data is related to the amount of softening already performed at the perturbative level. A leading or a next-to-leading description of the showering can therefore produce different values for ϵ, whose value is then not a "unique" and "true"one, but rather closely interconnected with the details of the description of the pQCD part of the problem.

In ref. [12] fits to D^* data taken by the ARGUS and OPAL experiments have been performed with NLO accuracy using a fragmentation description for the heavy quark production like the one described in Section 3, complemented with the inclusion of a non-perturbative component via the ansatz

$$D_k^{D^*}(\mu) = D_k^c(\mu) \otimes D_c^{D^*}, \tag{8}$$

represented by the convolution of a perturbatively calculable fragmentation function of the parton k into the heavy quark c and the non-perturbative form $D_c^{D^*}$ describing the $c \to D^*$ transition. This non perturbative form is taken to be scale independent, i.e. all scaling effects are assumed to be described by the Altarelli-Parisi evolution of the perturbative part $D_k^c(\mu)$. A similar approach had already been introduced in [9].

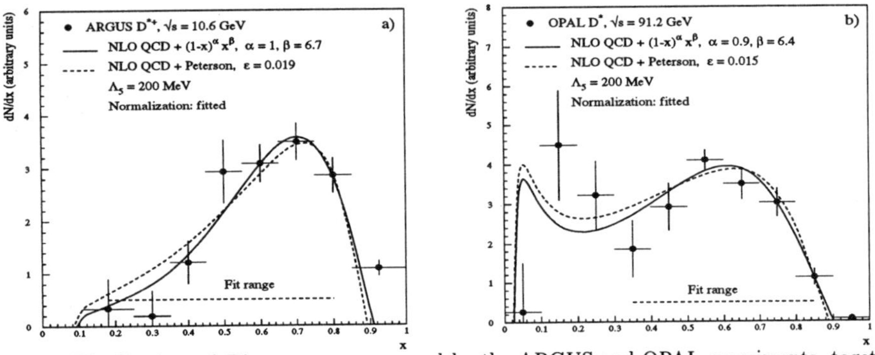

Fig. 5. Distributions of D^* mesons as measured by the ARGUS and OPAL experiments, together with the theoretical curves[12] fitted to the same data with the $(1-x)^\alpha x^\beta$ (full line) and the Peterson (dashed line) non-perturbative fragmentation functions.

Results for these fits are shown in fig. 5. The value for ϵ has been consistently found to be of order 0.02 rather than the customary 0.06 one, resulting instead from fits with leading order evolution. Recalling the previous discussion, this comes to no surprise: next-to-leading order evolution softens more the heavy quark spectrum, and a harder non-perturbative fragmentation function is therefore needed to provide a satisfactory description of the data (see [12] for a full discussion).

On the phenomenological side, and making use of the universality argument, one can now argue that the use of a "harder" Peterson form with $\epsilon = 0.02$ is probably more suited when combined with a NLO perturbative calculation like the FMNR one which, albeit only at fixed order, contains NLL gluon radiation. Decreasing ϵ means increasing the cross section at large p_T, being the p_T distribution steeply falling with increasing transverse momentum. This could help reconciling the HERA experimental data[13] with the perturbative NLO calculation, which was shown to underestimate them a little when convoluted with a Peterson with $\epsilon = 0.06$: fig. 6 shows, on the left, how the cross section for D^* photoproduction at HERA increases with decreasing ϵ and, on the right, a comparison of the H1 data with the fixed order prediction by FMNR ($\epsilon = 0.06$) and the fragmentation functions one with $\epsilon = 0.02$. One should notice that the p_T values involved are still pretty small: this means that the fixed order calculation is still reliable and the accuracy of the resummed one has to be assessed first by comparing with the former. In this case they are found to be in good agreement, the difference in the plot being mainly given by the different ϵ values.

Last but not least, it is worth mentioning how, going from LO to NLO analyses, a similar hardening of the non-perturbative fragmentation function is also expected for the b quark. The corresponding increase of the hadroproduction bottom p_T distributions[14] would be welcome in the light of the Tevatron data presently over-

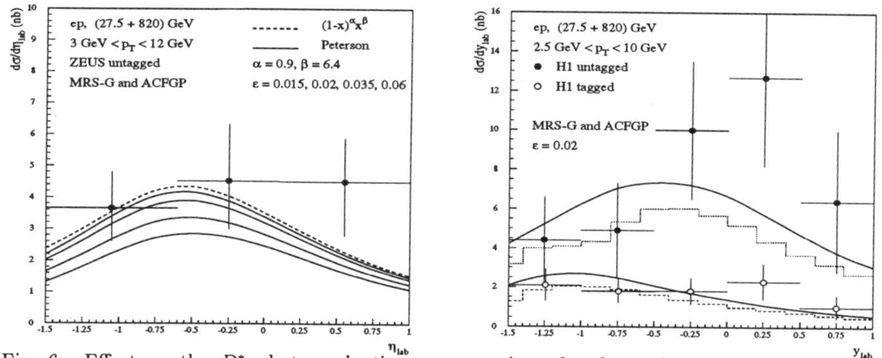

Fig. 6. Effect on the D^* photoproduction cross section of a decreasing value for ϵ (left), and comparison of H1 data with the fixed order calculation by FMNR (histograms, $\epsilon = 0.06$) and the fragmentation function approach (smooth lines, $\epsilon = 0.02$).

shooting the theoretical predictions by at least 30%.

Acknowledgements. I wish to thank the Organizers of this Workshop for the invitation to give this talk, Mario Greco for his collaboration and Paolo Nason for the many conversations on the heavy quarks physics items I've been reviewing here.

5. References

1. M. Ciafaloni, these Proceedings
2. J.C. Collins, D. Soper and G. Sterman, *Nucl. Phys.* **B263** (1986) 37
3. R.K. Ellis and P. Nason, *Nucl. Phys.* **B312** (1989) 551; J. Smith and W.L. van Neerven, *Nucl. Phys.* **B374** (1992) 36
4. P. Nason, S. Dawson and R.K. Ellis, *Nucl. Phys.* **B303** (1988) 607; *Nucl. Phys.* **B327** (1989) 49; W. Beenakker *et al.*, *Phys. Rev.* **D40** (1989) 54; *Nucl. Phys.* **B351** (1991) 507
5. S. Frixione, M.L. Mangano, P. Nason and G. Ridolfi, *Nucl. Phys.* **B412** (1994) 225; *Nucl. Phys.* **B431** (1994) 453; *Phys. Lett.* **B348** (1995) 633; *Nucl. Phys.* **B454** (1995) 3; hep-ph/9510253
6. S. Frixione, M.L. Mangano, P. Nason and G. Ridolfi, CERN-TH/97-16 (hep-ph/9702287)
7. M. Cacciari, M. Greco, *Zeit. Phys.* **C69** (1996) 459; *Nucl. Phys.* **B421** (1994) 530;
8. J.C. Collins and W.-K. Tung, *Nucl. Phys.* **B278** (1986) 934
9. B. Mele and P. Nason, *Nucl. Phys.* **B361** (1991) 626; G. Colangelo and P. Nason, *Phys. Lett.* **B285** (1992) 167
10. C. Peterson, D. Schlatter, I. Schmitt and P.M. Zerwas, *Phys. Rev.* **D27** (1983) 105
11. J. Chrin, *Zeit. Phys.* **C36** (1987) 163
12. M. Cacciari and M. Greco, *Phys. Rev.* **D55** (1997) 7134; M. Cacciari, M. Greco, S. Rolli and A. Tanzini, *Phys. Rev.* **D55** (1997) 2736
13. J. Breitweg et al. (ZEUS Collab.), preprint DESY 97-026; S. Aid et al. (H1 Coll.), *Nucl. Phys.* **B472** (1996) 32
14. M.L. Mangano, private communication

INELASTIC J/ψ PHOTOPRODUCTION AT HERA

Riccardo Brugnera
Representing the ZEUS and H1 Collaborations
Dipartimento di Fisica dell'Universitá di Padova and INFN Sezione di Padova
via Marzolo 8 Padova, Italy
E-mail: brugnera@vxdesy.desy.de

ABSTRACT

This article reviews the experimental results obtained by the ZEUS and H1 Collaborations at HERA in the study of the inelastic J/ψ photoproduction. The comparison between data and recent theoretical predictions is discussed.

1. Introduction

The inelastic reaction $e^+p \to e^+J/\psi X$ in the photoproduction regime ($Q^2 \approx 0$ GeV^2, where Q^2 is the photon virtuality) is thought to proceed via direct photon-gluon fusion, diffractive proton dissociation or resolved photon processes. These three possibilities are shown in Fig. 1. The data published by the ZEUS [1] and H1 [2] Collaborations are in a phase space region where the direct photon-gluon fusion process dominates all the other J/ψ production mechanisms. For this contribution, shown in Fig. 1a, full next-to-leading order (NLO) QCD calculations are available [3]. The predicted cross section is sensitive to the gluon density in the proton at an energy scale corresponding approximately to the heavy quark mass. Furthermore, perturbative QCD calculations using as input the parton densities extracted from other processes can provide a consistency test for QCD and give information on the mechanism of quarkonium formation.

The different processes in Fig. 1 can be distinguished by means of the inelasticity variable z, the fraction of the photon energy carried by the J/ψ in the proton rest frame [4]. The diffractive proton dissociation process dominates the high z region ($z > 0.9$), the photon-gluon fusion the intermediate z region, while the resolved photon process is dominant in the low z region. There are also differences in the p_T distribution, p_T being the J/ψ transverse momentum with respect to the incoming proton beam direction. Compared to photon-gluon fusion and resolved photon processes, the diffractive reaction produces J/ψ mesons with relatively low p_T ($p_T \lesssim 1$ GeV).

The direct photon-gluon fusion process is described in [3] in the framework of the colour-singlet model [5], in which the initial state photon and gluon interact giving a final state $c\bar{c}$ pair with the J/ψ quantum numbers through the emission of a hard gluon in the final state ($\gamma + g_1 \to J/\psi + g_2$). When a similar model was used to study J/ψ hadroproduction [6] at the Tevatron, the predictions, at lowest order in the strong coupling constant α_s, underestimated the data [7] by about one order of

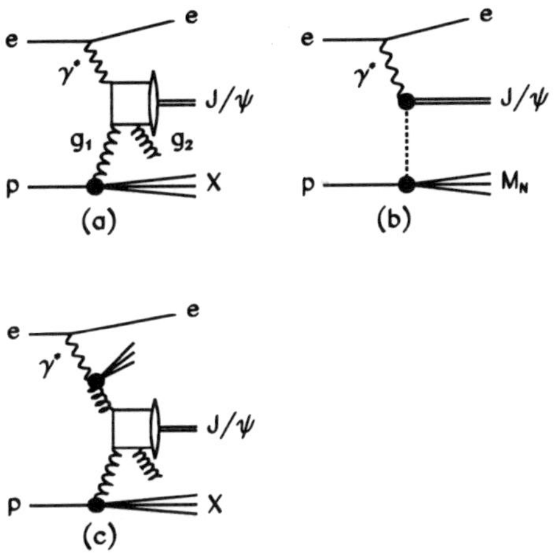

Fig. 1. Dominant inelastic J/ψ production mechanisms at HERA. Photon-gluon fusion is described by diagram (a). Diagrams (b) and (c) correspond to diffractive proton dissociation and resolved photon J/ψ production, respectively.

magnitude. The measured cross section could be explained in part by adding colour-octet contributions [8]. In this case the $c\bar{c}$ pair is produced in a colour-octet state (short distance process) and later binds to form a J/ψ (long distance process). While the short distance terms are calculable through perturbative QCD, the long distance terms are nonperturbative and have to be determined from the data themselves or from lattice calculations. It is therefore interesting to look for evidence of the octet mechanism at HERA, where it is expected to contribute at high z [9].

2. Kinematics

Schematic diagrams for the reaction:

$$e^+(k)p(P) \rightarrow e^+(k')\mathrm{J}/\psi(p_{J/\psi})X, \tag{1}$$

where each symbol in parentheses denotes the four-momentum of the corresponding particle, are shown in Fig. 1.

The kinematics of the inclusive scattering of unpolarised positrons and protons is described by the positron-proton centre of mass energy squared (s) and any two of the following variables:

- $Q^2 = -q^2 = -(k-k')^2$, the negative four-momentum squared of the exchanged photon;

- $y = (q \cdot P)/(k \cdot P)$, the fraction of the positron energy transferred to the hadronic final state in the rest frame of the initial state proton;
- $W^2 = (q+P)^2 = -Q^2 + 2y(k \cdot P) + M_p^2 \approx ys$, the centre of mass energy squared of the photon-proton system, where M_p is the proton mass.

The ZEUS [1] and H1 [2] measurements are restricted to photoproduction events, the Q^2 value ranging from the kinematic minimum $Q^2_{min} = M_e^2 y^2/(1-y) \sim 10^{-10}$ GeV^2, where M_e is the electron mass, to the value at which the scattered positron starts to be observed in the main detectors, $Q^2_{max} \sim 4$ GeV^2. The median Q^2 is approximately 10^{-4} GeV^2.

For reaction (1) the inelasticity variable, z, is defined by

$$z = \frac{P \cdot p_{J/\psi}}{P \cdot q} \simeq \frac{(E_{J/\psi} - pz_{J/\psi})}{2yE_e}, \tag{2}$$

where $E_{J/\psi}$ is the J/ψ energy and $pz_{J/\psi}$ its momentum component along the Z direction. In the proton rest frame, z is equal to $\frac{E_{J/\psi}}{E_\gamma}$, where E_γ is the photon energy.

3. Experimental Method

To obtain a sample of inelastic J/ψ events the ZEUS and H1 Collaborations have followed slightly different strategies. ZEUS selects J/ψ decaying into both leptonic decay modes (J/$\psi \to \mu^+\mu^-$ and e^+e^-) and to reject elastic J/ψ events requires an energy greater than 1 GeV in a cone along the incoming proton direction. H1, instead, uses only the muonic decay mode and requires at least an additional track (different from the two leptonic tracks) which comes from the interaction point. To remain in a region where the detectors are well understood and the signal to background ratio is high enough, cuts are imposed on W and z. The ZEUS data are in the ranges: $50 < W < 180$ GeV, $0.4 < z < 0.9$ (Fig. 2a and 2b show the invariant mass for $\mu^+\mu^-$ and e^+e^- samples, respectively), while H1 data are in the ranges: $30 < W < 150$ GeV, $0.45 < z < 0.9$.

The events so selected contain the following background processes:

- J/ψ from diffractive proton dissociation (Fig. 1a) ;
- J/ψ from resolved photon processes (Fig. 1b);
- J/ψ from ψ' decay.

The contributions of the first two processes have been estimated and have been subtracted from the data. The ψ' contribution has been estimated to be $\sim 15\%$ but has not been subtracted.

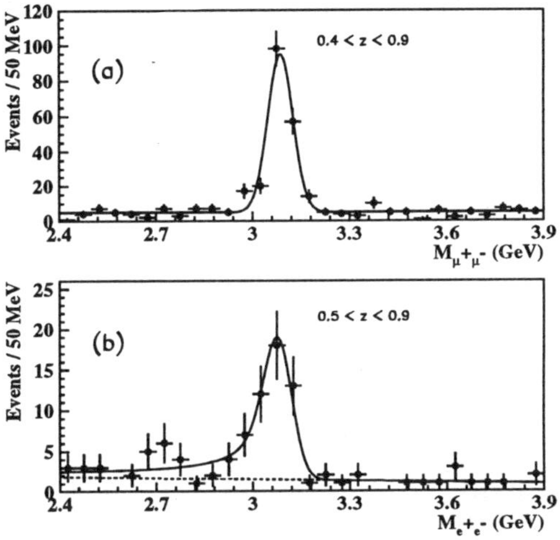

Fig. 2. Data from the ZEUS Collaboration [1]: (a) invariant mass spectrum for the muon pair sample ($0.4 < z < 0.9$ and $50 < W < 180$ GeV); (b) invariant mass spectrum for the electron pair sample ($0.5 < z < 0.9$ and $90 < W < 180$ GeV).

4. Integrated Cross Sections

Figure 3 shows photoproduction cross sections as a function of W for $z < 0.9$. The data come from ZEUS [1], H1 [2] and from low energy experiments [10,11,12]. The data of the ZEUS and H1 Collaborations are extrapolated down to $z = 0$, in order to be comparable with low energy experiments. The size of the extrapolation is small, $O(10\%)$, assuming the direct photon-gluon fusion as the only J/ψ production mechanism at low z. The curves in the plot correspond to the NLO calculation [3] computed with no cut on the p_T of the J/ψ , $z <$ 0.9, with a charm mass (m_c) of 1.4 GeV, $\Lambda_{QCD} = 300$ MeV and with renormalization and factorization scales of $\sqrt{2}m_c$. Dashed, continuous and dashed-dotted curves are obtained with different gluon distributions (MRSG [13], GRV [14] and CTEQ3M [15], respectively), compatible with those extracted from F_2 measurements at HERA [16,17]. The theoretical predictions were multiplied by a factor 1.15 to take into account ψ' production. The predictions are in qualitative agreement with the data, but the distinction among different parton densities is not possible because the NLO calculation is not well behaved in the limit $p_T \rightarrow 0$. Furthermore, there is a significant dependence on the values of the charm mass and Λ_{QCD}. The cross section varies as $1/m_c^3$ and α_s^2 as illustrated by the dotted line which is calculated with $m_c = 1.55$ GeV and $\Lambda_{QCD} = 215$ MeV and using the GRV gluon distribution.

A more quantitative comparison between data and theory can be made in the restricted kinematic range $p_T^2 >$ 1 GeV^2, where the calculation is much more reliable.

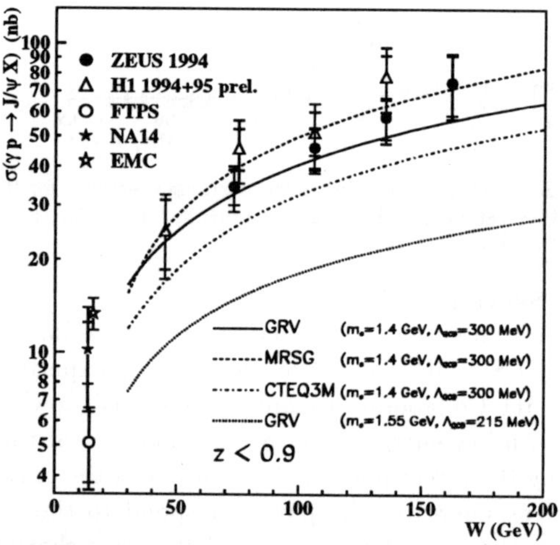

Fig. 3. The direct inelastic J/ψ photoproduction cross section as a function of W for $z < 0.9$. Data from ZEUS [1], H1 [2], FTPS [10], NA14 [11] and EMC [12] are shown. The lines correspond to the NLO prediction from [3] assuming the GRV [14] (continuous), MRSG [13] (dashed) and CTEQ3M [15] (dotted-dashed) gluon distributions with $m_c = 1.4$ GeV and $\Lambda_{QCD} = 300$ MeV, the dotted curve was obtained with GRV, $m_c = 1.55$ GeV and $\Lambda_{QCD} = 215$ MeV . The curves are scaled up by a factor of 1.15 to take into account the contribution from $\psi' \to J/\psi X$.

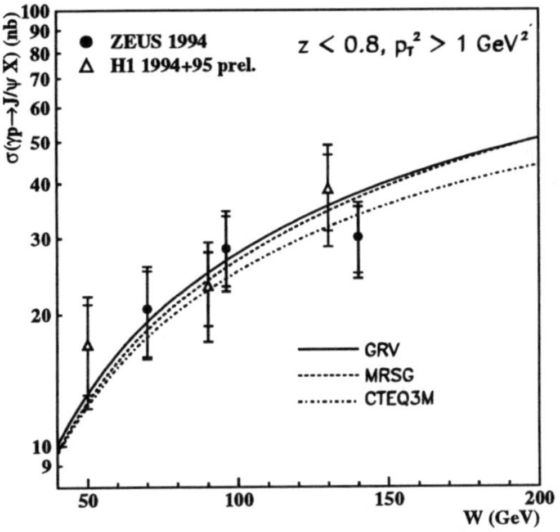

Fig. 4. The direct inelastic J/ψ photoproduction cross section as a function of W for $z < 0.8$ and $p_T^2 > 1$ GeV^2. Data from ZEUS [1] and H1 [2] are shown. The lines correspond to the NLO prediction from [3] assuming the GRV [14] (continuous), MRSG [13] (dashed) and CTEQ3M [15] (dotted-dashed) gluon distributions with $m_c = 1.4$ GeV and $\Lambda_{QCD} = 300$ MeV. The curves are scaled up by a factor of 1.15 to take into account the contribution from $\psi' \to$ J/ψX.

The NLO computation now allows an absolute comparison between data and models. The measured cross sections for $z < 0.8$ and $p_T^2 > 1$ GeV^2 are displayed in Figure 4.

The curves represent the NLO calculation using the different gluon distributions cited above. Data and theory are in good agreement using $m_c = 1.4$ GeV, $\Lambda_{QCD} = 300$ MeV and $\sqrt{2}m_c$ as energy scale. While for $z < 0.9$ the predictions are significantly different for different parametrizations of the gluon density, this is not so in the more restricted domain $z < 0.8$ and $p_T^2 > 1$ GeV^2. This is a consequence of the p_T^2 cut: the gluon distribution is probed at larger x_g values, where the differences between the various gluon densities are smaller; here x_g is the proton energy fraction carried by the incoming gluon.

5. Transverse Momentum Distribution

Figure 5 shows the differential cross section dσ/dp_T^2 for $z < 0.9$ in the range $50 < W < 180$ GeV for ZEUS and in the range $30 < W < 150$ GeV for H1. The curve shows the NLO prediction obtained with $m_c = 1.4$ GeV, $\Lambda_{QCD} = 300$ MeV, $\sqrt{2}m_c$ as energy scale and GRV as gluon distribution. For $p_T^2 > 1$ GeV^2 data and

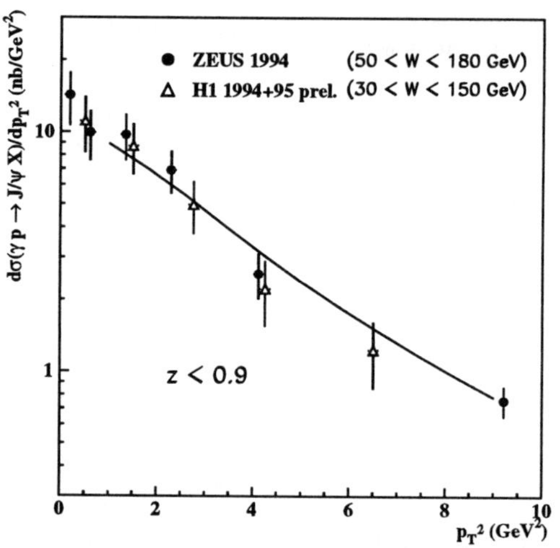

Fig. 5. Differential cross section $d\sigma/dp_T^2$ for the inelastic J/ψ for $z < 0.9$. Data from ZEUS [1] and H1 [2] are shown. The error bars indicate the quadratic sum of the statistical and systematic uncertainties. The NLO computation [3] with the GRV [14] structure function, $m_c = 1.4$ GeV and $\Lambda_{QCD} = 300$ MeV is shown as the solid line. The theoretical curve is drawn only for $p_T^2 > 1$ GeV^2 because in the low p_T region the calculation is not reliable. In the theoretical curve the 15% contribution of the ψ' has not been included.

theoretical calculation are in good agreement. A fit to the data using the function:

$$\frac{d\sigma}{dp_T^2} = Ae^{-bp_T^2} \tag{3}$$

gives:

$$b = 0.32 \pm 0.03 \text{ (stat)} \pm 0.01 \text{ (syst) GeV}^{-2} \tag{4}$$

for ZEUS and

$$b = 0.39 \pm 0.06 \text{ (stat)} \pm 0.03 \text{ (syst) GeV}^{-2} \tag{5}$$

for H1. A similar fit to the NLO calculation [3] yields a slope of $b = 0.3$ GeV^{-2} for $p_T^2 > 1$ GeV^2.

6. Distribution of z

Figure 6 shows the differential cross section $d\sigma/dz$ for $p_T^2 > 1$ GeV^2 in the range $50 < W < 180$ GeV for ZEUS and in the range $30 < W < 150$ GeV for H1. It is compared to the NLO calculation [3] discussed in section 4. Agreement in shape and normalization is found within the errors.

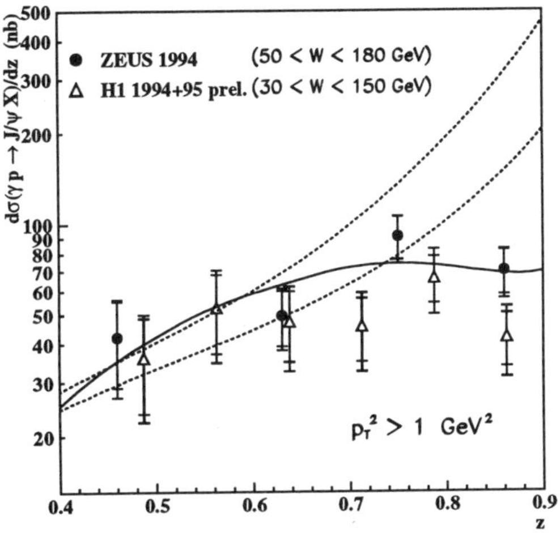

Fig. 6. Differential cross section $d\sigma/dz$ for the inelastic J/ψ for $p_T^2 > 1$ GeV^2. Data from ZEUS[1] and H1 [2] are shown. The inner error bars indicate the statistical uncertainties, the outer error bars the quadratic sum of the statistical and systematic uncertainties. The NLO computation [3] with the GRV [14] structure function, $m_c = 1.4$ GeV and $\Lambda_{QCD} = 300$ MeV is shown as a solid line. The two dashed lines are given by the sum of the colour-singlet and the colour-octet leading order calculations [9], using two different sets of nonperturbative terms. In the theoretical curves the 15% contribution of the ψ' has not been included.

Recently there has been theoretical activity attempting to solve the discrepancy between the J/ψ production cross section measurements in hadronic reactions and the colour-singlet model by invoking additional octet contributions [18]. A specific leading order calculation of J/ψ photoproduction at HERA has been carried out using values of the nonperturbative colour-octet terms determined from a fit [9] to the CDF data [7]. These calculations predict a cross section rising with z, which is not seen in the data. This is illustrated in Fig.6 where the two dashed lines show a sum of the colour-singlet and colour-octet contributions both calculated at leading order, using two different sets of values for the nonperturbative terms. The area between these two curves gives the size of the uncertainty in this type of calculation.

7. Conclusions

The ZEUS and H1 Collaborations have measured inelastic J/ψ photproduction in the range $30 < W < 180$ GeV and $z < 0.9$. The cross section rises with W. A NLO calculation for the photon-gluon fusion process agrees with the data both for the integrated cross section and the differential distributions over z and p_T^2 in the kinematic range $z < 0.8$ and $p_T^2 > 1$ GeV^2, using gluon distribution parametrizations compatible with those determined from the F_2 measurements performed at HERA. The predictions of a specific leading order colour-octet model, as formulated to fit the CDF data on J/ψ hadroproduction, are not consistent with the data.

1. ZEUS Collab., J. Breitweg et al., *DESY 97-147.*
2. H1 Collab., S. Aid et al., *Nucl. Phys.* **B472** (1996) 3;
   H1 Collab., Contribution to the Int. Conf. on HEP, Warsaw, 1996, ICHEP-PA02-085.
3. M. Krämer et al., *Phys. Lett.* **B348** (1995) 657;
   M. Krämer, *Nucl. Phys.* **B459** (1996) 3.
4. H. Jung, G.A. Schuler and J. Terron, *Int. Jour. of Mod. Phys.* **A7** (1992) 7955.
5. E.L. Berger and D. Jones, *Phys. Rev.* **D23** (1981) 1521.
6. R. Baier and R. Rückl, *Phys. Lett.* **B102** (1981) 364.
7. CDF Collab., F. Abe et al., *Phys. Rev. Lett.* **69** (1992) 3704;
   CDF Collab., A. Sansoni, *FERMILAB-CONF-95/263-E.*
8. G.T. Bodwin, E. Braaten and G.P. Lepage, *Phys. Rev.* **D51** (1995) 1125.
9. M. Cacciari and M. Krämer, *Phys. Rev. Lett.* **76** (1996) 4128.
10. FTPS Collab., B.H. Denby et al., *Phys. Rev. Lett.* **52** (1984) 795.
11. NA14 Collab., R. Barate et al., *Z. Phys.* **C33** (1987) 505.
12. EMC Collab., J. Ashman et al., *Z. Phys.* **C56** (1992) 21.
13. A. D. Martin, W. J. Stirling and R. G. Roberts, *Phys. Lett.* **B354** (1995) 155.
14. M. Glück, E. Reya and A. Vogt, *Z. Phys.* **C53** (1992) 127;
    M. Glück, E. Reya and A. Vogt, *Z. Phys.* **C53** (1992) 651.
15. H.L. Lai et al., *Phys. Rev.* **D51** (1995) 4763.
16. ZEUS Collab., M. Derrick et al., *Phys. Lett.* **B316** (1993) 412;
    ZEUS Collab., M. Derrick et al., *Z. Phys.* **C65** (1995) 379;
    ZEUS Collab., M. Derrick et al., *Z. Phys.* **C69** (1996) 607;
    ZEUS Collab., M. Derrick et al., *Z. Phys.* **C72** (1996) 399.
17. H1 Collab., I. Abt et al., *Nucl. Phys.* **B407** (1993) 515;
    H1 Collab., T. Ahmed et al., *Nucl. Phys.* **B439** (1995) 471;
    H1 Collab., S. Aid et al., *Nucl. Phys.* **B470** (1996) 3;
    H1 Collab., C. Adloff et al., *DESY 97-042.*
18. E. Braaten and S. Fleming, *Phys. Rev. Lett.* **74** (1995) 3327;
    P. Cho and A.K. Leibovich, *Phys. Rev.* **D53** (1996) 6203;
    B. Cano-Coloma and M.A. Sanchis-Lozano, *IFIC/97-29, hep-ph/9706270.*

INCLUSIVE J/ψ PHOTOPRODUCTION AT HERA

MICHAEL KRÄMER
Rutherford Appleton Laboratory
Chilton, OX11 0QX, UK
E-mail: Michael.Kraemer@rl.ac.uk

ABSTRACT

I discuss the impact of colour-octet contributions to inclusive J/ψ photoproduction at HERA. Emphasis is put on resolved photon processes and on J/ψ polarization, which will be experimentally accessible at HERA in the near future. Both analyses provide a powerful test of the NRQCD factorization approach to charmonium production.

1. Introduction

The production of heavy quarkonium states in high-energy collisions provides an important tool to study the interplay between perturbative and non-perturbative QCD dynamics. While the creation of heavy quarks in a hard scattering process can be calculated in perturbative QCD, the subsequent transition to a physical bound state introduces non-perturbative aspects. A rigorous framework for treating quarkonium production and decays has been developed only recently.[1] The factorization approach is based on the use of non-relativistic QCD (NRQCD)[2] to separate the short-distance parts from the long-distance matrix elements and explicitly takes into account the complete structure of the quarkonium Fock space. According to this formalism, the inclusive cross section for the production of a quarkonium state H can be expressed as a sum of terms, each of which factors into a short-distance coefficient and a long-distance matrix element:

$$\mathrm{d}\sigma(H+X) = \sum_n \mathrm{d}\hat{\sigma}(Q\overline{Q}\,[n]+X)\,\langle \mathcal{O}^H\,[n]\rangle \tag{1}$$

Here, $\mathrm{d}\hat{\sigma}$ denotes the short-distance cross section for producing an on-shell $Q\overline{Q}$-pair in a colour, spin and angular-momentum state labelled by n. The universal NRQCD matrix elements $\langle \mathcal{O}^H\,[n]\rangle$ give the probability for a $Q\overline{Q}$-pair in the state n to form the quarkonium state H. The relative importance of the various terms in (1) can be estimated by using NRQCD velocity scaling rules.[3] For $v \to 0$ (v being the average velocity of the heavy quark in the quarkonium rest frame) each of the NRQCD matrix elements scales with a definite power of v and the general expression (1) can be organized into an expansion in powers of the heavy quark velocity.

The general factorization formula (1) is believed to hold also for quarkonium production in hadron-hadron or photon-hadron collisions. In the case of inclusive open heavy flavour production it has been shown that the hadronic cross section factorizes into a partonic hard scattering cross section multiplied by light quark and gluon parton densities, with higher-twist corrections being suppressed by powers $\Lambda_{\rm QCD}/m_Q$.[4] The argument involves averaging over a sufficiently large range of the heavy quark transverse momentum p_T and uses the fact that the main contribution to the total inclusive cross section comes from $p_T \sim m_Q \gg \Lambda_{\rm QCD}$. Unfortunately, the situation is more involved for quarkonia production in hadronic collisions, since the region of small transverse $Q\overline{Q}$ momentum significantly contributes to the total cross section. At small transverse momentum however, the intermediate colour-singlet or colour-octet $Q\overline{Q}$ pair moves parallel with a remnant jet and might interact before the physical bound state has been formed.[a] Only in the heavy quark limit, where all scales involved in the bound state physics are much larger than $\Lambda_{\rm QCD}$, higher-twist terms are necessarily strongly suppressed. For charmonium production in hadronic collisions, it seems however reasonable to restrict the analysis to the region $p_T \gg \Lambda_{\rm QCD}$, where the general factorization formula (1) should safely be applicable. Excluding the small p_T region is also necessary to guarantee perturbative stability in higher-order QCD calculations of J/ψ photoproduction.[6,7]

The NRQCD formalism implies that colour-octet processes, in which the heavy-quark antiquark pair is produced at short distances in a colour-octet state and subsequently evolves nonperturbatively into a physical quarkonium, must contribute to the cross section. As discussed extensively in the literature[8,9], colour-octet contributions appear as the most plausible explanation of the large direct ψ production cross section observed at the Tevatron[10]. The NRQCD approach is certainly the correct theory for quarkonium production in the heavy quark limit. It is however not clear whether the charm quark mass is sufficiently large to allow for a reliable expansion in the heavy quark velocity. In order to establish the phenomenological significance of the colour-octet mechanism and the universality of the NRQCD matrix elements for charmonium production it is thus necessary to identify colour-octet contributions in different production processes. The analysis of J/ψ photoproduction at HERA appears to be a very powerful tool to constrain the colour-octet matrix elements and to test the picture of quarkonium production as developed in the context of the NRQCD factorization approach.

The production of J/ψ particles in high energy ep collisions at HERA is dominated by photoproduction events where the electron is scattered by a small angle producing photons of almost zero virtuality. The measurements at HERA provide information on the dynamics of J/ψ photoproduction in a wide kinematical region, 30 GeV $\lesssim \sqrt{s_{\gamma p}} \lesssim$ 200 GeV, corresponding to initial photon energies in a fixed-target experiment of 450 GeV $\lesssim E_\gamma \lesssim$ 20,000 GeV. The production of J/ψ particles in photon-proton collisions proceeds predominantly through photon-gluon fusion, where the photon interacts directly with the gluon from the proton. Besides the direct photoproduction channel, J/ψ production at HERA can also take place via resolved

[a]Higher-twist corrections to colour-singlet J/ψ photoproduction have been analysed recently.[5]

photon contributions where the photon behaves as a source of partons which interact with the partons in the proton. Resolved processes are expected to contribute significantly to the lower endpoint of the J/ψ energy spectrum and might be probed at HERA for the first time in the near future.[b]

2. Direct Photon Contributions

For J/ψ production and at leading order in the velocity expansion, the general factorization formula (1) reduces to the standard expression of the colour-singlet model[12]. The short-distance cross section for J/ψ photoproduction through direct photons is given by the subprocess

$$\gamma + g \to c\bar{c}\,[^3S_1, \underline{1}] + g \tag{2}$$

with $c\bar{c}$ in a colour-singlet state (denoted by $\underline{1}$), with zero relative velocity, and spin/angular-momentum quantum numbers $^{2S+1}L_J = {}^3S_1$. Relativistic corrections to the colour-singlet channel due to the motion of the charm quarks in the J/ψ bound state enhance the small-p_t region, but can be neglected for $p_T \gtrsim 1$ GeV.[13] The calculation of the higher-order perturbative QCD corrections to the short-distance cross section (2) has been performed recently.[6,7] Inclusion of the NLO corrections reduces the scale dependence of the theoretical prediction and increases the cross section significantly, depending in detail on the γp energy and the choice of parameters.[7] Details of the calculation and a comprehensive analysis of total cross sections and differential distributions for the energy range of the fixed-target experiments and for J/ψ photoproduction at HERA can be found elsewhere.[7]

The leading colour-octet configurations which contribute to J/ψ photoproduction at $p_T \neq 0$ are produced through the subprocesses[14]

$$\begin{aligned} \gamma + g &\to c\bar{c}\,[^1S_0, {}^3S_1, {}^3P_{0,1,2}, \underline{8}] + g \\ \gamma + q(\bar{q}) &\to c\bar{c}\,[^1S_0, {}^3S_1, {}^3P_{0,1,2}, \underline{8}] + q(\bar{q}) \end{aligned} \tag{3}$$

The transition of the colour-octet $c\bar{c}\,[^{2S+1}L_J, \underline{8}]$ pairs into a physical J/ψ state through the emission of non-perturbative gluons is described by the long-distance matrix elements $\langle \mathcal{O}^{J/\psi}\,[^{2S+1}L_J, \underline{8}]\rangle$. They have to be obtained from lattice simulations or measured directly in some production process. According to the velocity scaling rules of NRQCD, the colour-octet matrix elements associated with S-wave quarkonia should be suppressed by a factor of $v^4 \sim 10^{-2}$ compared to the leading colour-singlet matrix element. However, explicit calculation of the processes (3) shows that the short-distance factors of the $[^1S_0, \underline{8}]$ and $[^3P_{0,2}, \underline{8}]$ channels are strongly enhanced as compared to the colour-singlet term and compensate the $\mathcal{O}(10^{-2})$ suppression of the

[b]Elastic mechanisms only contribute to the region of small J/ψ transverse momentum and can be suppressed by an appropriate cut in p_T.[11]

corresponding non-perturbative matrix elements. This short-distance enhancement is due to t-channel gluon exchange which contributes to the $[{}^1S_0, {}^3P_{0,2}, \underline{8}]$ cross sections, but not to the leading-order colour-singlet channel.

3. Resolved photon contributions

Inclusive J/ψ photoproduction can also take place via resolved photon interactions, where the photon first splits into a flux of light quarks and gluons which then may fuse with a gluon or light quark from the proton to form a $c\bar{c}$ pair. This process has been extensively analyzed in the past in the framework of the colour-singlet model, where the cross section is given by gluon-gluon fusion into a ${}^3S_1$ colour-singlet $c\bar{c}$ pair, which subsequently hadronizes into a J/ψ:

$$g + g \to c\bar{c}\,[{}^3S_1, \underline{1}] + g \tag{4}$$

The process (4) contributes to the overall cross section only marginally except near the lower endpoint of the J/ψ energy spectrum.[15]

Within the NRQCD factorization approach, however, more resolved-photon channels have to be considered, where J/ψ production proceeds via colour-octet $c\bar{c}$ pairs.[16] The leading colour-octet contributions are:

$$\begin{aligned} g + g &\to c\bar{c}\,[{}^1S_0, {}^3S_1, {}^3P_{0,1,2}, \underline{8}] + g \\ g + q(\bar{q}) &\to c\bar{c}\,[{}^1S_0, {}^3S_1, {}^3P_{0,1,2}, \underline{8}] + q(\bar{q}) \\ q + \bar{q} &\to c\bar{c}\,[{}^1S_0, {}^3S_1, {}^3P_{0,1,2}, \underline{8}] + g \end{aligned} \tag{5}$$

They are in every respect analogous to the ones which have been argued to strongly increase the J/ψ production cross section in $p\bar{p}$ collisions at the Tevatron.[8,9] Colour-octet contributions to resolved J/ψ photoproduction thus provide an important direct test of the NRQCD explanation of the Tevatron data.

4. The J/ψ Energy Distribution

The J/ψ energy distribution $\mathrm{d}\sigma/\mathrm{d}z$ offers unique possibilities to study the relative importance of the different mechanisms that contribute to J/ψ photoproduction. The scaling variable z is defined by $z = p \cdot k_\psi \,/\, p \cdot k_\gamma$, $p, k_{\psi,\gamma}$ being the momenta of the proton and J/ψ, γ particles, respectively. In the proton rest frame, z is the ratio of J/ψ to γ energy, $z = E_\psi/E_\gamma$. Adopting NRQCD matrix elements consistent with those extracted from the fits to prompt J/ψ data at the Tevatron[9] (see Table 1), one finds significant colour-octet contributions near the upper and lower endpoint of the J/ψ energy spectrum. This is demonstrated in Fig. 1, where I plot the colour-singlet and colour-octet contributions to $\mathrm{d}\sigma/\mathrm{d}z$ at a typical HERA energy of $\sqrt{s_{\gamma p}} = 100$ GeV in the restricted range $p_\perp \geq 1$ GeV via direct and resolved photon

Table 1. Values of the NRQCD matrix elements used in the numerical analysis, with the velocity and mass scaling.

$\langle \mathcal{O}^{J/\psi}\,[^3S_1,\underline{1}]\rangle$	$1.16\ \mathrm{GeV}^3$	$m_c^3 v^3$
$\langle \mathcal{O}^{J/\psi}\,[^1S_0,\underline{8}]\rangle$	$10^{-2}\ \mathrm{GeV}^3$	$m_c^3 v^7$
$\langle \mathcal{O}^{J/\psi}\,[^3S_1,\underline{8}]\rangle$	$10^{-2}\ \mathrm{GeV}^3$	$m_c^3 v^7$
$\langle \mathcal{O}^{J/\psi}\,[^3P_0,\underline{8}]\rangle/m_c^2$	$10^{-2}\ \mathrm{GeV}^3$	$m_c^3 v^7$

processes.[c] The colour-octet processes are predicted to exceed the colour-singlet ones by more than an order of magnitude for $z \gtrsim 0.6$ and $z \lesssim 0.2$. Resolved contributions are negligible for $z \gtrsim 0.3$, but colour-octet terms are expected to become significant at the lower end of the energy spectrum.[d] The theoretical predictions are compared to experimental results obtained by the H1[18] and ZEUS[19] collaborations at HERA. From Fig. 1 one can conclude that the large colour-octet enhancement in the region $z \gtrsim 0.6$ is not supported by the experimental data and that the J/ψ energy spectrum is adequately described by the colour-singlet channel. Taking into account the large uncertainty of the cross section normalization due to the choice for the charm quark mass and the QCD coupling, the current experimental data on the J/ψ energy distribution can be accounted for by the colour-singlet channel alone. This can be judged by comparing the experimental data with the shaded band in Fig. 1 which indicates how much the colour-singlet cross section is altered if m_c and $\Lambda^{(4)}$ vary in the range $1.35\ \mathrm{GeV} < m_c < 1.55\ \mathrm{GeV}$ and $200\ \mathrm{MeV} < \Lambda^{(4)} < 300\ \mathrm{MeV}$.[e]

One should be careful to interpret the discrepancy between the theoretical predictions for the colour-octet contributions and the HERA data shown in Fig. 1 as a failure of the factorization approach. As pointed out recently, the NRQCD expansion breaks down close to the phase space boundary $z \to 1$, and no prediction can be made for $z \gtrsim 0.75$ without resumming large corrections in the velocity expansion.[21] Therefore, the shape of $d\sigma/dz$ near the endpoint region cannot be used to access the importance of colour-octet contributions. Nevertheless, the analysis of the the J/ψ energy distribution provides a crucial constraint on the colour-octet matrix elements if one averages over a sufficiently large interval $\Delta z \gg v^2 \sim 0.25$ containing the endpoint $z = 1$, as shown in Fig. 2. Colour-octet mechanisms should contribute significantly to the region $0.6 \lesssim z \leq 1$ and exceed the colour-singlet cross section by more than an order of magnitude. While the overall size of the cross section is subject to large uncertainties due to m_c and α_s, the distinctive modification of the J/ψ energy spectrum by colour-octet processes in the low and large z region should clearly be

[c] Note that a cut in the J/ψ transverse momentum, $p_t \gtrsim 1$ GeV, is sufficient to exclude the elastic contributions; no additional cut in z is required.

[d] Similarly, colour-octet processes enhance the low-z region in J/ψ photoproduction at large p_T via fragmentation mechanisms.[17]

[e] To study the α_s dependence of the cross section, I use consistently adjusted sets of parton densities.[20]

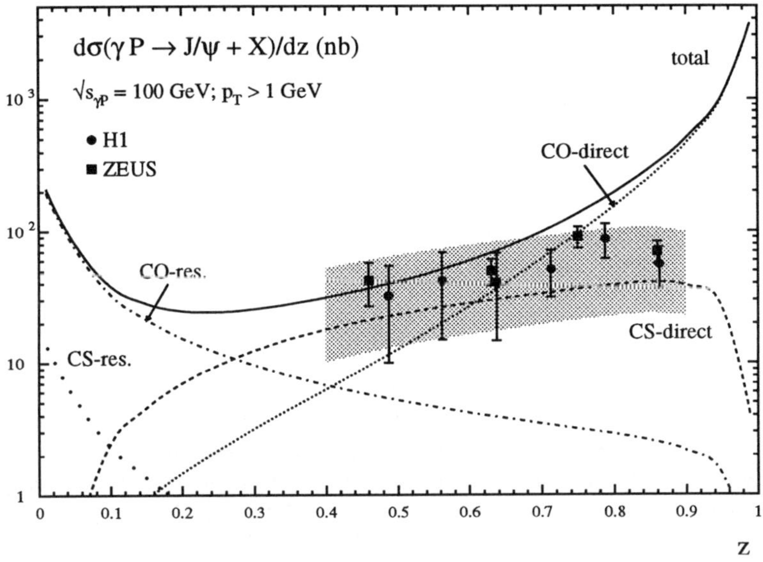

Fig. 1. Colour-singlet (CS) and colour-octet (CO) contributions to the J/ψ energy distribution $d\sigma/dz$ at the photon-proton centre of mass energy $\sqrt{s_{\gamma p}} = 100$ GeV integrated in the range $p_T \geq 1$ GeV, compared to experimental data[18,19]. Both direct and resolved photon processes are included. Parameters: $m_c = 1.5$ GeV, renormalization/factorization scale $\mu = 2m_c$, GRV parton distribution functions[20] with $\Lambda^{(4)} = 200$ MeV. The shaded band reflects the uncertainty in the normalization of the direct colour-singlet contributions due to variation of m_c and α_s, as described in the text.

visible in the experimental data. Moreover, the low-z region of the energy spectrum is not expected to be sensitive to higher order terms in the velocity expansion and can thus reliably be predicted within leading-twist NRQCD.

The results at HERA seem to indicate that the values of the colour-octet matrix elements $\langle \mathcal{O}^{J/\psi}[^1S_0, \underline{8}]\rangle$ and $\langle \mathcal{O}^{J/\psi}[^3P_0, \underline{8}]\rangle$ are considerably smaller than suggested by the fits to the CDF data at the Tevatron. This does however not necessarily imply the non-universality of the NRQCD matrix elements for J/ψ production. The determination of the $\langle \mathcal{O}^{J/\psi}[^1S_0, \underline{8}]\rangle$ and $\langle \mathcal{O}^{J/\psi}[^3P_0, \underline{8}]\rangle$ matrix elements from the Tevatron data is very sensitive to all effects that modify the shape of the J/ψ transverse momentum distribution at $p_t \lesssim 5$ GeV and might yield much too large values.[22] The theoretical uncertainties include the small-x behaviour of the gluon distribution, the evolution of the strong coupling, next-to-leading order contributions, intrinsic transverse momentum of the partons in the proton, and also systematic effects inherent in NRQCD, such as the inaccurate treatment of the energy conservation in the hadronization of

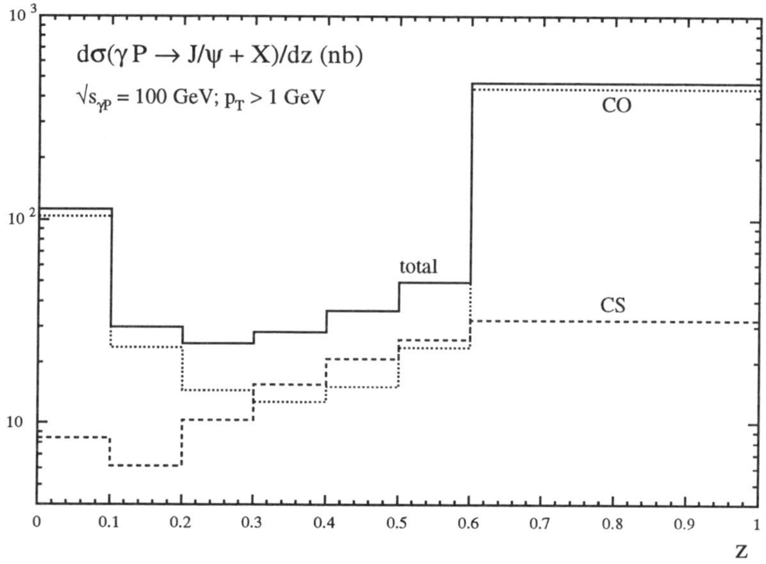

Fig. 2. Same as Fig. 1, averaged over large intervals in z.

the colour-octet $c\bar{c}$ pairs. With higher statistics data at HERA it will be possible to extract more detailed information on the colour-octet matrix elements, in particular from the analysis of the J/ψ energy distribution.

5. J/ψ Polarization

The NRQCD factorization approach allows for unambiguous predictions of the quarkonium polarization.[23] Transverse polarization of J/ψ and ψ' particles, produced directly at large transverse momentum in $p\bar{p}$ collisions at the Fermilab Tevatron, has in fact emerged as the most prominent test of colour-octet contributions and spin symmetry in charmonium production.[24] Polarization signatures in J/ψ photoproduction have been analysed in detail in the past within the colour-singlet model.[25] Recently, a comprehensive calculation has been performed in the context of the NRQCD factorization approach, including colour-singlet and colour-octet contributions to both direct as well as resolved photon processes.[26] Results are available for the most general decay angular distribution in any given reference frame. Here, I only consider the polar-angle distribution in the s-channel helicity frame, in which the polarization axis in the J/ψ rest frame is defined as the direction of the J/ψ three-momentum in the photon-proton cms frame. The polar angular distribution in the decay $J/\psi \to l^+l^-$ is given by

$$\frac{\mathrm{d}\Gamma}{\mathrm{d}\cos\theta} \propto 1 + \alpha\cos^2\theta, \tag{6}$$

with θ the angle between the lepton three-momentum in the J/ψ rest frame and the polarization axis. Fig. 3 displays colour-singlet and colour-octet contributions to the polar asymmetry α, defined in (6), as function of the J/ψ energy variable z. It is worth

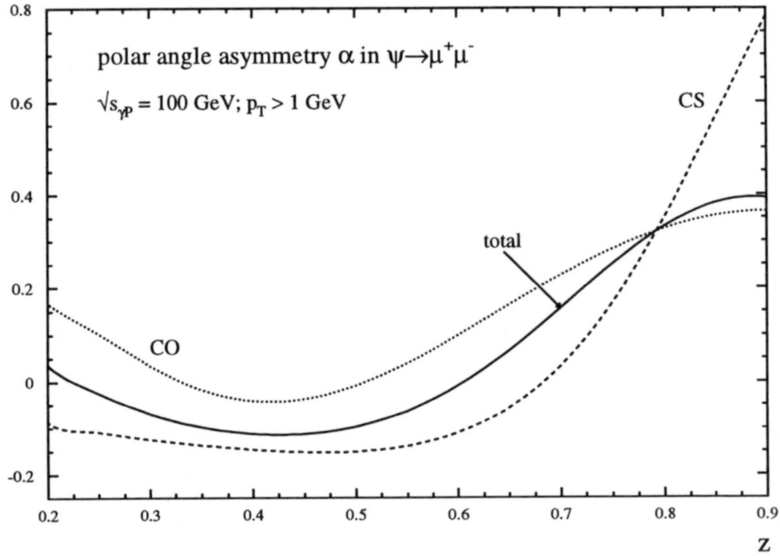

Fig. 3. Colour-singlet (CS) and colour-octet (CO) contributions to the polar angle asymmetry α in (6) at the photon-proton centre of mass energy $\sqrt{s_{\gamma p}} = 100$ GeV integrated in the range $p_T \geq 1$ GeV. Both direct and resolved photon processes are included.

pointing out that the asymmetry predictions involve ratios of cross sections and are thus to a large extent insensitive to uncertainties in the charm quark mass, the QCD coupling or parton distribution functions. Therefore, polarization measurements are a rather clean test of the underlying quarkonium production mechanisms and will allow to access the relative importance of colour-singlet and colour-octet contributions.

6. Conclusions

I have discussed colour-singlet and colour-octet contributions to J/ψ photoproduction at HERA. The high-statistics data to be expected in the near future will allow the analysis of resolved photon processes and J/ψ polarization to further test the phenomenological significance of colour-octet contributions. Various other channels and final states can be accessed in the future at HERA, like photoproduction of ψ', Υ and χ_c particles[16,27], associated $J/\psi + \gamma$ production[16,28,29], fragmentation contributions at large p_T[30,17] as well as deep inelastic J/ψ production[31]. The future analyses of inclusive quarkonium production at HERA offer unique possibilities to test the NRQCD factorization approach to charmonium production.

7. Acknowledgements

I would like to thank Martin Beneke, Matteo Cacciari and Mikko Vänttinen for their collaboration on quarkonium production.

8. References

1. G.T. Bodwin, E. Braaten and G.P. Lepage, Phys. Rev. **D51** (1995) 1125
2. W.E. Caswell and G.P. Lepage, Phys. Lett. **B167** (1986) 437
3. G.P. Lepage, L. Magnea, C. Nakhleh, U. Magnea and K. Hornbostel, Phys. Rev. **D46** (1992) 4052
4. J.C. Collins, D.E. Soper and G. Sterman, Nucl. Phys. **B263** (1986) 37
5. J.P. Ma, Nucl. Phys. **B498** (1997) 267; see also S.J. Brodsky, W.-K. Tang and P. Hoyer, Phys. Rev. **D52** (1995) 6285
6. M. Krämer, J. Zunft, J. Steegborn and P.M. Zerwas, Phys. Lett. **B348** (1995) 657
7. M. Krämer, Nucl. Phys. **B459** (1996) 3
8. E. Braaten and S. Fleming, Phys. Rev. Lett. **74** (1995) 3327
9. M. Cacciari, M. Greco, M.L. Mangano and A. Petrelli, Phys. Lett. **B356** (1995) 560; P. Cho and A.K. Leibovich, Phys. Rev. **D53** (1996) 150 and *ibid.* **53** (1996) 6203; M. Beneke and M. Krämer, Phys. Rev. **D55** (1997) 5269
10. F. Abe et al. [CDF Collab.], Phys. Rev. Lett. **69** (1992) 3704; A. Sansoni, [CDF Collab.], FERMILAB-CONF-95/263-E
11. W. Koepf, P.V. Landshoff, E.M. Levin and N.N. Nikolaev, Proc. "Future Physics at HERA", Eds. A. de Roeck, G. Ingelman and R. Klanner, DESY, Hamburg, 1996 and references therein; M. Diehl, these proceedings
12. E.L. Berger and D. Jones, Phys. Rev. **D23** (1981) 1521; R. Baier and R. Rückl, Phys. Lett. **102B** (1981) 364
13. W.Y. Keung and I.J. Muzinich, Phys. Rev. **D27** (1983) 1518; H. Jung, D. Krücker, C. Greub and D. Wyler, Z. Phys. **C60** (1993) 721; H. Khan and P. Hoodbhoy Phys. Lett. **B382** (1996) 189
14. M. Cacciari and M. Krämer, Phys. Rev. Lett. **76** (1996) 4128; P. Ko, J. Lee and H.S. Song, Phys. Rev. **D54** (1996) 4312

15. H. Jung, G.A. Schuler and J. Terrón, Proc. "Physics at HERA", Eds. W. Buchmüller and G. Ingelman, DESY, Hamburg, 1992 and Int. J. Mod. Phys. A7 (1992) 7955
16. M. Cacciari and M. Krämer, Proc. "Workshop on Future Physics at HERA", Eds. G. Ingelman, A. De Roeck and R. Klanner, DESY Hamburg, 1996
17. B.A. Kniehl and G. Kramer, DESY-97-036 (hep-ph/9703280) and DESY-97-110 (hep-ph/9706369)
18. S. Aid et al. [H1 Collab.], Nucl. Phys. **B472** (1996) 3
19. M. Derrick et al. [ZEUS Collab.], presented by L. Stanco at the International Workshop on Deep Inelastic Scattering, Rome, April 1996; R. Brugnera, these proceedings
20. M. Glück, E. Reya and A. Vogt, Z. Phys. **C67** (1995) 433; A. Vogt, Phys. Lett. **B354** (1995) 433
21. M. Beneke, I.Z. Rothstein and M.B. Wise, CERN-TH-97-086 (hep-ph/9705286)
22. M. Beneke and M. Krämer[9]; B. Cano Coloma, M.A. Sanchis Lozano, IFIC-97-29 (hep-ph/9706270)
23. M. Beneke and I.Z. Rothstein, Phys. Lett. **B372** (1996) 157 and Phys. Rev. **D54** (1996) 2005 [erratum: *ibid.* **D54** (1996) 7082]; E. Braaten and Y.-Q. Chen, Phys. Rev. **D54** (1996) 3216; M. Beneke and M. Krämer[9]
24. P. Cho and M.B. Wise, Phys. Lett. **B346** (1995) 129
25. R. Baier and R. Rückl, Nucl. Phys. **B201** (1982) 1; J.G. Körner, J. Cleymans, M. Kuroda and G.J. Gounaris, Nucl. Phys. **B204** (1982) 6 [erratum: *ibid.* **B213 (1983) 546**] and Phys. Lett. **114B** (19195) 82
26. M. Beneke, M. Krämer and M. Vänttinen, to be published
27. J.P. Ma, Nucl. Phys. **B460** (1996) 109
28. T. Mehen, Phys. Rev. **D55** (1997) 4338
29. M. Cacciari, M. Greco and M. Krämer, Phys. Rev. **D55** (1997) 7126
30. R. Godbole, D.P. Roy and K. Sridhar, Phys. Lett. **B373** (1996) 328
31. S. Fleming, MADPH-96-966 (hep-ph/9610372), talk at "Quarkonium Physics Workshop: Experiment Confronts Theory", Chicago, IL, 13-15 Jun 1996; S. Fleming and T. Mehen, JHU-TIPAC-96022A (hep-ph/9707365)

Chapter 8

Elastic and Diffractive *ep* Scattering

Elastic and Proton-dissociative Vector Meson Production at Small and Large Q^2 at HERA

Robert KLANNER
University of Hamburg
E-mail: klanner@desy.de

presented for the H1- and ZEUS-Collaborations.

ABSTRACT

This talk presents recent HERA results on the elastic and proton-dissociative production of the vector mesons ρ^0, ω, ϕ, and J/ψ from photons in the Q^2-range between 0 and 30 GeV^2 and the negative momentum transfer squared to the proton, $|\,t\,|$, between 0 and 3 GeV^2. The data are compared to results from lower energy experiments and various predictions.

1. Introduction

Elastic lepto production of vectors mesons (VM) as shown in Fig. 1 gives access to the reaction $\gamma^* p \rightarrow VMp$, which, similar to elastic scattering in hadron-hadron interactions, represents the simplest photon-hadron diffractive reaction. At the electron-proton collider HERA elastic VM-production is studied in a new range of $\gamma^* p$ center of mass energy $W_{\gamma^* p}$, up to $\sim$ 200 GeV, for photon virtualities Q^2 between 10^{-10} and 30 GeV^2 and, in the case of quasi-real photoproduction, for negative momentum transfers squared between the incoming and outgoing hadronic system $|\,t\,|$ up to 5 GeV^2. Thus it is possible to investigate the transition from the well studied soft QCD-regime of photoproduction of light vector mesons at low momentum transfers into the perturbative QCD (p-QCD) regime for three different energy scales:

- the mass m_{VM} of the vector meson,
- the square root of the photon virtuality Q,
- the momentum transfer to the outgoing baryonic system $\sqrt{|\,t\,|}$.

The expectations for the production characteristics are : the Vector-Dominance-Model (VDM) and the soft Pomeron picture for the soft, non-perturbative domain, but p-QCD if at least one of the energy scales is hard. The signatures of "soft" are a weak energy dependence of the cross-section, the validity of the additive quark model, when taking into account the γ-VM couplings, a sharp forward diffractive peak in the t-distribution and a Q^2-dependence given by the VM-propagator term. The signatures of p-QCD, where VM-production can be described by the exchange of a gluon ladder[1], are: a strong increase of the cross-section with energy, reflecting the strong rise of the

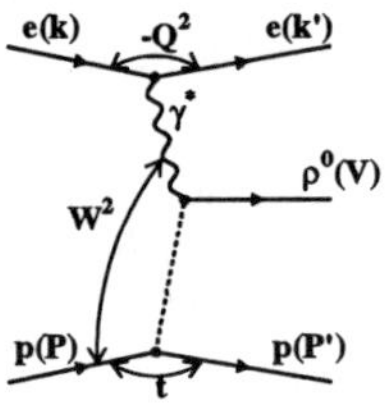

Figure 1: Diagram of elastic electro-production of ρ^0-mesons.

gluon distribution in the proton with decreasing $x_{Bjorken}$-value observed at HERA, a significant contribution from longitudinal photons, and production cross section ratios for the vector mesons with values not too different from $\rho^0 : \omega : \phi : J/\psi \sim 1 : 1/9 : 2/9 : 8/9$. These values are obtained from the charges of the valence quarks of the vector mesons.

To measure $\gamma^* p \to VMp$ at HERA, H1 and ZEUS use the following experimental techniques:

- The decay products of the vector mesons are measured in the central detector.
- Untagged low $Q^2 < 4$ GeV^2 and low $|t| < 1$ GeV^2: Neither the scattered positron nor the scattered proton are detected. $W_{\gamma^* p}$ is obtained from energy-longitudinal momentum $(E - p_z)$ conservation, and the transverse momentum p_t of the vector meson is used to estimate the momentum transfer to the proton. The background from events with diffractively excited protons is subtracted statistically. In a subsample of the untagged events ($\sim 5\%$) the ZEUS experiment measures the proton in the leading proton spectrometer LPS (proton-tagged events). Both H1 and ZEUS use forward (in the proton beam direction) detectors to identify events with diffractively excited protons.
- Tagged low $Q^2 < 0.01$ GeV^2 and high $|t| > 1$ GeV^2: The scattered electron, with an energy corresponding to $W_{\gamma^* p} = 70 - 120$ GeV, is tagged in a small angle electromagnetic calorimeter (the 44 m tagger). The value of $W_{\gamma^* p}$ is obtained from $(E - p_z)$ conservation, and the value of t from transverse momentum balance. The results presented contain both events with elastic and diffractively excited protons - the latter dominating at high $|t|$.
- Tagged intermediate Q^2 (0.25 to 0.85 GeV^2) and low $|t| < 1$ GeV^2: The electron is measured using the small angle beam-pipe-calorimeter (ZEUS-BPC). The kinematic quantities are reconstructed using the precisely measured angle of the scattered electron and the momenta of the decay products of the vector meson. Events with diffractively excited protons are subtracted statistically.

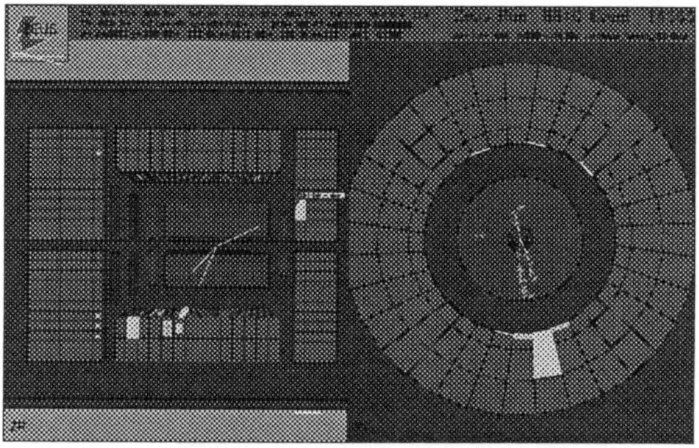

Fig. 2. A DIS ρ^0-event from the ZEUS experiment.

- High $Q^2 > 1$ GeV^2 (DIS) and low $|t| < 1$ GeV^2: Both H1 and ZEUS measure the positron in their main calorimeters. The reconstruction of the kinematic variables and the subtraction of the diffractive background is identical to tagged intermediate Q^2 vector meson production.

As an illustration Fig. 2 shows a $\rho^0 \to \pi^+\pi^-$- DIS-event measured in the ZEUS-detector, with $W_{\gamma^* p} \sim 100$ GeV and $Q^2 = 120$ GeV^2.

2. Photoproduction

2.1. Results at small $|t|$

Fig.3 shows the measured integrated cross-sections for untagged elastic photoproduction of ρ^0-, ω-, ϕ- and J/ψ-mesons as function of $W_{\gamma p}$ for $|t| < 0.5$ GeV^2 together with the total photoproduction cross-section[2]. To guide the eyes, curves $\sigma \propto W_{\gamma p}^a$ are drawn through the HERA data points, where from VDM and soft Pomeron exchange $a \sim 0.16$ is expected for the total cross-section and $a \sim 0.22$ for elastic vector meson production.[a] It can be concluded, that for $W_{\gamma p} > 10$ GeV the VDM prediction gives a good description of the elastic production of light vector mesons. J/ψ-production however, has a much steeper $W_{\gamma p}$-dependence with a value $a \sim 0.8$, as expected from p-QCD taking into account the steeply rising gluon density in the proton with decreasing $x_{Bjorken}$.

[a]The value is derived from the Regge formula $d\sigma/d\,|t| \propto (W_{\gamma p}^2)^{2\alpha(t)-2}$, with $\alpha(t) = 1.08 + \alpha' \cdot t$ and the measured slope of $d\sigma/d\,|t|$

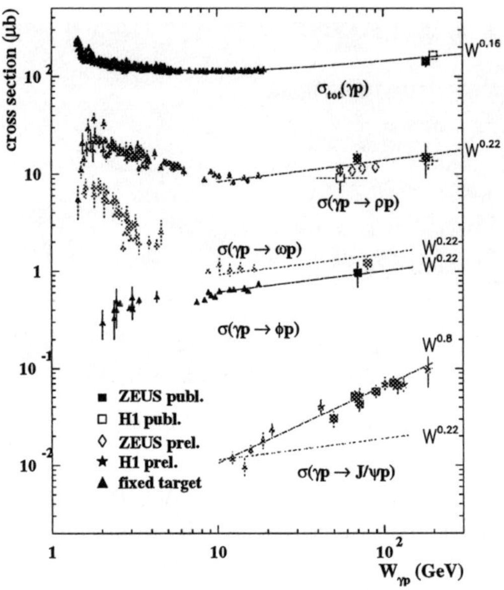

Fig. 3. The total and elastic VM-cross sections for the vector mesons ρ^0, ω, ϕ and J/ψ as a function of $W_{\gamma p}$ in the photoproduction regime.

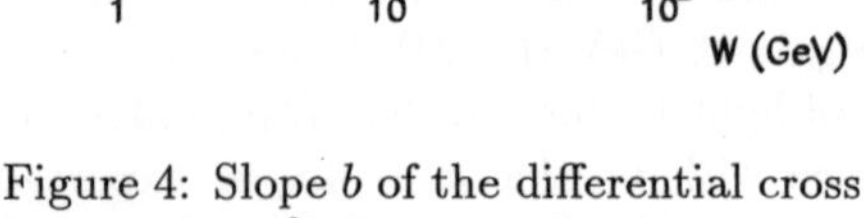

Figure 4: Slope b of the differential cross section for ρ^0-photo production versus energy.

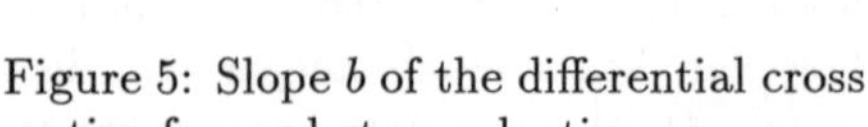

Figure 5: Slope b of the differential cross section for ω-photo production versus energy.

For the same reactions the slopes b of the differential cross-section $d\sigma/dt \sim e^{-b\cdot t}$ have been determined[2] as shown in Figs. 4, 5, 6 and 7. To guide the eyes, lines $b(W_{\gamma p}) = b_0 + 2\alpha' \cdot \ln(W_{\gamma p}^2)$ with a slope of the soft Pomeron trajectory $\alpha' = 0.25$, are drawn. They provide a good description of all the data. A fit to the ZEUS-data on ρ^0-production gives a value[b] of $\alpha' = 0.23 \pm 0.15^{+0.07}_{-0.10}$ GeV^{-2} and to the H1-data on J/ψ-production of $\alpha' = 0.36 \pm 0.21$ GeV^{-2}. Data with significantly higher accuracy are required to determine if the value of α' changes between the soft-QCD and the p-QCD regime. At a given $W_{\gamma p}$ the values of b decrease with increasing VM-mass, reflecting a decrease of the corresponding VMp interaction radius.

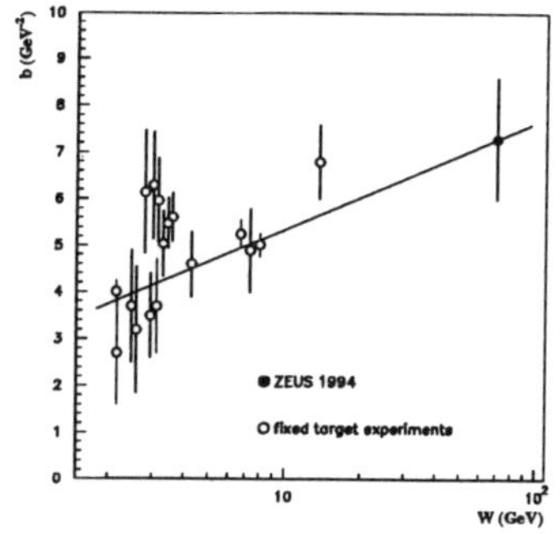

Figure 6: Slope b of the differential cross section for ϕ-photo production versus energy.

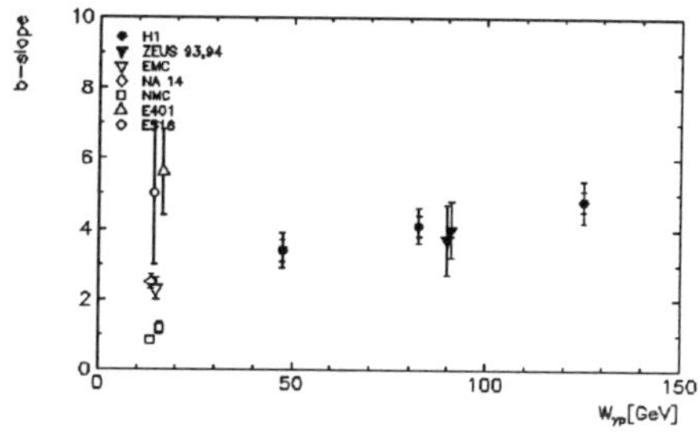

Figure 7: Slope b of the differential cross section for J/ψ-photo production versus energy.

Within the framework of the VDM and the additive quark model, the elastic VM-photoproduction is related to the total $VM\ p$ cross-sections, which in turn is related to the total πp- and Kp-cross-sections:

$$\sigma_{\gamma p \to VMp} = \frac{1}{16\pi} \cdot \frac{\alpha}{\frac{f_{VM}^2}{4\pi}} \cdot \frac{\sigma_{tot,VMp}^2}{b} \quad ,$$

where f_{VM} is the VM decay constant. From the ZEUS-1994 data[3] we find values $\sigma_{tot,\rho^0 p} = 28.0 \pm 1.2 \pm 2.8$ mb and $\sigma_{tot,\omega p} = 26.0 \pm 2.5^{+3.0}_{-4.2}$ mb, to be compared to the value from the additive quark model of $1/2(\sigma_{tot,\pi^+ p} + \sigma_{tot,\pi^- p}) = 28$ mb. The corresponding values for ϕ-production are: $\sigma_{tot,\phi p} = 19.0 \pm 7.0$ mb and $\sigma_{tot,K^+ p} + \sigma_{tot,K^- p} - \sigma_{tot,\pi^+ p}$ = 19.9 mb. Thus, within experimental errors, VDM and the additive quark model work at HERA for the photoproduction of ρ^0-, ω- and ϕ-mesons.

The H1 and ZEUS-experiments have also measured the spin density matrix elements of the $\rho^0, \omega\phi$, and the J/ψ in the s-channel helicity system[2]. Within the

[b] Here and in the following, the first error is the statistical and the second systematic; in case only one error is given, the square root of the quadratic sum of both is given.

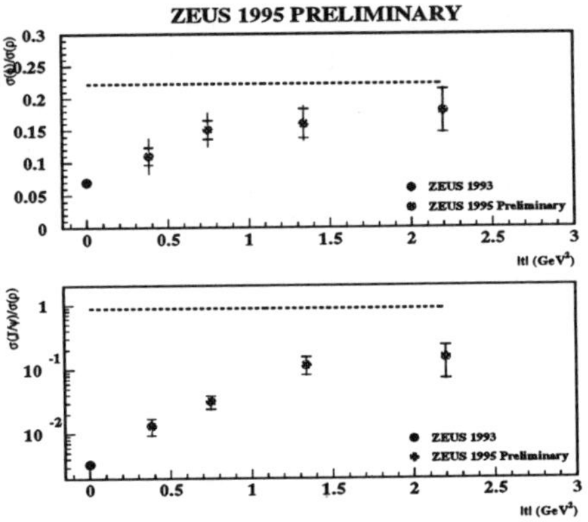

Fig. 8. Cross-section ratios of ϕ to ρ and J/ψ to ρ as funtion of $|t|$.

experimental errors (which are as small as ± 0.01 for the latest ZEUS measurement for the ρ^0), it is found that in photo production the helicity of the photon is transferred to the ρ^0 in the s-channel helicity system, also known as s-channel helicity conservation.

2.2. *Results at large* $|t|$

Using the 44m-tagger, a small tungsten-scintillator calorimeter positioned approximately 2 cm from the electron beam 44 m downstream of the interaction point in the positron beam direction, ZEUS has measured[4] the photo-production of ρ^0-, ϕ- and J/ψ-mesons up to $|t|$-values of 3 GeV^2 in the $W_{\gamma p}$-range 70 - 120 GeV^2. The data contain events with recoiling protons and diffractively excited protons. Fig. 8 shows the cross-section ratios ϕ to ρ^0 and J/ψ to ρ^0 as function of t. As expected from p-QCD models, the ratios rise (a factor 100 in the latter case !) towards the values 2/9 and 8/9 (shown as dashed lines in the figures), respectively, obtained from the quark-charges squared and assuming a flavour independent production mechanism.

3. Intermediate and high Q^2 production

3.1. ρ^0 *and* ϕ*-production*

The results on the $W_{\gamma^* p}$ dependence of $\sigma_{\gamma^* p \to \rho^0 p}$ [5] are shown in Fig. 9, with lines

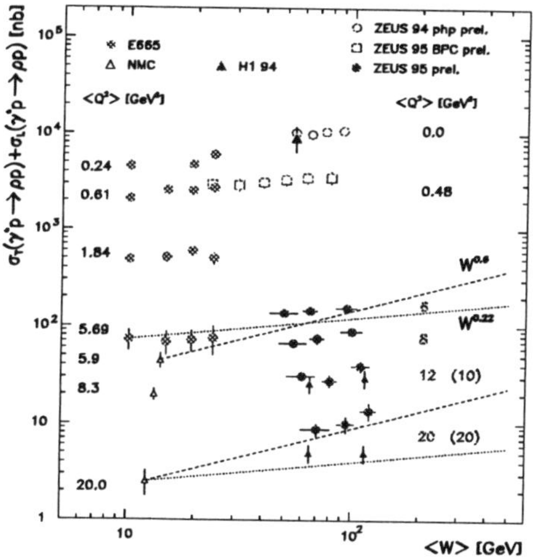

Fig. 9. Energy dependence of the cross-sections for ρ^0- production as function of $W_{\gamma p}$ for different values of Q^2.

$W^a_{\gamma^* p}$ for $a = 0.22$ and $a = 0.6$ drawn to guide the eyes. Given are the measured cross-sections $\sigma_T + \epsilon \cdot \sigma_L$, where the values for ϵ, the relative flux of longitudinal to transverse photons, are close to 1 for the HERA experiments, and between 0.59 and 0.99 for the low energy data. For $Q^2 < 1$ GeV^2 the soft-QCD exponent $a = 0.24$ provides a good description of both HERA and low energy data. At higher $Q^2 > 1$ GeV^2, the HERA data on their own are not precise enough to distinguish the soft-QCD from the p-QCD exponent[c]. There is a tendency for an increase of a with increasing Q^2 from the HERA data, but the precision is not sufficient to draw conclusions. At lower energies, there is a significant disagreement between the results from the NMC and the E665 experiments. Thus the situation needs clarification.

The Q^2-dependence[5] of the cross-section $\sigma_T + \epsilon \cdot \sigma_L$ can be described by the ansatz $\sigma \propto (m_\rho^2/(Q^2 + m_\rho^2))^m$, with values of m[d] close to 2. Also the E665 collaboration finds a value compatible with 2. It is worth noting, that fits as suggested by the VDM-model $\sigma \propto (m_\rho^2/(Q^2 + m_\rho^2))^m \cdot (1 + \epsilon \cdot R(Q^2))$, where $R(Q^2)$ is obtained from the measured spin-density matrix as discussed later, gives values $m \sim 2.5$, incompatible with the VDM-value of $m = 2$.

[c] The values obtained by the ZEUS collaboration for the exponents at different values of Q^2 are: $a = 0.18 \pm 0.05 \pm 0.13$ for $Q^2 = 0.47$ GeV^2, $a = 0.00 \pm 0.33 \pm 0.27$ for $Q^2 = 5.9$ GeV^2, $a = 0.22 \pm 0.38 \pm 0.26$ for $Q^2 = 8.3$ GeV^2, $a = 0.86 \pm 0.46 \pm 0.40$ for $Q^2 = 12.0$ GeV^2, and $a = 1.15 \pm 0.92 \pm 0.56$ for $Q^2 = 20.0$ GeV^2.

[d] ZEUS for $Q^2 = 0.25 - 0.84$ GeV^2 : $m = 1.94 \pm 0.12 \pm 0.18$; ZEUS for $Q^2 = 5 - 20$ GeV^2 : $m = 2.52 \pm 0.28 \pm 0.24$; H1 for $Q^2 = 5 - 20$ GeV^2 : $m = 1.9 \pm 0.30$.

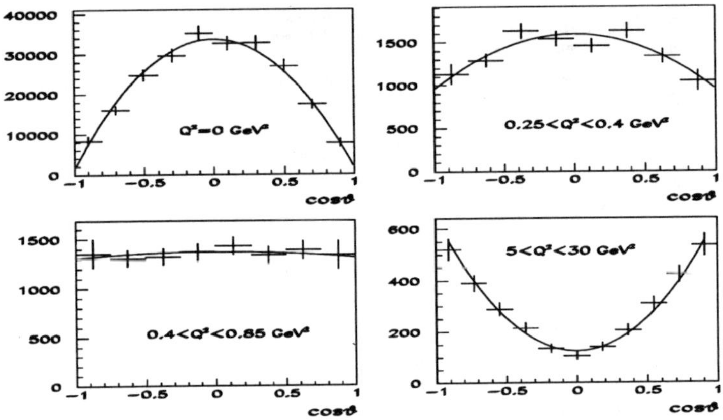

Fig. 10. Distribution of $\cos\theta$ for ρ^0-production for different Q^2-values.

The Q^2-dependence of elastic ϕ-production has also been measured[5] and can be described with $\sigma \propto (m_\phi^2/(Q^2+m_\phi^2))^m$ with a value of m compatible 2.

The value of the slope b of the differential cross-section $d\sigma/dt$ for ρ^0 production decreases from $10.9 \pm 0.3^{+1.0}_{-0.5}$ GeV^{-2} (ZEUS $Q^2 = 0$) to 4.1 ± 0.8 GeV^{-2} (ZEUS $Q^2 = 20$ GeV^2) indicating that the effective ρ^0p interaction radius decreases with increasing Q^2. The data are not sufficiently accurate to determine if the value of b depends on energy.

Fig. 10 , which shows the distribution of the cosine of the polar angle of the π^+ in the s-channel system for different Q^2-intervals, directly demonstrates the transition from the dominance of transverse to longitudinal polarisation of the ρ^0. The expected decay distribution is: $dN/d\cos\theta \propto (1 - r_{00}^{04} + (3r_{00}^{04} - 1)\cdot\cos^2\theta)$, where the density matrix element r_{00}^{04} gives the probability to find the ρ^0 in a longitudinally polarised state. Fits to the angular distributions have been performed and up to 9 ρ^0 density matrix have been determined[2]. They are compatible with the hypothesis of s-channel helicity conservation. Assuming the latter, the ratio of longitudinal to transverse cross-section $R = \sigma_L/\sigma_T$, shown in Fig. 11 has been determined. At low Q^2 we find $R \propto Q^2$, as expected from the VDM-model. At higher Q^2-values a tendency of R to flatten is observed, which can be explained within p-QCD[6]. Similar results, although with bigger errors, are obtained in ϕ-production, where a value of $r_{00}^{04} = 0.77 \pm 0.13 \pm 0.02$, corresponding to $R = 3.3$ has been observed for $Q^2 = 20$ GeV^2.

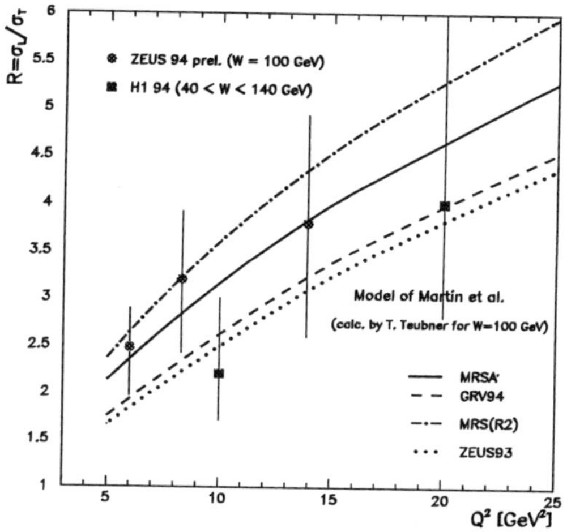

Fig. 11. Dependence of $R = \sigma_L/\sigma_T$ for ρ^0-production as function of Q^2 and comparison to QCD-calculations.

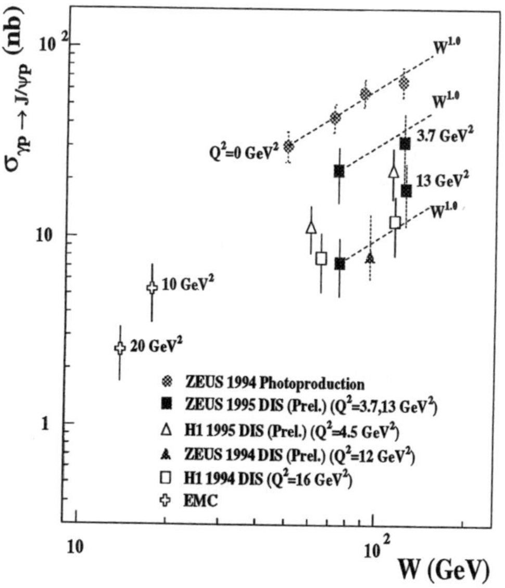

Fig. 12. Dependence of J/ψ-production on $W_{\gamma^* p}$.

3.2. J/ψ-production and particle ratios

Fig. 12 shows the energy dependence of J/ψ-production[5] for different values of Q^2, with the line $W^a_{\gamma^*p}$ with $a = 1$, drawn to guide the eyes. Such an energy dependence, which is expected from p-QCD, is compatible with the data.

H1 has also measured the Q^2-dependence[5] in the interval from 0 to 30 GeV^2 of the cross section for W_{γ^*p} between 30 and 150 GeV and finds a value of $m = 1.9 \pm 0.3$ for the parametrisation $\sigma \propto (m^2_{J/\psi}/(Q^2 + m^2_{J/\psi}))^m$. Thus there appears to be a "universal" Q^2-dependence for the elastic production of ρ^0-, ϕ-, and J/ψ-mesons.

Fig. 13 shows the ratio of ϕ to ρ^0 and Fig. 14 the J/ψ to ρ^0 cross-sections as function of Q^2. We observe a significant increase of this ratio, towards the values of 2/9 and 8/9 given by the quark charges. The values predicted[7] are: $\rho^0 : \phi : J/\psi \simeq (1 \cdot 0.8) : (2/9 \cdot 1.2) : (8/9 \cdot 3.5)$.

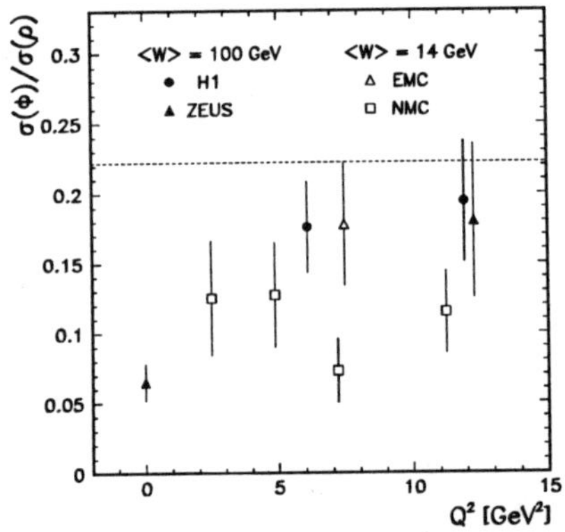

Figure 13: Ratio of ϕ to ρ^0-cross-section as function of Q^2.

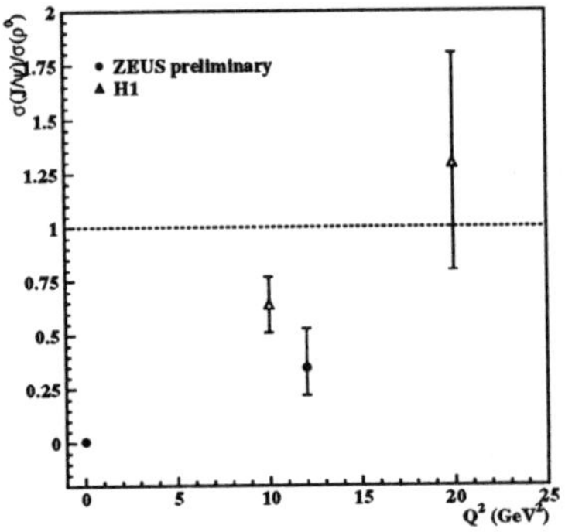

Figure 14: Ratio of J/ψ to ρ^0-cross-section as function of Q^2.

4. Summary and conclusions

To summarise:

- The reaction $\gamma^*p \to VMp$ is a good place to study diffraction in the soft-QCD and the p-QCD regime for different energy scales.

- Vector Meson Dominance and soft Pomeron phenomenolgy gives a successful description of photoproduction of light vector mesons at low values of $|t|$.
- For intermediate values of the photon virtuality the data on light vector meson production qualitatively agree with VDM-ideas - there are however discrepancies at the quantitative level.
- The question if the p-QCD description is valid for light vector mesons for high photon virtualities Q^2 has not yet been answered by the experiments: the measurement of the energy dependence of the cross-section by the HERA experiments is not yet sufficiently precise and there is a discrepancy for the low energy data. Polarisation and cross-section ratios are consistent with the p-QCD predictions.
- Photoproduction of light vector mesons at high momentum-transfers to the outgoing hadronic system exhibits the features predicted by p-QCD.
- The reaction $\gamma^* p \to J/\psi p$ shows the properties expected by p-QCD for all values of Q^2 studied.

5. Acknowledgements

I would like to thank the organisers, the sponsors of this meeting, the conference secretary and the staff of the Ringbergschloß for the excellent organisation and the most pleasant and stimulating atmosphere of this workshop.

6. References

1. S.J.Brodsky et al., *Phys. Rev.* **D 50** (1994) 3134.
2. Talks presented at the Workshop DIS-97, Chicago, April 1997:
   F. Gaede, Study of exclusive vector meson production with H1,
   T. Monteiro, Elastic ρ^0-production at intermediate and high Q^2 from ZEUS,
   L. Adamczyk, Vector mesons in photoproduction from ZEUS,
   L. Bellagamba, Elastic J/ψ-production in γp and DIS from ZEUS, *and*
   A. Wegner, Inelastic J/ψ-production and determination of the gluon density in the proton;
   ZEUS Collab., M. Derrick et al., *Z. Phys.* **C 63** (1994) 391;
   H1 Collab., S. Aid et al., *Z. Phys.* **C 69** (1995) 27;
   ZEUS Collab., M. Derrick et al., *Z. Phys.* **C 69** (1995) 39;
   H1 Collab., T. Aid et al., *Nucl. Phys.* **B 463** (1996) 3;
   H1 Collab., C.Arndt et al., Proton Dissociative ρ and Elastic Φ Electroproduction at HERA, DESY 97-082 (1997), subm. to Z. Phys. C;

ZEUS Collab., J. Breitweg et al., Elastic and Proton-Dissociative ρ^0 Photoproduction at HERA, to be subm. to *Z. Phys* (1997);
ZEUS Collab., M. Derrick et al., *Z. Phys.* **C 73** (1996) 73;
ZEUS Collab., M. Derrick et al., *Phys. Lett.* **B 377** (1996) 259;
H1 Collab., T. Ahmed et al., *Phys. Lett.* **B 338** (1994) 507;
ZEUS Collab., M. Derrick et al., *Phys. Lett.* **B 350** (1995) 120;
H1 Collab., S. Aid et al, *Nucl. Phys.* **B 472** (1996) 3;
ZEUS Collab., J. Breitweg et al., *Z. Phys.* **C 75** (1997) 215.

3. ZEUS Collab., M. Derrick et al., *Z. Phys.* **C 73** (1996) 73.
4. *L. Adamczyk*, Vector mesons in photoproduction from ZEUS, presented at the Workshop DIS-97, Chicago, April 1997.
5. NMC Collab., M. Arneodo et al., *Nucl. Phys.* **B 429** (1994) 503;
E665 Collab., M. R. Adams et al, *Z. Phys.* **C 72** (1996) 237;
H1 Collab., S. Aid et al, *Nucl. Phys.* **B 468** (1996) 3;
ZEUS Collab., M. Derrick et al., *Phys. Lett.* **B 380** (1996) 220;
ZEUS Collab., M. Derrick et al., *Phys. Lett.* **B 356** (1995) 601;
and talks presented at the Workshop DIS-97, Chicago, April 1997:
F. Gaede, Study of exclusive vector meson production with H1,
T. Monteiro, Elastic ρ^0-production at intermediate and high Q^2 from ZEUS,
L. Bellagamba, Elastic J/ψ-production in γp and DIS from ZEUS.
6. A.D. Martin, M.G. Ryskin, T. Teubner, *Phys. Rev.* **D 55** (1997) 4329.
7. L. Frankfurt, W. Koepf and M. Strikman, *Phys. Rev.* **D 54** (1996) 3194.

THEORY OF ELASTIC VECTOR MESON PRODUCTION: SOME RECENT DEVELOPMENTS

MARKUS DIEHL
CPT, Ecole Polytechnique,*
91128 Palaiseau, France
E-mail: diehl@pth.polytechnique.fr

ABSTRACT

I review some recent developments in the QCD description of elastic vector meson production, focusing on issues like the meson wave function, the gluon density in the proton, factorisation, the use of parton-hadron duality and the nonperturbative vacuum.

1. Introduction

I will discuss some recent developments in the description of elastic vector meson production at high energy within QCD, restricting myself to the cases where either the photon virtuality Q^2 or the vector meson mass M_V is large while the squared momentum transfer t from the proton is small. I will highlight progress and problems in the theory of these processes, concerning issues such as the meson wave function, the gluon density in the proton and factorisation. I shall also spend some time on new approaches, namely the use of parton-hadron duality[1] and a nonperturbative description of the scattering[2]. My aim is not to attempt a detailed comparison of theory predictions with data but rather to indicate where and why the predictions of various authors are different. I will not have time to speak about phenomenological models based on Regge theory.

2. Simple approach

The common picture of most QCD models for elastic vector meson production at high energy is shown in Fig. 1, where I also define the kinematics. At large $\gamma^* p$ c.m. energy W the dominating exchange is the pomeron, which is described by the exchange of two gluons coupling on one side to the proton and on the other to a quark loop to which are attached the virtual photon and the vector meson. The vector meson couples to the quark line of the loop by its wave function $\psi(z, k_T)$. Here z and k_T are defined by a Sudakov decomposition for the quark momentum in the meson, $k = zq' + \zeta p + k_T$.

As a baseline let me present a very simple version of this model, recently used by

*Unité propre 14 du Centre National de la Recherche Scientifique.

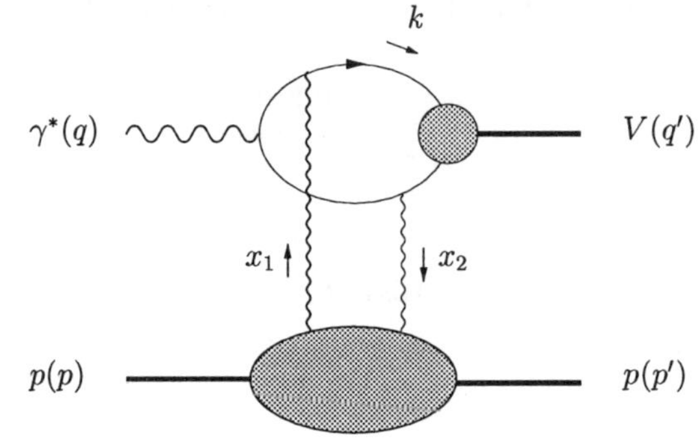

Fig. 1. One of the Feynman diagrams for $\gamma^* + p \to V + p$ with a vector meson $V = \rho$, ρ', ω, ϕ, J/ψ, etc. More diagrams are obtained by attaching the gluons to the quark loop in different ways.

Cudell and Royen[3]. For the meson it takes a nonrelativistic, constituent quark description, namely a wave function peaked at $z = \frac{1}{2}$ and $k_T = 0$, so that the quark and antiquark with mass $M_V/2$ share the four-momentum of the meson equally. It makes the approximation that the exchanged gluons do not interact and couple directly to the three constituent quarks of the proton, following the work of Low, Nussinov, Gunion and Soper[4].

With these assumptions Cudell and Royen find the Q^2- and t-dependence of the cross sections for ρ, ϕ and J/ψ production in fair agreement with data from HERA, EMC and NMC, except for problems with the fixed target J/ψ data. Let me remark that the cross sections for longitudinal and transverse photons respectively behave like $\sigma_L \sim 1/Q^6$ and $\sigma_T \sim 1/Q^8$ for $Q^2 \gg M_V^2$, while for nonasymptotic Q^2 the behaviour in this variable is more complicated, even in this simple model where a fixed strong coupling α_s was taken. One finds however a ratio $\sigma_L/\sigma_T = Q^2/M_V^2$ for all $Q^2 \geq 0$, which is in disagreement with the data—a point I will come back to in Sec. 4. As the approximation of noninteracting gluons does not allow to describe the energy dependence of the cross section the authors introduce an overall phenomenological factor in σ_L and σ_T, and find consistency with the data if this factor is allowed to depend on W, but not on Q^2 or M_V. Little change in the results is obtained when the propagators of the exchanged gluons are taken as nonperturbative, as was done in earlier work of Donnachie, Landshoff and Cudell[5]. Finally, the authors observe that their model agrees with the data down to $Q^2 = 0$ even for the light mesons ρ and ϕ. Notice that in this case one has no large virtuality in the quark loop to justify the perturbative treatment of the quarks and their coupling to the photon.

3. Sophistication of the model

Having seen that even the simplest form of the model depicted in Fig. 1 does not fare badly in describing data we will now take a look at several of its refinements.

3.1. Meson wave function

The meson wave function $\psi(z, k_T)$ turns out to be a major source of uncertainty in the predictions of the model. It determines which virtualities dominate the integration over the quark loop and in particular influences the overall normalisation of the cross section, its Q^2-dependence and the production ratios for different mesons. Various choices have been made in the literature[2,6,7,8,9,10,11], sometimes leading to appreciable differences in numerical results. For J/ψ photoproduction, as an example, Frankfurt et al.[7] find a suppression of the cross section due to finite k_T in the wave function which is significantly stronger than the one estimated by Ryskin et al.[6] For ρ electroproduction a wide range of suppression factors has been obtained by Frankfurt et al.[8] according to the choice of $\psi(z, k_T)$, with a particular sensitivity to its large-k_T tail.

On the theory side it is known that as the renormalisation scale for the wave function tends to infinity, corresponding to very large Q^2 in the process, the k_T-integrated wave function $\int dk_T^2\, \psi(z, k_T)$ behaves like $z(1-z)$ due to QCD evolution, both for longitudinal and transverse meson polarisation. This is very different from the $\delta(z-\frac{1}{2})$ in the nonrelativistic wave function, but the difficult question is of course to assess how far one is from the asymptotic regime for the values of Q^2 one has in experiment. Investigations of the wave function using the operator product expansion and QCD sum rules have been made by Ball and Braun[12] and Halperin and Zhitnitski[11]. The latter find that at $z \to 0$ and $z \to 1$ the wave function should only depend on $\frac{k_T^2}{z(1-z)}$, which in particular excludes a factorising ansatz $\psi(z, k_T) = \phi(z) \cdot \chi(k_T)$. They also argue that from a conceptual point of view there should be no perturbative tail like $1/k_T^2$ in the wave function and that the corresponding corrections to the meson-quark vertex should be explicitly treated as α_s-corrections to the leading order result.

In the operator product expansion framework the effects of finite k_T in $\psi(z, k_T)$ are of higher twist. It is important to realise that there are other higher twist contributions from diagrams where one or more gluons enter the wave function blob in Fig. 1. They correspond to higher fock states of the meson such as $q\bar{q}g$ and $q\bar{q}gg$, which have respectively been considered by Halperin[13] and Hoodbhoy[10]. Moreover, taking only into account a $q\bar{q}$ wave function $\psi(z, k_T)$ is not gauge invariant. To see this consider that a finite k_T involves the transverse component of the operator $i\partial_\mu$

between quark fields in a matrix element. To recover gauge invariance one must pass to the covariant derivative, $i\partial_\mu + A_\mu$, whose transverse component involves physical gluon degrees of freedom.

Finally let me mention a problem raised by Frankfurt et al.[7] concerning the J/ψ wave function. Calculating the transition amplitude of a J/ψ to a timelike photon via a charm quark loop with wave functions $\psi(z, k_T)$ obtained in various nonrelativistic potential models they find that the region of k_T larger than the charm mass contributes up to 30% of the loop integral, thus calling into question the consistency of a nonrelativistic description. Note that the transition $J/\psi \to \gamma^* \to e^+e^-$ is commonly used to fix the normalisation of the wave function from the semileptonic J/ψ decay width.

3.2. Quark mass

Which quark mass should be chosen in the loop of Fig. 1 is another source of uncertainty. Surprisingly this even holds for heavy mesons. Ryskin et al.[6] point out that the cross section scales like the eighth power of $M_{J/\psi}/(2m_c)$, resulting in huge changes of its overall normalisation even if one allows for a modest change in the charm mass m_c. Incidentally, Frankfurt et al.[7] advocate to use the running quark mass rather than the pole or constituent mass.

In the case of light quarks both the choice of current and constituent masses has been made in the literature[6,8,5,9]. Considering the limit of zero quark mass m_q reveals that light meson production from transverse photons is a higher twist effect[6]: the cross section σ_T is proportional to m_q^2 in the collinear approximation for the quarks, i.e. if one sets their transverse momentum k_T to zero in the calculation of the loop and uses the k_T-integrated meson wave function, corresponding to what one does when using collinear parton densities in DIS. At least one of the quantities m_q or k_T must be nonzero to obtain a nonvanishing σ_T from diagrams as shown in Fig. 1.

3.3. Off-diagonal gluon density

We now focus on the lower blob in Fig. 1, the coupling of the two gluons to the proton, including their interaction. If the outgoing proton had the same momentum as the incoming one this blob would be given by the gluon density in the proton. In our reaction this is however not the case: even if the proton has zero transverse momentum after the scattering it has lost a fraction $x = (M_V^2 + Q^2)/(W^2 + Q^2)$ of its longitudinal momentum, which is necessary to make a timelike vector meson out of a spacelike photon. Thus the longitudinal momentum fractions x_1 and x_2 of the gluons with respect to p (cf. Fig. 1) are not equal, their difference being $x_1 - x_2 = x$. One can argue that in the leading $\ln Q^2 \cdot \ln(1/x)$ approximation it is indeed the gluon density $g(x)$ that describes the blob[14,15], as the typical values of x_1 and x_2 in the loop integration are of order x and to leading $\ln(1/x)$ all such values are equivalent.

The most prominent phenomenological consequence of this description is the strong rise of the cross section in W when either of the scales Q^2 or M_V^2 is hard. At this point I should remark that only the imaginary part of the $\gamma^* p \to Vp$ amplitude is actually calculated from the Feynman diagrams of Fig. 1, cutting them in the s-channel. The contribution of the real part to the cross section is negligible if the W-dependence is weak as in the case of soft pomeron exchange, otherwise it can be accounted for in an approximate way[14,6].

Beyond leading $\ln(1/x)$ the quantity describing the blob in Fig. 1 is certainly different from the gluon distribution that contributes to inclusive DIS, and there has been much theoretical interest recently in such so-called "off-diagonal parton densities". Strictly speaking they are not "densities" by the way, since they correspond to matrix elements of parton fields between *different* proton states and do not have a probability interpretation. Let me note that their evolution equations differ from the DGLAP equations by their splitting functions[16,17]. Some studies have been performed[10,17] to estimate the difference between the off-diagonal and the usual, diagonal gluon densities, finding no big difference at small x, but much remains to be done in this field. An obvious remark is that the dependence of the off-diagonal gluon density on t is specific of its asymmetric kinematics and cannot be predicted by approximating it with the diagonal density measured in other processes.

In the same way as in DIS at small x one can also go beyond the leading $\ln Q^2$ approximation and instead of $g(x)$ consider the unintegrated gluon density $f(x, l_T)$ with a finite transverse momentum l_T of the gluons[6].

The question of which factorisation scale is appropriate in the gluon density is numerically very important since we know from DIS that parton distributions at small x change rapidly with this scale. Generically it is set by the typical virtualities in the quark loop, i.e. by Q^2 and the quark mass, but various concrete choices have been advocated in the literature: for J/ψ production the scale $\frac{1}{4}(Q^2 + M_{J/\psi}^2)$ used by Ryskin et al.[6] is smaller than that of Frankfurt et al.[7], whereas Nemchik et al.[9] choose $Q^2 + M_V^2$ times a factor between 0.07 and 0.2 depending on the meson and its polarisation. Let me however emphasise that *any* choice of factorisation scale can only be an educated guess of which value will make higher order corrections small. As in other reactions such as for instance jet production experiment should feel free to try different scales $const \cdot (Q^2 + M_V^2)$, and also $const \cdot Q^2$ for light mesons.

3.4. Factorisation theorem

For the asymptotic limit where Q^2 is much larger than all masses in the process a factorisation theorem has been proven by Collins et al.[18] within perturbative QCD. By an analysis of Feynman diagrams and power counting arguments they find that to all orders in α_s the amplitude for $\gamma^* p \to Vp$ factorises into a hard scattering part,

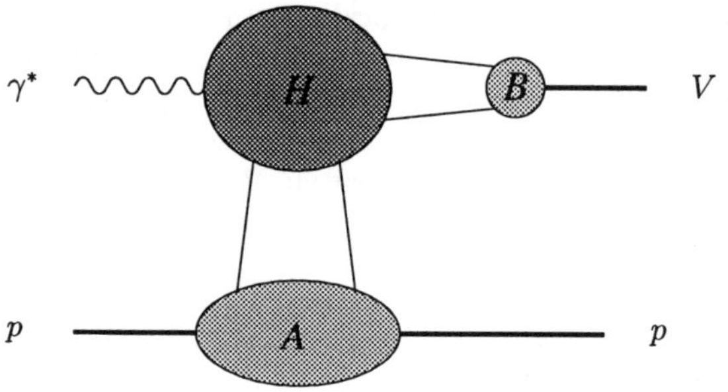

Fig. 2. Factorisation of the amplitude for $\gamma^* p \to Vp$ with a longitudinal photon into a hard scattering part H, an off-diagonal quark or gluon density A and a meson wave function B.

a collinear, off-diagonal quark or gluon distribution in the proton, and a collinear $q\bar{q}$ wave function for the vector meson (see Fig. 2). The corresponding amplitude has a power behaviour like $1/Q$.

This power behaviour and factorisation property holds only for a *longitudinal photon.* If the γ^* is transverse the authors find that the amplitude only behaves like $1/Q^2$, and that there is no factorisation like in Fig. 2: soft, "wee" partons can bypass the hard scattering and directly go from the proton blob A to the meson blob B. The power behaviour in Q confirms from a different point of view the statement that light meson production from transverse photons is a higher twist process.

It is worth noting that these findings are not restricted to the small-x region but valid for general x, where the quark distribution in the proton is relevant in addition to that of the gluons.

Finally it is clear that the direct phenomenological application of these results requires "sufficiently large" values of Q^2, and we have already seen that higher twist effects such as the transverse momentum in the meson wave function can be numerically important in the HERA regime. Also one should keep in mind that the Q^2-behaviour obtained by power counting can be strongly masked by logarithmic scaling violations which are strong at small x.

3.5. Rescattering effects

Let me just briefly mention that several papers[6,7,9,19] have investigated the effects of rescattering of the $q\bar{q}$-pair on the proton, in other words shadowing corrections or multiple-pomeron exchange, and found that they can be important in kinematic situations accessible at HERA. An observable that is naturally sensitive to such effects is the t-dependence of the cross section, which has been investigated in detail by

Gotsman et al.[19]

In this context one may remark that an exponential parametrisation $d\sigma/dt \propto \exp(bt)$ at small and moderate values of t is no more than a fit to a simple function: from theory one does not expect such a behaviour to be exact. Remember for instance that ordinary elastic form factors of hadrons are not exponentials. One should be aware that just comparing an experimentally fitted slope parameter b with the logarithmic slope of $d\sigma/dt$ at $t = 0$ calculated in theory can be misleading. To compare the full t-dependence is of course the most thorough way to proceed, but if one looks for convenient "handy" parameters other choices than b might be useful, such as the mean value $\langle t \rangle$ proposed by Cudell and Royen[3].

4. The ratio $R = \sigma_L/\sigma_T$

As I have already emphasised the physics of light meson production is quite different with transverse and longitudinal photon polarisation. It turns out that σ_T is more sensitive to small quark virtualities and thus to infrared physics than its counterpart for longitudinal photons[1,8,9]. This can be understood from the properties of the transition $\gamma^* \to q\bar{q}$: while for a longitudinal photon configurations are preferred where the longitudinal momenta of quark and antiquark are comparable, a transverse photon likes to split into a $q\bar{q}$-pair where the quark or antiquark carries only a small fraction of the photon momentum and is soft. This is the aligned jet configuration, which is of prime importance for the physics of diffraction. The interaction with the gluons hardly changes the longitudinal momenta of q and $\bar{q}$, so that this configuration corresponds to small z or $1-z$ in the meson wave function. To which extent σ_T is dominated by the soft region thus depends on how strongly small z or $1-z$ are suppressed in $\psi(z, k_T)$.

In the simple model presented in Sec. 2, where a meson wave function peaked at $z = \frac{1}{2}$ is used, the ratio R comes out as Q^2/M_V^2, which is far too big to describe the data for Q^2 in the range of some GeV^2. In the model of Nemchik et al.[9] the greater infrared sensitivity of σ_T is reflected in a lower factorisation scale of the gluon density than in σ_L, and the authors find that R grows less fast than linearly in Q^2 and that its rate of growth in Q^2 depends on x.

4.1. Using parton-hadron duality

To circumvent the uncertainties associated with the meson wave function in this delicate context, Martin et al.[1] invoke parton-hadron duality: to obtain the cross section for ρ-production they calculate the diffractive production of an open $q\bar{q}$-pair by two-gluon exchange, project out the appropriate $q\bar{q}$ partial wave to make a vector meson of given helicity, and integrate the corresponding cross section over the $q\bar{q}$ invariant mass $M_{q\bar{q}}$ in an interval around M_ρ. The authors stress that in this mass range partonic final states other than $q\bar{q}$ are heavily suppressed and that the dominating hadronic final state is a pair of pions. In the ratio R several uncertainties are expected to cancel to a large extent, e.g. uncertainties about the appropriate interval of $M_{q\bar{q}}$ and about radiative corrections (the authors estimate that there should be a large K-factor). As in the calculation by Nemchik et al.[9] σ_T is dominated by lower quark virtualities than σ_L, now in the context of open $q\bar{q}$-production, and through the different factorisation scales in the gluon density the authors obtain a ratio R that grows less than linearly in Q^2 and is in fair agreement with HERA data for R in the range 5 $\text{GeV}^2 < Q^2 < 20$ GeV^2. Again this ratio is predicted to depend on x.

5. A nonperturbative model

To finish let me present an approach from a perspective of nonperturbative physics. The model of Dosch et al.[2] starts from the high-energy scattering of a quark or antiquark, which is described as the scattering in an external gluon field[20], like in the semiclassical model of diffraction of Buchmüller and Hebecker[21]. This gluon field is then averaged over in the sense of a path integral, according to the stochastic model of the nonperturbative QCD vacuum[22]. High-energy scattering thus becomes related with nonperturbative parameters such as the SVZ gluon condensate.

The nonabelian character of the gluon field is crucial in this model: not single quarks or antiquarks are scattered but rather colour singlet configurations of $q\bar{q}$ or qqq, their transverse separation having a strong influence on the strength of the scattering. The physical picture for vector meson production is then as shown in Fig. 3. For the three-quark wave function of the proton and the $q\bar{q}$ wave function of the vector meson an ansatz has to be made, whereas the splitting of the virtual photon into $q\bar{q}$ is calculated perturbatively. As in the simple two-gluon exchange model discussed in Sec. 2 the W-dependence of the cross section cannot be predicted in the present form of the model.

Its authors find fair agreement of their results with EMC and NMC data for Q^2 between 2 and 10 GeV^2 and W between 10 and 20 GeV, looking at observables such as the Q^2- and t-dependence of the cross section and its normalisation, except that their cross section for the ϕ is about a factor 2 above the data. They also describe the ratio R for ρ-production measured by NMC and the t-dependence of J/ψ photoproduction. They conclude by emphasising the need to incorporate nonperturbative aspects of the photon wave function for small Q^2 and M_V, and perturbative gluon contributions which they expect to become important as W increases from fixed target to HERA energies.

6. Conclusions

The model depicted in Fig. 1 is a good candidate for the description of elastic

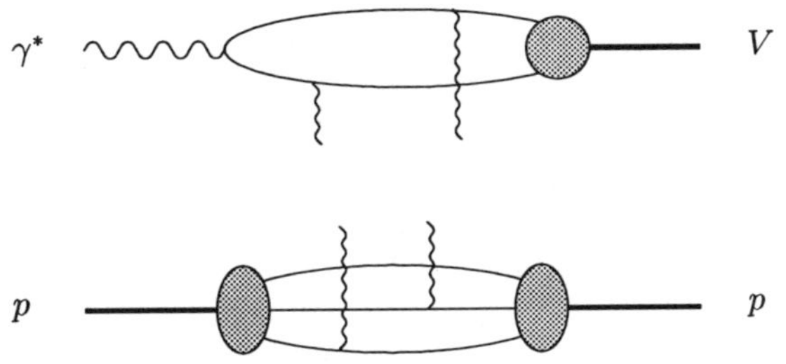

Fig. 3. The process $\gamma^* p \to Vp$ in the model of Dosch et al.[2] Quarks and antiquarks are scattered in the gluon field of the nonperturbative QCD vacuum.

vector meson production at large Q^2 or meson mass in QCD. Even a very simple version of this model as presented in Sec. 2 can describe quantitative features of the data.

Progress has been made in the theory of meson wave functions and of the off-diagonal gluon distribution in the proton. For the asymptotic regime of very large Q^2 a factorisation theorem has been worked out which further elucidates the different physics of σ_L and σ_T.

On the phenomenological side various difficulties or uncertainties present themselves if one is to make quantitative predictions in the nonasymptotic regime, in particular for light meson production from transverse photons, or if in turn one aims to extract information from the data on the meson wave function or the gluon distribution. An important task is to identify observables which are sensitive to only a few theoretical effects.

Finally, as I showed in Sec. 4.1 and 5, there are promising alternative approaches, which highlight the connection of this reaction with open $q\bar{q}$-production, and with nonperturbative physics and the QCD vacuum.

7. Acknowledgements

It is a pleasure to acknowledge conversations with T. Gousset, R. Klanner, P. V. Landshoff and A. D. Martin. Special thanks go to T. Gousset for carefully reading the manuscript. Finally, I would like to thank B. Kniehl, G. Kramer and A. Wagner for organising this very fruitful workshop.

8. References

1. A D Martin, M G Ryskin, and T Teubner, *Phys. Rev.* **D55** (1997) 4329

2. H G Dosch, T Gousset, G Kulzinger and H J Pirner, *Phys. Rev.* **D55** (1997) 2602
3. J R Cudell and I Royen, *Phys. Lett.* **B397** (1997) 317
4. F E Low, *Phys. Rev.* **D12** (1975) 163;
   S Nussinov, *Phys. Rev. Lett.* **34** (1975) 1286;
   J F Gunion and D E Soper, *Phys. Rev.* **D15** (1977) 2617
5. A Donnachie and P V Landshoff, *Nucl. Phys.* **B311** (1988/89) 509;
   J R Cudell, *Nucl. Phys.* **B336** (1990) 1
6. M G Ryskin, R G Roberts, A D Martin and E M Levin, hep-ph/9511228
7. L Frankfurt, W Koepf and M Strikman, hep-ph/9702216
8. L Frankfurt, W Koepf and M Strikman, *Phys. Rev.* **D54** (1996) 3194
9. J Nemchik, N N Nikolaev and B G Zakharov, *Phys. Lett.* **B341** (1994) 228
10. P Hoodbhoy, *Phys. Rev.* **D56** (1997) 388
11. I Halperin and A Zhitnitsky, *Phys. Rev.* **D56** (1997) 184
12. P Ball and V M Braun, *Phys. Rev.* **D54** (1996) 2182
13. I Halperin, hep-ph/9704265
14. S J Brodsky et al., *Phys. Rev.* **D50** (1994) 3134
15. J Bartels, H Lotter and M Wüsthoff, *Phys. Lett.* **B379** (1996) 239
16. I I Balitskii and V M Braun, *Nucl. Phys.* **B311** (1988/89) 541;
    X Ji, *Phys. Rev.* **D55** (1997) 7114
17. A V Radyushkin, *Phys. Lett.* **B385** (1996) 333; hep-ph/9704207;
    L Frankfurt, A Freund, V Guzey and M Strikman, hep-ph/9703449
18. J C Collins, L Frankfurt and M Strikman, hep-ph/9611433
19. E Gotsman, E Levin and U Maor, hep-ph/9701236
20. O Nachtmann, *Ann. Phys. (N.Y.)* **209** (1991) 436
21. W Buchmüller and A Hebecker, *Nucl. Phys.* **B476** (1996) 203
22. H G Dosch, E Ferreira and A Krämer, *Phys. Rev.* **D50** (1994) 1992

DIFFRACTION AT HERA

RESULTS FROM THE H1 AND ZEUS EXPERIMENTS

R.ROOSEN

Inter University Institute for High Energies, Vrije Universiteit Brussel, Pleinlaan 2,1050,Brussels, Belgium

E-mail: roosen@hep.iihe.ac.be

ABSTRACT

An experimental overview of photon dissociative diffraction at large Q^2, based on results from the ZEUS and H1 experiments at HERA, is given. The different methods to tag photon dissociative events are reviewed. The analysis of the $F_2^{D(3)}$ structure function is discussed. The knowledge extracted from this diffractive analysis, implemented in a Monte Carlo model, is subsequently used for predicting various aspects of the diffractive final states. It is shown that, at present, the predictions of this model explain most of the diffractive final state features remarkably well.

1. Diffraction: Definition and Kinematics

Elastic and total cross sections of hadron-hadron and photon-hadron interactions are well described by the Donnachie Landshoff [1] formula:

$$\sigma_{tot} = Xs^{0.0808} + Ys^{-0.4525} \tag{1}$$

where s is the centre of mass energy in the reaction and X and Y are two parameters. Eq. (1) shows that at large $\sqrt{s}$ these cross sections are dominated by the first term on the r.h.s. in Eq. (1). In terms of Regge theory Eq. (1) can also be interpreted as being due to the exchange of two trajectories: a reggeon and a pomeron trajectory, the latter being described by the equation:

$$\alpha_{I\!P} = \alpha_{I\!P}(0) + \alpha' t \tag{2}$$

with $\alpha_{I\!P}(0) = 1.0808$ and $\alpha' = 0.25 GeV^{-2}$. If we adopt the definition that *diffraction* equals *pomeron exchange* and combine this with Feynman's definition that: *The probability there are no hadrons in a sufficiently large gap of rapidity Δy goes as $exp(-a\Delta y)$ where the parameter a depends on the quantum numbers carried by the gap*, one comes to an experimental definition once one can relate a to the Regge trajectory. In Regge theory, the relation between a and the Regge trajectory is given by:

$$a = 2(1 - \alpha_{trajectory}) \tag{3}$$

Eq. (3) shows that only in the case of the pomeron the probability for finding hadrons in a large rapidity gap is not exponentially surpressed in contrast to pion or reggeon exchange. Hence Bjorken's definition of diffraction : *A process is diffractive if and only if there is a large rapidity gap in the final phase space which is not exponentially surpressed.*

Diffractive events can be pictured as shown Fig. 1. In case the proton remains intact after the interaction, one has either quasi elastic vector meson production ($\gamma\ p \rightarrow Vp$) (V=ρ, ϕ, ω,..) or single photon dissociation ($\gamma\ p \rightarrow Xp$). If the proton breaks up in a sytem Y, the process is called single proton dissociation ($\gamma\ p \rightarrow VY$) or double dissociative diffraction ($\gamma\ p \rightarrow XY$). In the remainder of the paper we will be only concerned with photon dissociative diffraction, the quasi elastic vector meson production will be discussed in a different session at this conference. In order to describe the diffractive deep inelastic event dynamics, the following kinematical deep inelastic variables x, y, Q^2 and W are commonly introduced (Fig. 1). In addition, because of the presence of the gap two other variables can be defined i.e. β and $x_{\mathbb{P}}$:

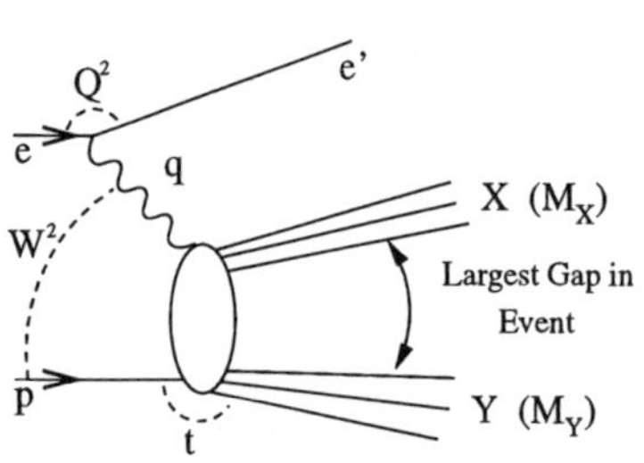

Fig. 1. Definition of the kinematic variables.

$$\begin{aligned} \beta &= \frac{Q^2}{2q.(P-Y)} \approx \frac{Q^2}{Q^2+M_x^2} \\ x_{\mathbb{P}} &= \frac{q(P-Y)}{q.P} \approx \frac{Q^2+M_x^2}{Q^2+W^2} \\ t &= (P-Y)^2 \end{aligned} \quad (4)$$

where P,Y and q stand for the four vectors of the proton, the system Y and the photon and t is the four momentum transfer at the proton vertex. The variables β and $x_{\mathbb{P}}$ are related via the expression $x = \beta.x_{\mathbb{P}}$. These definitions can be applied to any process once a gap has been defined and is not dependent on the definition of what is a diffractive event. If one furthermore assumes that the exchanged object is a pomeron one can interprete $x_{\mathbb{P}}$ as being the three-momentum fraction of the pomeron in the proton and β as being the momentum fraction of the parton in the pomeron which interacts with the photon.

2. Diffractive event tagging $\gamma^* p \rightarrow Xp$

2.1. H1-experiment

The diffractive event selection[a] in the H1 experiment relies entirely on the absence of any activity in the forward detectors which cover a range in rapidity of $3.4 \leq \eta \leq 7.5$. This ensures that the photon diffracted system X is fully contained in the

[a]Details on experimental results in diffractive scattering can be found in the Proceedings of the 5th International Workshop on Deep Inelastic Scattering and QCD, Chicago, 1997.

main detector. Because in this selection the scattered proton is not tagged, double dissociative processes in which $M_Y \leq 1.6$ will be included in the photon dissociative event sample. This contribution is estimated to be of the order of 5%. Requiring furthermore that $x_{\mathbb{P}} \leq 0.05$ ensures that the Y system carries more than 95% of the the scatered proton.

2.2. ZEUS-experiment

The Zeus collaboration uses two different methods to select diffractive events. In the first method, the rapidity gap is defined by the most forward energy deposition of 400 MeV in the calorimeter, defining a rapidity gap of $\eta \leq 1.8$. This small η_{max} value limits the acceptance for large M_X values and a Monte Carlo method is required to subtract the non-diffractive background.

The second method is based on the difference in the behaviour of the invariant mass distributions in the diffractive and non-diffractive event samples. This difference is expressed by the two formulas:

$$\frac{dN^{\textbf{Non-Diffractive}}}{d\ln M_X^2} \propto exp(b \cdot \ln M_X^2) \qquad \frac{dN^{\textbf{Diffractive}}}{d\ln M_X^2} \propto C^{ste} \tag{5}$$

Hence extrapolation of the M_X distribution at large values of $\ln(M_X)$ allows to define the excess of events at lower M_X as rapidity gap events, avoiding a rapidity gap cut.

2.3. Forward Proton Spectrometers: LPS (ZEUS), FPS(H1)

Both experiments are equipped with a forward proton spectrometer. The FPS is equipped with 4 Roman pots in which the main detection elements consists of scintillating fibers. The ZEUS LPS spectrometer consists of 6 silicon detector. These spectrometers offer in contrast to the previously described methods, the possibility of tagging the scattered proton. As the distribution of the quantity

$$x_L = p_{proton}^{scattered} / p_{proton}^{incident} = 1 - x_{\mathbb{P}}$$

peaks very strongly at $x_L \approx 1$ for diffractive events this quantity allows a clear event tag. This method of tagging diffractive events is less subject to background ($\leq 5\%$) and in addition allows the measure the t-distribution in the process.

3. Measurement of the diffractive Structure function $F_2^{D(3)}$

The diffractive structure function, as measured by H1, is based on a deep inelastic event sample in which the scattered electron has an energy above 8 GeV and for which no signals are detected in the forward detectors ($7.5 \leq \eta \leq 3.4$). The cross section from which the structure function is extracted is measured for $x_{\mathbb{P}} \leq 0.05$ and for

$M_Y \leq 1.6$ GeV. With these selections the background contained in the event sample resulting from events with $x_{\mathbb{P}} \geq 0.05$ is small and taken into account in the Monte Carlo smearing corrections. In order to correct for data losses and smearing effects a Monte Carlo simulation has been used.
The diffractive structure function $F_2^{D(3)}$ is defined via the differential cross section:

$$\frac{d^4\sigma_{ep}^D \to e'XY}{dQ^2 d\beta dx_{\mathbb{P}} dt} = \frac{4\pi\alpha^2}{\beta Q^4}\left(1 - y + \frac{y^2}{2(1+R)}\right) F_2^{D(4)}(Q^2,\ \beta,\ x_{\mathbb{P}}, t) \tag{6}$$

Eq. (6) contains 2 unknowns: first, the variable t is not measured and therefore it will be integrated over; second, the variable R is as yet not measured and theoretical estimates lead to very different values. Therefore R is set to zero, leading to the following expression after integration over t:

$$\frac{d^3\sigma_{ep}^D \to e'XY}{dQ^2 d\beta dx_{\mathbb{P}}} = \frac{4\pi\alpha^2}{\beta Q^4}\left(1 - y + \frac{y^2}{2}\right) F_2^{D(3)}(x_{\mathbb{P}}, Q^2, \beta) \tag{7}$$

The kinematic domain over which the diffractive structure function is measured is:

$$\begin{aligned} 2.5 \leq\ & Q^2 \leq 65\ \text{GeV}^2 \\ 0.01 \leq\ & \beta \leq 0.9 \\ 1.10^{-4} \leq\ & x_{\mathbb{P}} \leq 0.05 \end{aligned} \tag{8}$$

From the 1993 data analysis the conclusion was reached that the $F_2^{D(3)}$ structure function factorises in $x_{\mathbb{P}}$ i.e could be written as:

$$F_2^{D(3)}(x_{\mathbb{P}}, Q^2, \beta) = f_{\mathbb{P}/p}(x_{\mathbb{P}}) \cdot F_2^{D(3)} = x_{\mathbb{P}}{}^{-n} \cdot A(\beta, Q^2) \tag{9}$$

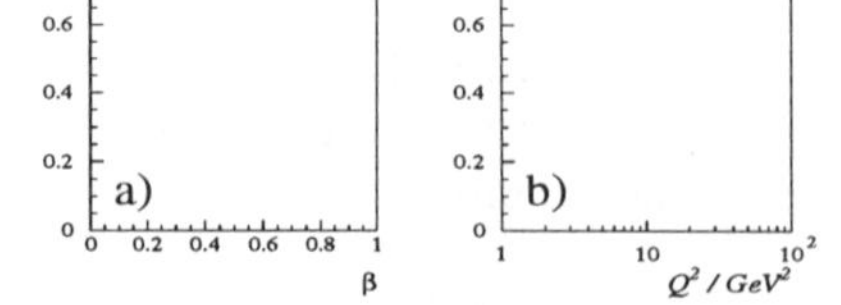

Fig. 2. $x_{x_{\mathbb{P}}}.F_2^{D(3)}(x_{\mathbb{P}}, Q^2, \beta)$ and the result of the a fit of the form $F_2^{D(3)} = A(\beta, Q^2)/x_{x_{\mathbb{P}}}^{n(\beta))}$. The errors represent the statistical and systematic errors in quadrature.

where $f_{\mathbb{P}/p}(x_{\mathbb{P}})$ stands for the pomeron flux factor. The exponent n is, in the framework of Regge theory, connected to the pomeron trajectory exchanged in the t-channel. With the increase in statistics in the 1994 data, it turned out that the n value could no longer be considered constant. A fit of the data to a polynomial dependence of $n(\beta)$ in the expression for $x_{\mathbb{P}}{}^{-n}$, assuming a dependence of n on β but not on Q^2 showed that the parameter n varied with β [5] (Fig. 2a). Fig. 2b also shows the result of a similar fit, but this time allowing n to vary with Q^2 and not with β, indicating no evidence for a $\ln Q^2$ dependence of n. Fig. 2a clearly indicates that n is not constant and that for $\beta \leq 0.2$ its value drops below unity. A possible scenario which could explain this factorisation breaking is that in addition to pomeron exchange, a small contribution from a subleading meson trajectory may be present (e.g. $f_2(1270)$) for which $n \approx 0$ and which may have a softer β dependence. A fit to the data of the prediction of such a model, in which the pomeron is described by $F_2^{I\!P}(\beta,Q^2).x_{I\!P}{}^{-n_{I\!P}}$ and the meson by $C_M.F_2^M(\beta,Q^2)\cdot x_{I\!P}{}^{-n_M}$ shows that the meson contribution is important at small β values and large $x_{I\!P}$ values. ($F_2^M(\beta,Q^2)$ is taken from the GRV parametrisation of the pion [2])

From this fit the following values for the $n_{I\!P}$ and n_M are obtained:

$$n_{I\!P} = 1.29 \pm 0.6\ (sys) \pm 0.03\ (model) \quad (10)$$

$$n_M = 0.3 \pm 0.3\ (stat) \pm 0.6\ (sys) \pm 0.3\ (model) \quad (11)$$

The uncertainties quoted take into account a full systematic error analysis as well as the meson structure model dependences. Integrating over t, assuming a t-dependence of type $\propto e^{b_{I\!P}}, e^{b_M}$ and a Regge trajectory of the form $\alpha_{I\!P,M} = \alpha_{I\!P,M}(0) + \alpha'_{I\!P,M}t$, one can calculate the meson and pomeron intercepts. For the meson one finds, assuming $b_M = 6$ and $\alpha'_M = 1$,

$$\alpha_M = 0.6 \pm 0.1\ (stat) \pm 0.3\ (sys) \quad (12)$$

in good agreement with the standard value. Similarly for the pomeron, assuming $b_{I\!P} = 6$ and $\alpha'_{I\!P} = 0$, the result is

$$\alpha_{I\!P} = 1.15 \pm 0.02\ (stat) \pm 0.04\ (sys) \quad (13)$$

Assuming a value for $\alpha'_{I\!P} = 0.3\ \mathrm{GeV}^{-2}$, the estimate for $\alpha_{I\!P}$ becomes:

$$\alpha_{I\!P} = 1.18 \pm 0.02\ (stat) \pm 0.04\ (sys) \quad (14)$$

The Zeus Collaboration, using the alternative $\ln(M_X^2)$-method, binning the data in 4 Q^2 and 2 M_X bins, obtain for the pomeron intercept, averaged over t, the values shown in Fig. 3. These values are in good agreement with those obtained by the H1 Collaboration as indicated by the dashed lines (the 1-σ error band). The full line indicates the Donachie Landshoff soft pomeron intercept value.

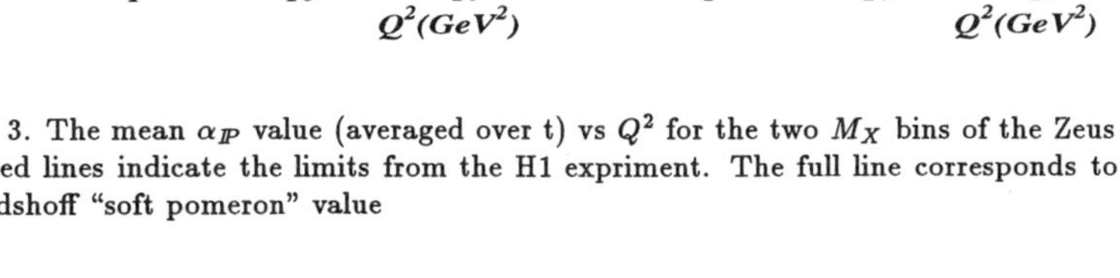

Fig. 3. The mean $\alpha_{I\!P}$ value (averaged over t) vs Q^2 for the two M_X bins of the Zeus analysis. The dotted lines indicate the limits from the H1 expriment. The full line corresponds to the Donachie Landshoff "soft pomeron" value

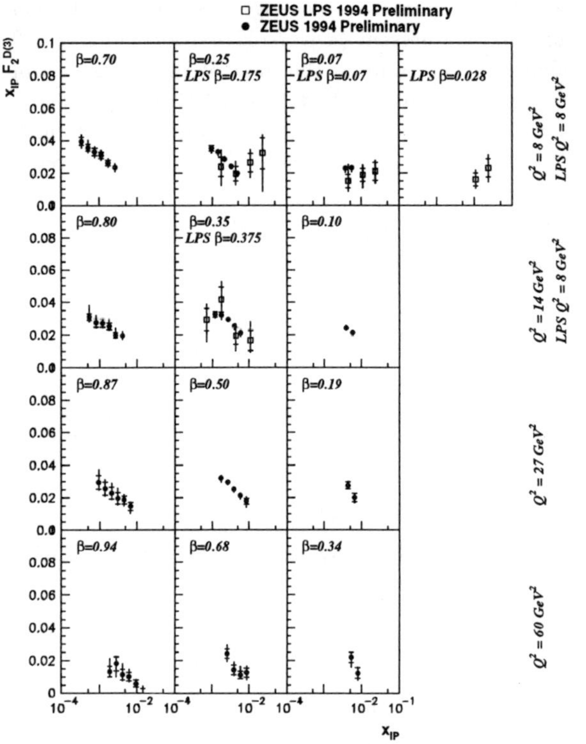

Fig. 4. $x_{I\!P}.F_2^{D(3)}$ obtained with two different analysis methods: the $\ln(M_X^2)$-method and tagging the leading proton with the LPS

An alternative way to extract the diffractive structure function as done by the ZEUS Collaboration is by tagging the leading proton. This spectrometer is able to tag and measure the diffracted proton with an acceptance $0.6 \leq x_L \leq 1.0$ and $p_\perp^2 \leq 1$ GeV^2. Because diffractive events are characterised by large values of x_L, a clean event sample can be obtained by requiring $x_L \geq .97$ in which the non-diffractive background is less than 5%. Fig. 4 shows the measurement of the diffractive structure function at large Q^2 as obtained by the Zeus Collaboration using two different methods: tagging the leading proton and using the $\ln(M_X^2)$ method. The results of both methods agree well in the overlapping region.

4. Partonic Structure of Diffractive Exchange

For a factorisable structure function in $x_{I\!P}$, $\int dx_{I\!P} F_2^{D(3)}(\beta, Q^2, x_{I\!P})$ describes the partonic structure of the diffractive exchange. But even in the case of non-factorisation, the integral can be shown [6] to be a measurement of the deep inelastic structure of the diffractive exchange. The integrated structure function is closely related to the fracture function for a semi-inclusive hadronic final state with a leading isospin-1/2 baryon [6] and can as such be interpreted in the context of the QCD evolution of parton densities of the Regge pole exchange coupling to the isospin-1/2 baryon. $\tilde{F}_2^D(\beta, Q^2)$ defined as:

$$\tilde{F}_2^D(\beta, Q^2) = \int_{x_{I\!P}=0.0003}^{x_{I\!P}=0.05} F_2^D(\beta, Q^2) dx_{I\!P}$$

and averaged over t ($|t| \geq 1$ GeV^2), is shown in Fig. 5a,b. At fixed Q^2 there is no evidence for a decreasing parton momentum fraction as observed in hadrons.

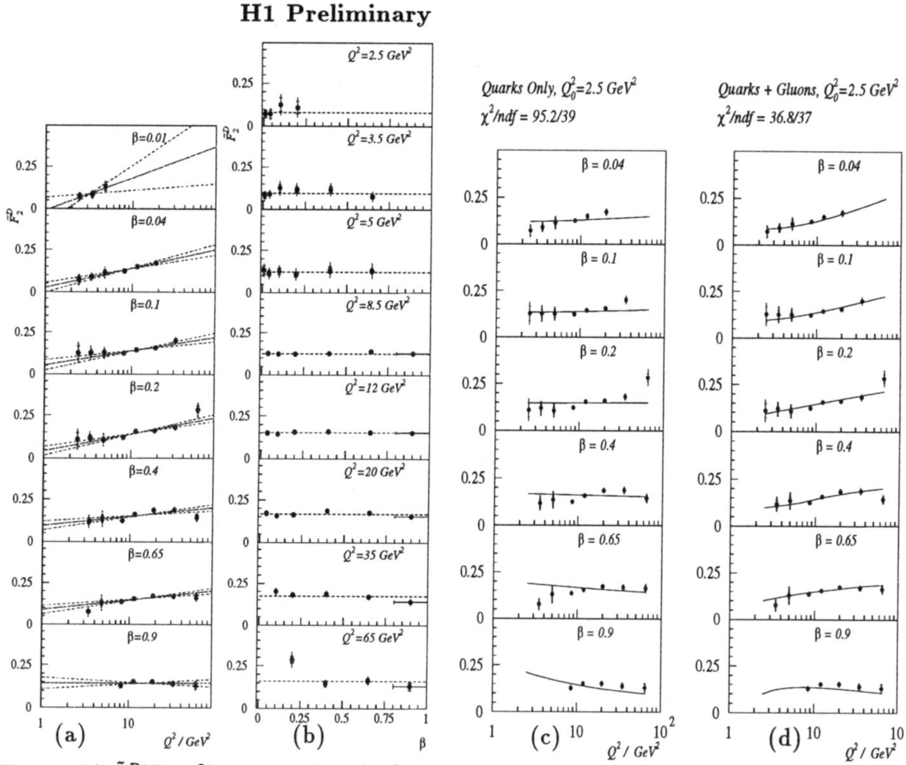

Fig. 5. a,b) $\tilde{F}_2^D)(\beta, Q^2)$ as a function of Q^2 for different β values (left) and as a function of β for different Q^2 values (right); the superimposed lines correspond to the best fit to a linear dependence in $\ln Q^2$ (continous line) in the Q^2 dependencies and to a constant dependence in β dependencies. c) The $\tilde{F}_2^D$ dependence on β and Q^2 assuming only quarks at the starting scale $Q_0^2 = 2.5$ GeV^2; d) The $\tilde{F}_2^D$ dependence on β and Q^2 assuming both quarks and gluons at the starting scale $Q_0^2 = 2.5$ GeV^2

Although the variation of $\tilde{F}_2^D(\beta, Q^2)$ with Q^2 is moderate, the violation of scale invariance is clearly present. Peculiar is that even at large values of β this rise in $\ln Q^2$ still persists, supporting the hypothesis of a substantial gluon component in the diffractive exchange. A QCD analysis performed by the H1-collaboration has confirmed this qualitative picture. A $u + d + s$ flavour singlet quark and a gluon distribution defined at a Q_0^2 starting scale are parametrised as

$$x_{i/\mathbb{P}} \cdot f_i(x_{i/\mathbb{P}}) = A_i x_{i/\mathbb{P}}^{B_i} (1 - x_{i/\mathbb{P}})^{C_i}$$

where i= q,g and $x_{i/\mathbb{P}}$ is the fraction of the exchanged momentum carried by the parton. These parton distrubutions are then evolved using the DGLAP [7] evolution equations to values of $Q^2 \geq Q_0^2$, with the parameters $A_{q,g}$, $B_{q,g}$ and $C_{q,g}$ to be determined from the fit. The result of this calculation is shown in Fig. 5c,d. Fig. 5c clearly shows that a diffractive exchange only consisting of quarks ($A_q = 0$) is unable to describe the data. In contrast, the inclusion of the gluons leads to an excellent fit as shown by Fig. 5d. These results imply that at a starting scale of $Q_0^2 = 5$ GeV^2 the diffractive exchange is characterised by a hard gluon behaviour in which the gluons carry 80% of the momentum.

5. Diffractive final states

If diffraction is due to the exchange of a pomeron which is dominated by gluons as the $F_2^{D(3)}$ analysis has shown, then a cruxial test as to the validity of these results would be to model these mechanisms and compare the diffractive final state properties with these model predictions.

5.1. *Monte Carlo model*

The Monte Carlo which was developped (RAPGAP [4]) to simulate diffractive final states is based on the Ingelman-Schlein [3] model in which the pomeron has a partonic structure. Inelastic $e\mathbb{P}$ interactions are simulated based on the elementary processes:

$$eq \rightarrow eq \qquad O(\alpha) \tag{15}$$
$$eg \rightarrow eq\,\bar{q} \qquad BGF \quad O(\alpha.\alpha_s.g) \tag{16}$$
$$eq \rightarrow eqg \qquad QCDC \quad O(\alpha.\alpha_s) \tag{17}$$

To avoid divergencies in the matrixelement calculations with massless quarks a $\hat{p}_t^2$-cutoff was imposed, $\hat{p}_t^2 \leq 2$ GeV^2. Two sets of parton distributions were used to describe the partonic structure of the pomeron: the first set corresponds to the situation were only u,d and s quarks are contributing in the partonic processes, while in the second set a leading gluon distribution was used where the gluons carry aproximately 80% of the diffractive exchange as obtained from the $F_2^{D(3)}$ analysis [5]. Both sets are defined at a scale $Q_0^2 = 2.5\ GeV^2$ and evolve according the DGLAP evolution equations. Hereafter, we will refer to the two sets resp. as RGQ and RGG. Subsequent to the "hard" scattering, two different fragmentation schemes have been used namely the leading log parton showers and the colour dipole model. In both cases the hadronisation is done using the LUND string model as implemented by JETSET [8].

5.2. *Corrections and systematic errors*

As the RAPGAP model describes well the overall features of the hadronic final state, this Monte Carlo was used to correct for limited acceptance and finite resolution corrections. The distributions shown in the following have all been corrected for acceptance and resolution effects.

5.3. Results

From the processes (15),(16) and (17) one can already anticipate which characteristics the final state will exhibit when the pomeron is either dominated by quarks or by gluons. If the pomeron has mainly a quark content, processes (15),(17) will be important. The quark interacting with the photon has a small virtuality leading to a small intrinsic k_t in the two body partonic reaction. Hence in the $\gamma I\!P$ system the struck quark and its remnant will be strongly aligned with the $\gamma^* I\!P$ axis. Transverse energy will be characterised by hadronisation and soft bremstrahlung. Events with large transverse momentum may also occur, due to process (17) but only to order α_s. This picture changes when one assumes a "gluon dominated pomeron". Then the dominant process is the BGF-process leading to a 3 body partonic final state in which the intrinsic k_t can be finite and therefore one expects more transverse energy.

5.3.1. Energy flow, p_T and x_F

Fig. 6a shows the energy flow $\frac{1}{N}\frac{dE}{d\eta^*}$ in the $\gamma^* I\!P$ frame for three intervals of the invariant mass M_X^2 , where η^* stands for the pseudo-rapidity w.r.t to the direction of the exchanged boson. In the figure, also the RAPGAP predictions RGQ and RGG are shown for two different fragmentation schemes. Comparision between data and Monte Carlo clearly shows that RGQ prediction lacks energy in the central region $\eta^* \approx 0$ and this is independent of the fragmentation sheme. In contrast, and as expected, the RGG predicts more energy centrally and fits quite well to the data.

Fig. 6.b shows the ${p^*}_T^2$ distribution for charged particle tracks for both the H1 data and the EMC data. For the H1 data, the value for $< M_X >$=10 GeV is slightly less then the corresponding $< W >$=10 GeV value for EMC. The EMC data at higher Bjorken x-values resulting from valence quarks are seen to fall below the H1 data. Also the RGQ prediction fails to describe the data whereas RGG prediction is in good agreement with the H1 data.
The "Seagull" plot in Fig. 6c, showing the H1 and EMC data, exhibits the correlation between the $< p_T^{*2} >$ and x_F, Feynmann-x. Although at a slightly lower centre of mass energy, the H1 data have a significantly larger $< p_T^{*2} >$ than the EMC data except for the highest x_F data point, indicating that in a substantial fraction of the events a high p_T hard process is involved. The forward/backward asymmetry in x_F observed for the EMC data can be understood as result of less radiation emanating from the proton because it is a more extended object than the quark. By the same token, the symmetry in the case of the H1 data can be interpreted as the interacting quark and the remnant having similar sizes and therefore produce a similar amount

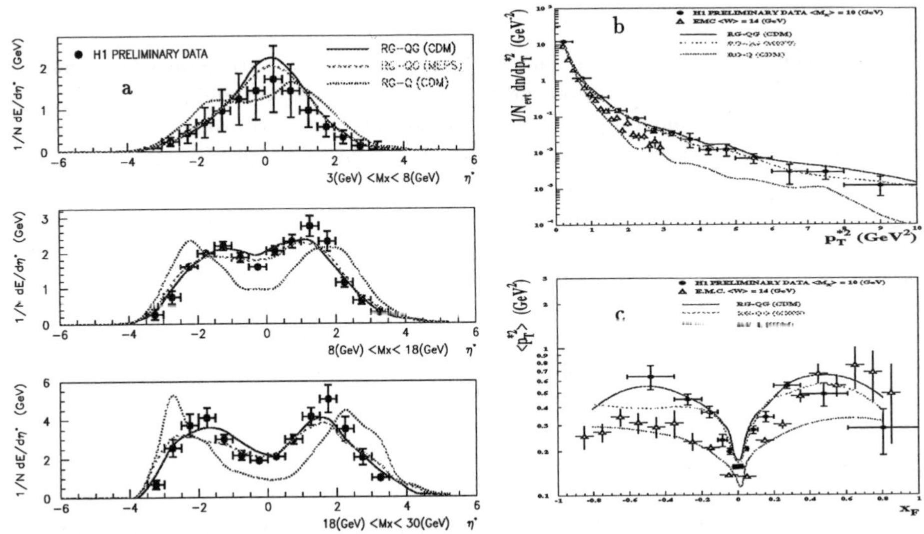

Fig. 6. a) Energy flow $\frac{1}{N}\frac{dE}{d\eta^*}$ in the $\gamma^* I\!P$ frame as a function of the invariant mass M_X^2 in the kinematic region $7.5 \leq Q^2 \leq 100$ Gev^2, $0.05 \leq y \leq 0.6$ and $x_{I\!P} \leq 0.025$. b) $< p_T^{*^2} >$ distribution for the H1 data in the $\gamma^* I\!P$ system and for the EMC μp data in the $\gamma^* p$ frame. c) "Seagull" plot for the H1 data in the $\gamma^* I\!P$ system and for the EMC μp data in the $\gamma^* p$ frame. The curves are the RAPGAP model predictions

of radiation. Again the RGG prediction fits the data well.

5.3.2. Thrust Analysis

A different approach to investigate the partonic composition of the pomeron is to apply a thrust analysis to the hadronic final state in the $\gamma^* I\!P$ frame. Thrust is defined as:

$$T = \left(1/\sum_i^N \mid \vec{p_i} \mid\right) . \max_{\vec{a}} \sum_i^N \mid \vec{p_i}.\vec{a} \mid \tag{18}$$

where the $\vec{p_i}$ are the final state momenta and $\vec{a}$ is a direction along which the projected momentum flow is maximal. This axis permits, with respect to the $\gamma^* I\!P$ axis to define a p_T^{thrust} variable. The value of thrust reaches its minimum value for a pencil like back to back topology, and its maximal value for a spherical final state with infinite multiplicity. Fig. 7.a shows the average thrust as a function of $1/M_X$ for the H1 and e^+e^- data. The increase of the average thrust value with M_X indicates that the events at large M_X develop a more back to back structure. Fig. 7b is similar to Fig. 7a but the events are split in a high and a low p_T event sample. Higher p_T^{thrust} values correspond to lower average event thrust i.e. there seems to be significant more gluon radiation.

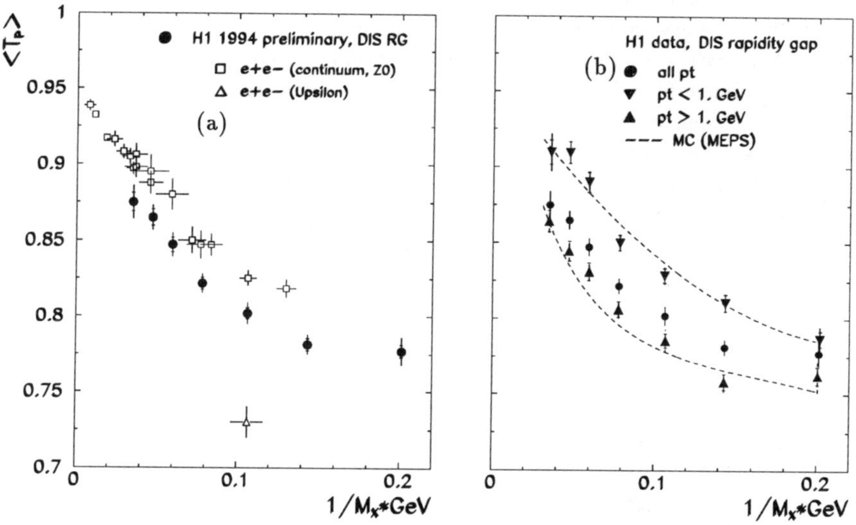

Fig. 7. a) The average thrust as a function of $1/M_X$. Both H1 diffractive data and e^+e^- data are shown. b) same as a) but for the event samples with $p_T^{thrust^2} \geq 1(3)$ GeV^2

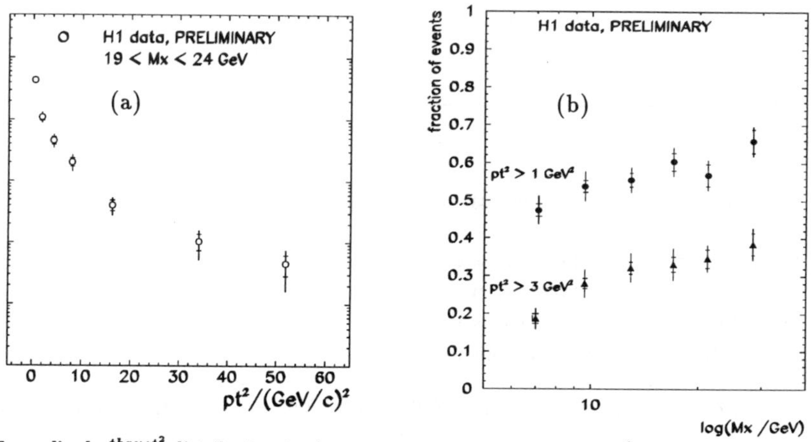

Fig. 8. a) Normalised $p_t^{thrust^2}$ distribution in the mass range $19 \leq M_X \leq 24$ Gev^2. b)Event fractions as a function of $\log(1/M_X)$ for the event sample $p_T^{thrust^2} \geq 1(3)$ GeV^2

This evidence is further corroborated by Fig. 8a and Fig. 8b which give resp. the inclusive p_T^{thrust} distribution and the fraction of p_t^{thrust} events with $p_T^{thrust} \geq 1$ GeV and $p_T^{thrust} \geq 3$ GeV. Although the p_T^{thrust} distribution in Fig. 7c peaks at small p_T^{thrust} values, there exists a substantial fraction of events at large p_T^{thrust}. All this can be taken as evidence for a large gluon content in the diffractive exchange. However the difference between in $< T >$ between e^+e^- and large rapidity gap events (Fig. 7a) indicate that the latter cannot be understood as simple $q\bar{q}$ events and that higher parton multipicties in large rapidity gap events are important. As shown in Fig. 7b the predictions of RGG are in good agreement with the data.

5.4. Particle correlations

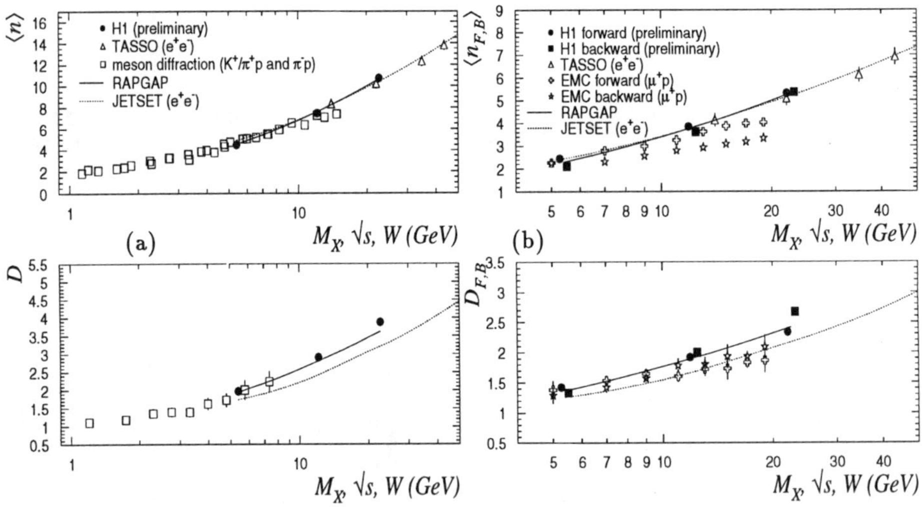

Fig. 9. a) Full event multiplicity moments. b) Single hemisphere multiplicity moments

The differences in the processes underlying the diffractive exchange will also reflect in the event multiplicity structure. As a reference we will compare to e^+e^- data in which the partonic final state is dominated by $q\bar{q}$ pairs. Fig. 9a shows for an event sample which covers the kinematic region

$$7.5\ GeV^2 \leq Q^2 \leq 100\ Gev^2\ ,\ 0.05 \leq y \leq 0.6\ ,\ x_{I\!P} \leq 0.025\ ,\ M_X \geq 3\ GeV$$

the average charged particle multiplicity $< n >$ and the dispersion $D = \sqrt{< n^2 > - < n >^2}$ for the global X system as a function of M_X. The H1 data points regarding the average multiplicity are in agreement with the meson diffraction data at lower energy and indicate a slightly faster rise than the e^+e^- data. The dispersion D in the H1 data is significantly larger than the JETSET prediction which adequately describes the e^+e^- event features. Fig. 9b is similar to Fig. 9a but the forward and backward hemispheres are taken separately. The mean multiplicity and the dispersion in the forward/backward hemispheres are very similar. This result is in agreement with the conclusion also drawn from Fig. 6c. The curves shown in Fig. 9a and Fig. 9b are the RGG prediction which describe the data fairly well.

5.4.1. Jets

Jets are observed in diffractive interaction both in photoproduction and D.I.S. Due to the high p_T involved, these events can also shed light on the the partonic content of the pomeron. Using a sample of 399 events in the kinematic region:

$$Q^2 \leq 4\ GeV^2 \quad , \quad 0.2 \leq y \leq 0.8 \quad , \quad -1.5 \leq \eta^{jet} \leq 1.5 \quad and \quad \eta_{max}^{hadron} \leq 1.8$$

the Zeus collaboration has measured the di-jet cross sections $d\sigma/d\eta^{jet}$ and $d\sigma/dE_T^{jet}$ (Fig. 10) using a cone algorithme and imposing an energy cut $E_T^{jet} \geq 6$ GeV. The

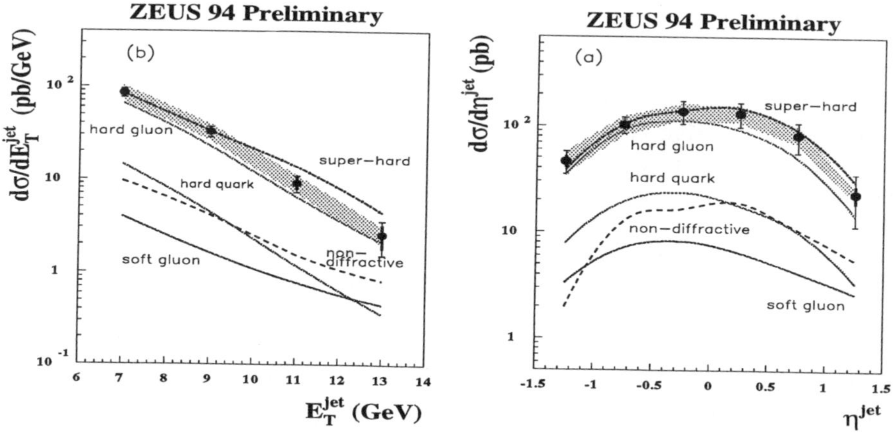

Fig. 10. Photoproduction di-jet cross section as function of η^{jet} (a) and as function of E_T^{jet}. The curves are the Pompyt predictions assuming various parton distributions in the pomeron

curves shown are the model predictions from the POMPYT Monte Carlo with different parton densities for the quarks and gluons in pomeron. As clearly indicated by Fig. 10, the data exclude a soft gluon distribution and require a hard ($\beta f_{g/\mathbb{P}} = 6\beta \cdot (1-\beta)$) or a superhard ($\beta f_{g/\mathbb{P}} = 0.1/(1-\beta)$) gluon. A similar H1 analysis [10] confirms these conclusions. Also di-jet DIS data lead to the same results.

5.5. Charm production

If the pomeron has a large gluon content, then the BGF process at the partonic level should be dominating, leading to an enhanced charm production. This has been confirmed by both the H1 and Zeus collaborations using the reaction :
$e + p \to D^{*\pm} + X + p$ in which the D^* are reconstructed through their decay mode $D^{*\pm} \to \bar{D}^{*0}\pi^\pm \to (K^\mp\pi^\pm)\pi^\pm$. The measured cross sections for $\mid \eta^{D^*} \mid \leq 1.5$ and $p_T^{D^*} \geq 1$ GeV^2 are:

$$\sigma(e^+p \to e^+D^{*\pm}Xp) = \left(380 \pm \begin{matrix}150\\120\end{matrix}\,(stat) \pm \begin{matrix}140\\110\end{matrix}\,(sys)\right) pb \quad (H1) \qquad (19)$$

$$\sigma_{(diff)}(D^{*\pm}) = \left(875 \pm 248\,(stat) \pm \begin{matrix}395\\199\end{matrix}\,(sys)\right) pb \quad (Zeus)$$

For the ZEUS measurement the double dissociation contribution is not subtracted.

These values are compatible with the RGG prediction of $\sigma_{diff}^{RGG} \approx 200pb$ and disagree with RGQ prediction $\sigma_{diff}^{RGQ} \approx 10pb$.

6. Conclusions

The analysis of the $x_{I\!P}$-dependence of the $F_2^{D(3)}$ structure function is shown to depend on β but not on Q^2 at the present level of experimental precision. This factorisation breaking in β is consistent with a contribution from meson exchange. The pomeron trajectory is found to have a intercept slightly higer than that obtained in soft hadronic physics. The observed scaling violations in the $\ln Q^2$-dependence of $F_2^{D(3)}$ at fixed β are interpreted as a substantial contribution of gluons in the partonic structure of the diffractive exchange. A QCD analysis of $F_2^{D(3)}$ using the DGLAP evolution shows that parton distributions in the pomeron are dominated by a leading gluon at a Q_0^2 starting scale of 2.5 GeV^2. The Monte Carlo program RAPGAP which treats the diffractive interaction as inelastic $eI\!P$ interactions, in which the pomeron is assumed to have a partonic substructure and which uses the parton distributions obtained from the previous analysis, is at the present level of experimental accuracy capable of explaining most of the different features of the diffractive final state such as energy flow, thrust distributions, multiplicity moments and jet properties.

7. Acknowledgements

The author would like to thank the Organising Committee of the Conference for the invitation to the Ringberg Workshop. He also takes the opportunity to thank his H1 and Zeus colleagues for providing him with the information and for the fruitful discussions in the preparation of the manuscipt.

8. References

1. A. Donnachie, P. V. Landshoff, *Phys. Lett.* **B296** (1992),227.
2. M. Gluck, E. Reya, A. Vogt, Z.Phys. **C53** (1992) 651-656.
3. G. Ingelman and P. Schlein, Phys. Lett. **B152** (1985) 256.
4. H. Jung, *Hard Diffractive scattering in high energy ep collisions and the Monte Carlo generator RAPGAP Comp.Phys. Comm.* 86 (1995) 147.
   H. Jung *RAPGAP2.02 Program Manual* 1996 unpublished.
5. H1 Collaboration: *A Measurement and QCD Analysis of the Diffractive Structure Function* $F_2^{D(3)}$, Proc. of the 28th Int. Conf. on HEP (ICHEP '96), Warsaw, 1996.
6. L. Trentadue and G. Veneziano, Phys. Lett. **B323** (1994) 201.
7. Yu. Dokshizer, JETP **46** (1977) 641.
   V.N. Gribov and L.N. Lipatov, Sov. Journ. Nucl. Phys. **15** (1972) 78.
   G. Altarelli and G.Parisi, Nucl. Phys. **B126** 1977 298.
8. T. Sjöstrand, Comp. Phys. Comm. **82** (1994) 74.
9. EMC Collaboration, Comparison between hadronic final states produced in μp and e^+e^- interactions. *Zeit Physik* **C35** (1987) 417.
10. P. Marage, *Diffractive Di-Jet Production in H1* $5^t h$ International Workshop on Deep Inelastic Scattering and QCD, Chicago, 1997.

HARD DIFFRACTIVE SCATTERING

ZOLTAN KUNSZT
Institute of Theoretical Physics, Eidgenossische Technische Hochschule, Zurich
Ch-8093 Zurich, Switzerland
E-mail: kunszt@itp.phys.ethz.ch

ABSTRACT

I discuss hard diffractive scattering in the framework of perturbative QCD and Regge-parametrization.

1. Diffractive deep inelastic scattering

1.1. Kinematics

Measurements of deep inelastic diffractive scattering at HERA [1,2,3] provided us with some of the most interesting experimental results which allow one to test new theoretical ideas on diffractive processes in an unexplored regime. The data triggered much theoretical work. The concept of diffractive parton number densities (or differential fracture functions)[4,5] with standard DGLAP evolution appears to be the key ingredient of the QCD description. It was conjectured that these new types of parton number densities factorize similarly to the parton number densities of non-diffractive deep inelastic scattering and satisfy the same Altarelli-Parisi evolution equation. The Ingelman-Schlein model[6] emerges subsequently by assuming Regge-behaviour[7] for the $x_{\mathcal{P}}$ and t dependence.

Diffractive deep inelastic scattering is defined by restricting our studies into final states when the hadrons can be classified into a high mass (M_X) and a low mass (M_Y) group with large rapidity gap between them. Denoting the four-momenta of virtual photon, proton and electron with q, p and k and the four-momenta of the high mass and low mass hadronic system with p_X and p_Y, respectively, the measured differential cross-section can be written in terms of the diffractive structure function F_2^D as

$$\frac{d\sigma^D}{dx dQ^2 dx_{\mathcal{P}} d\,|t|} = \frac{4\pi\alpha^2}{xQ^4}\left(1 - y + \frac{y^2}{2}\right)\frac{dF_2^D}{dx_{\mathcal{P}} d|t|}(x, Q^2, x_{\mathcal{P}}) \qquad (1)$$

where

$$Q^2 \;=\; -q^2\,, \quad y = \frac{q{\cdot}p}{k.p}\,, \quad x = \frac{Q^2}{2p.q}\,, \quad x_{\mathcal{P}} = \frac{q{\cdot}p - q{\cdot}p_y}{q{\cdot}p} \approx \frac{Q^2 + M_X^2}{Q^2 + W^2}\,,$$

$$\beta = \frac{x}{x_{\mathcal{P}}} = \frac{Q^2}{2q.p - 2q.p_y} \approx \frac{Q^2}{Q^2 + M_X^2}, \quad W^2 = (q+p)^2, \quad t = (p - p_Y)^2 \quad (2)$$

Requiring small M_Y and $x_{\mathcal{P}}$ the rapidity gap is inevitable. For example the H1 collaboration required $M_Y < 1.6\,\mathrm{GeV}^2$ and $x_{\mathcal{P}} < 0.05$. In the H1 measurement, the momentum transfer from the proton to the Y-system $t = (p - p_Y)^2$ is not measured but is restricted to the small interval $|t_{\min}| < |t| < 1.0\,\mathrm{GeV}^2$. The ZEUS collaboration collected also data when the Y system was identified with a fast forward proton with a measured value of t. The longitudinal cross-section was assumed to be zero.

1.2. Diffractive parton number densities

The factorization theorem of deep inelastic scattering allows the decomposition

$$F_2(x, Q^2) = \sum_a \int dx' f_{a/p}(x', \mu) \hat{F}_{2a}(x/x', Q^2, \mu^2) \quad (3)$$

It is natural to ask whether similar decomposition remain valid also for diffractive deep inelastic scattering giving

$$\frac{dF_2}{dx_{\mathcal{P}} d|t|}(x, Q^2, x_{\mathcal{P}}, t) = \sum_a \int dx' \frac{df_{a/p}}{dx_{\mathcal{P}} d|t|}(x', \mu, x_{\mathcal{P}}, t) \hat{F}_{2a}(x/x', Q^2, \mu^2) \quad (4)$$

where in both equations $\hat{F}_2$ denotes the finite hard scattering contribution. The dependence on t and $x_{\mathcal{P}}$ is completely absorbed in the diffractive parton number densities. Berera and Soper[5] pointed out that the operator definition of the parton number densities can be generalized for diffractive processes. The quark parton number density for example defined as

$$f_{q/p}(x, \mu) = \frac{1}{4\pi} \sum_{s,X} \int dy^- e^{ixp^+y^-}$$
$$< p, s|\bar{\tilde{q}}(0, y^-, \vec{0})|X > \gamma^+ < X|\tilde{q}(0,0,0)|p, s > \quad (5)$$

is modified in the case of diffractive densities as

$$\frac{df_{q/p}}{dx_{\mathcal{P}} d|t|}(\beta, \mu, x_{\mathcal{P}}, t) = \frac{1}{4\pi} \sum_{s,s',X} \int dy^- e^{i\beta p^+ y^-}$$
$$< p, s|\bar{\tilde{q}}(0, y^-, \vec{0})|p', s'; X > \gamma^+ < p', s'; X|\tilde{q}(0,0,0)|p, s > \quad (6)$$

where X indicates all possible final states with a proton of momentum p' and polarization s' and $\tilde{q}$ is the color singlet quark operator given in terms of quark and gluon fields as

$$\tilde{q}(0, y^-, \vec{0}) = \mathcal{P} e^{ig \int_{y^-}^{\infty} dx^- A_c^+(0, x^-, \vec{0}) T_c} q(0,0,0) \quad (7)$$

where T_c is the color matrix and $\mathcal{P}$ demands path ordering in color space. In fig.1 the factorization in DIS is shown graphically. The crucial observation is that the ultraviolet structure of the bilocal operators defining the non-diffractive and diffractive parton number densities are the same therefore they fulfill the same renormalization group equation. As a result, the diffractive parton number densities at fixed values of $x_{\mathcal{P}}$ and t (assuming t is small) also have to fulfill the DGLAP evolution equation. The apparent difference in the Q^2 behaviour of the DIS and diffractive DIS data must have its origin in the dramatically different initial parton distributions. Factorization

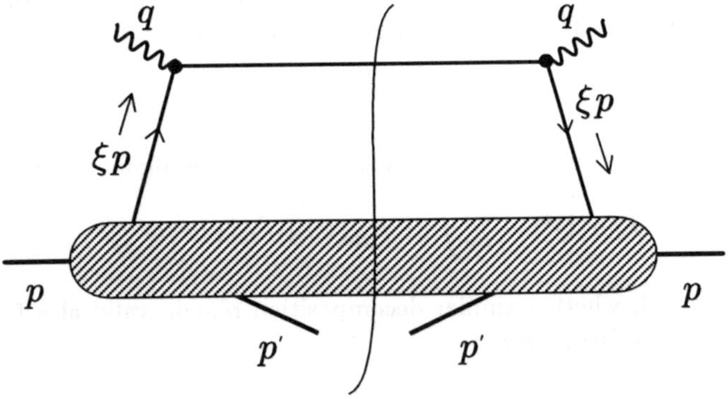

Fig. 1. Diffractive factorization in DIS.

theorem with evolution equation has been suggested by Trentadue and Veneziano[4] in the context of deep inelastic scattering with one observed final state hadron allowing also target fragmentation region. In the QCD improved parton model collinear singularities given by low p_T gluons emitted into the target fragmentation region can not be absorbed into the standard fragmentation and structure functions and a new non–perturbative function, the so called fracture function, had to be introduced. In diffractive deep inelastic scattering the observed forward proton is in the target fragmentation region. Therefore the fracture function formalism applies. Berera and Soper pointed out[5] that if the forward proton transverse momentum remains unintegrated and is required to be small the diffractive parton number densities can be identified with the fracture functions. With explicit calculation Graudenz[8] has shown that the factorization theorem for fracture functions is fulfilled in next-to-leading order accuracy. It is generally believed that it remains valid for diffractive deep inelastic scattering to all orders in perturbation theory.

1.3. Regge factorization and parametrization

As we have seen above both the standard parton number densities and the diffractive parton number densities are defined as matrix element of bilocal operator between proton states and they are given by 'soft physics', they can not be calculated in perturbation theory. The parametrization of the initial distribution at some Q_0^2 scale, however, can be motivated by phenomenological models. For diffractive scattering Regge theory is the most successful framework. Assuming the dominance of the Pomeron trajectory one obtains

$$\frac{df_{q/\mathcal{P}}}{dx_{\mathcal{P}}d\,|t|}(\beta,\mu,x_{\mathcal{P}},t) = \frac{|\gamma(t)|^2}{8\pi^2} x_{\mathcal{P}}^{-2\alpha_{\mathcal{P}}(t)} f_{a/\mathcal{P}}(\beta,t;\mu) \tag{8}$$

where $\alpha_{\mathcal{P}}(t)$ is the trajectory function and $\gamma(t)$ is its coupling to the proton. From fits to hadronic cross-sections[7] one obtains $\alpha_{\mathcal{P}} \approx 1.08$. The function $f_{a/\mathcal{P}}(\beta,t;\mu)$ is called the parton distribution of the Pomeron since formally the cross-section of diffractive deep inelastic scattering is given by folding the hard scattering cross-section with this function. This description was suggested by Ingelman and Schlein using the notion of 'parton constituents of the Pomeron'. The latter concept should be treated with care since the Pomeron can not be interpreted as a particle emitted by the proton before the parton distribution was probed. One of the most significant qualitative consequences of the assumption of the Pomeron parametrization is the factorization of the $x_{\mathcal{P}}$ and t dependence. The HERA diffractive structure function data, however, show a modest amount of factorization breaking [2,1,3]. It could, however, be accommodated by invoking a sum over Regge trajectories, each with a different intercept and structure function:

$$F_2^D(\beta,Q^2;x_{\mathcal{P}},t) = \sum_{\mathcal{R}} F_{\mathcal{R}}(t) x_{\mathcal{P}}^{-2\alpha_{\mathcal{R}}(t)+1}\, F_2^{\mathcal{R}}(\beta,Q^2)\,, \tag{9}$$

which would yield an effective power of the $x_{\mathcal{P}}$ dependence which depends on β but is approximately independent of Q^2. Integrating over t and small bins of $x_{\mathcal{P}}$ what is measured is a linear combination of $F_2^{\mathcal{R}}$ structure functions or, equivalently, the parton distributions in an effective colour singlet target:

$$\int dx_{\mathcal{P}}dt\; F_2^D(\beta,Q^2;x_{\mathcal{P}},t) = \sum_{\mathcal{R}} A_{\mathcal{R}}\, F_2^{\mathcal{R}}(\beta,Q^2) = \beta \sum_q e_q^2 \sum_{\mathcal{R}} q_{\mathcal{R}}(\beta,Q^2)\,, \tag{10}$$

where the coefficients $A_{\mathcal{R}}$ are independent of β and Q^2. Since the DGLAP equations are *linear* in the parton distributions, the Q^2 evolution of the integrated structure function F_2^D should also be calculable perturbatively. The $x_{\mathcal{P}}$ dependence was fitted by treating $\alpha_{\mathcal{P}}(0)$ as a free parameter and the H1 and ZEUS collaboration have found that its value is somewhat larger than the soft Pomeron values. The value of a recent fit obtained by H1 is $\alpha_{\mathcal{P}} \approx 1.20 \pm 0.04$. The values obtained by H1 and ZEUS are consistently between the values of the intercept of the soft and hard (BFKL) Pomeron.

Various fits for the parton distributions in the 'Pomeron' $f_{q/\mathcal{P}}(x,\mu^2)$ have been proposed, ranging from the two extremes of mainly gluons to mainly quarks [2,9,10,11]). Already the early data favoured the hard gluon parametrization. From the recent

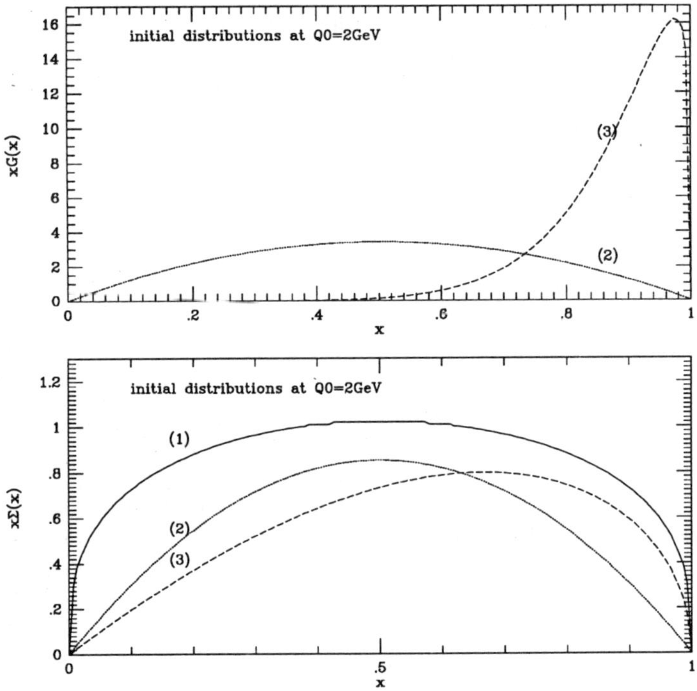

Fig. 2. Starting parton distributions and their evolved values

H1 analysis of the 1994 data[3] one can conclude that the parametrization which only assumes quarks at $Q^2 = 3\,\mathrm{GeV}^2$ is excluded. Since the diffractive parton number densities do not have momentum sum rule the overall normalization of the quark singlet $\Sigma(x, \mu^2)$ and the gluon $g(x, \mu^2)$ densities have to also be fitted from the data. The normalization, however, will not change with Q^2 evolution. It is the *product* of $f_{q/\mathcal{P}}$ and $f_{\mathcal{P}}$ that appears in the expression for the structure function, therefore it is convenient to simply impose a momentum sum rule on the parton distributions and absorb an overall normalization $\mathcal{N}$ into $f_{\mathcal{P}}$. In fig. 2 the initial parton distributions are shown at $\mu^2 = 4\,\mathrm{GeV}^2$ for hard gluon (3) no-gluon (1) and soft gluon (2) fits.

There has been some warning[14] that one can not accept the fitted gluon distribution if it is peaked at $x = 1$ since the higher twist contributions become important in the range $(1 - x)\mu^2 < 1\,\mathrm{GeV}^2$. This criticism, however, should concern only the measured data points. If the data data points in this range are excluded from the fit the obtained densities can be meaningfully calculated using the DGLAP evolution equation also in this range. One can easily see that the densities in this range can not be directly related to measurable quantities without adding also higher twist terms.

Finally I note that a number of models have been investigated in the literature to calculate the diffractive deep inelastic structure function at some fixed scale. Interesting semi-classical description has been developed by Buchmueller and Hebbeker[12,13] consistent with the aligned jet parton picture[13,15]. The Pomeron is soft and in leading order the intercepts is $\alpha_{\mathcal{P}}(0) = 1.0$. Models with simple two gluon exchanges have also been considered [14].

2. Hard diffractive processes in hadron-hadron scattering

Assuming the validity of factorization theorem also for diffractive hard scattering processes using the DIS value of the densities the cross-section values in hadron-hadron collisions can be calculated by using the standard formula

$$\frac{d\Delta\sigma^{\rm SD}}{dx_{\mathcal{P}}dt} = \sum_{a,b}\int f_{a/p}(x_a,\mu)\frac{df_{a/p}}{dx_{\mathcal{P}}dt}(x_b,\mu,x_{\mathcal{P}},t)\Delta\hat{\sigma}(x_a x_b s,..) \qquad (11)$$

Here the label SD makes reference to single diffractive production. The generalization of this formula for double diffractive scattering is trivial. Data on single diffractive jet production has been obtained first by UA8[17]. Its comparison with the theory was not clear because of high values of t. The validity of the factorization theorem, however, is debated[16]. In particular it is expected that it may be strongly violated. The soft exchanges before the hard scattering do not cancel if hadrons are observed in the target fragmentation region because not all the final states containing the hard-scattered object (jet or W-boson or Higgs boson etc.) are summed over. Interaction between spectator partons can lead to extra particles to fill the rapidity gap. It has been suggested to correct the prediction of the naive application of the factorization theorem with the so called survival probability of the rapidity gap[21,22]. The parton model is predicted to overestimate the cross-sections of hard diffractive scattering by an amount given by such a survival probability. To gain experimental information on this relevant question one may proceed by comparing the naive predictions with the data. The cleanest process to study would appear to be weak boson production at the Tevatron $p\bar{p}$ collider. The naive prediction was first calculated before the HERA data in Ref. [18] neglecting possible effects of Q^2 evolution. It has been found that the single diffractive component of the total W cross section could be as large as 20%. It is convenient to study integrated 'single diffractive' events by $x_{\mathcal{P}} < 0.1$, integrating over all t and to normalize the result to the total W-production cross-section. In practice, of course, the events are defined by rapidity gaps of a certain minimum size, and therefore the observed diffractive cross section must be corrected to the theoretical prediction based on, say, $x_{\mathcal{P}} < 0.1$ using a Monte Carlo simulation [19]. Using the various fits discussed above, it is straightforward to reevaluate the single diffractive

W cross section with Q^2 evolution and with updated values for the Pomeron flux factors. Two recent analysis[10,11] have found values in the range

$$\frac{\Delta\sigma_W^{\mathrm{SD,th}}}{\sigma_W} \approx 4\%\text{—}6\%. \tag{12}$$

The important point to note here is that the predictions coming from different fits are quite similar. The single diffractive W cross section at the Tevatron samples the quarks at $\langle x_{q/P}\rangle \sim 0.4$. At low Q^2 the quark distributions at this x are constrained by the HERA F_2^D data to be roughly the same. As Q^2 increases the distributions diverge, reflecting the quantitatively different gluon contributions to the DGLAP evolution. However at $Q^2 \sim 10^4$ GeV^2, the relevant value for W production, the difference between the quarks in the three models is still not very large, and the predictions for $\sigma^{SD}(W)$ are correspondingly similar since they are directly related to the HERA data. In recent measurements the CDF Collaboration has found a smaller ratio

$$\frac{\Delta\sigma_W^{\mathrm{SD,th}}}{\sigma_W} \approx 1.55 \pm .55 \tag{13}$$

The surviving probability of the factorized result is about $S \approx 20 - 30\%$. Similar analysis for inclusive jet production gave even smaller value $S \approx 10\%$. These results indicate that the use of the (invalid) factorization theorem very likely strongly overestimates the cross-section values of diffractively produced heavy quarks and Higgs bosons at LHC. It would be important to develop some theoretical model in which the survival probability could be estimated.

3. Acknowledgments

I thank Dave Soper and James Stirling for helpful discussions. I also thank F. Hautmann for reading the manuscript and Bernd Kniehl for organizing an interesting workshop.

4. References

1. ZEUS Collaboration: M. Derrick et al., *Phys. Lett.* **B315** (1993) 481; **B332** (1994) 228; **B338** (1994) 483; *Z. Phys.* **C68** (1995) 569.
2. H1 Collaboration: T. Ahmed et al., *Nucl. Phys.* **B429** (1994) 477. T. Ahmed et al., *Phys. Lett.* **B348** (1995) 681.
3. H1 Collaboration: Adloff et.al., DESY 97-158, hep-ex/9708016 (1997).
4. L. Trentadue and G. Veneziano, *Phys. Lett.* **B323** (1994) 201.
5. A. Berera and D.E. Soper, *Phys. Rev.* **D53** (1996) 6162.
6. G. Ingelman and P. Schlein, Phys. Lett. **B152** (1985) 256.

7. A. Donnachie and P.V. Landshoff, *Phys. Lett.* **B191** (1987) 309; **B198** (1987) 590(E).
8. D. Graudenz, preprint, CERN-TH-96-052 (1996), hep-ph 9701334.
9. T. Gehrmann and W.J. Stirling, *Z. Phys.* **C70** (1996) 89.
10. Z. Kunszt and W.J. Stirling, in the Proceedings of the Workshop Deep Inelastic Scattering, 1996, Rome, e-print archive hep-ph/9609245.
11. L. Alvero, J. Collins, J. Terron and J. Whitemore preprint CTEQ-701 and hep-ph/9701374.
12. W. Buchmueller and A. Hebecker, *Nucl. Phys.* **B476** (1996) 203; W. Buchmueller, M F. McDermott and A. Hebecker, *Nucl. Phys.* **B487** (1997) ; preprint DESY 97-099.
13. M. F. McDermott, these Proceedings.
14. J. Bartels, these Proceedings.
15. J. D. Bjorken, preprint, SLAC-PUB-7096.
16. J. C. Collins, L. Frankfurt amd M. Strikman, *Phys. Lett.* **B307** (1993) 161.
17. UA8 collaboration: R. Bonino et al., *Phys. Lett.* **B211** (1988) 239; A. Brandt et al., *Phys. Lett.* **B297** (1992) 417.
18. P. Bruni and G. Ingelman, *Phys. Lett.* **B311** (1993) 317.
19. K. Goulianos for the CDF Collaboration, Fermilab-Conf-96/233-E; P. L. Melese for the CDF Collaboration, Fermilab-Conf-96/231-E.
20. A. G. Brandt, for D0 Collaboration, Fermilab-Conf-96/185-E.
21. J. D. Bjorken, *Phys.Rev.* **D47** (1993) 101.
22. S. E. Soper, preprint hep-ph/9707384.
23. M. Heyssler, Durham preprint DTP/96/10 (1996).
24. A. Schäfer, O. Nachtmann and R. Schöpf, *Phys. Lett.* **B249** (1990) 331; D. Graudenz and G. Veneziano, *Phys. Lett.* **B365** (1996) 302; M. Heyssler, Z. Kunszt and W. J. Stirling, *Phys. Lett.* **B406**(1997) 95.

Are charm and high-$p_\perp$ jets the keys to understanding diffraction in DIS

M.F. McDermott
DESY, Theory Group, Notkestrasse 85,
Hamburg, 22607, Germany
E-mail: mcdermot@mail.desy.de

ABSTRACT

Following an introduction which explains some basic ideas about diffraction, the semiclassical approach to diffraction in deep inelastic scattering (DIS) is outlined. Some phenomenological tests of this picture, concerning the $p_\perp$ and mass spectra of open charm and jet production are given.

1. Introduction

To introduce the main topic of this talk, I would like, first of all, to recall some basic ideas concerning diffraction that it is useful to have at the back of one's mind when thinking about diffraction in high energy processes such as DIS at small x. I will then discuss predictions for diffractive events containing either high $p_\perp$-jets or charm quarks in the final state, for two-gluon exchange models and the semiclassical approach.

1.1. The optical model

Let us begin by reminding ourselves of the phenomena of diffraction in optics. Consider a broad beam of plane polarized light incident on a small piece of Polaroid, which has its axis of polarization at an angle, θ, to the polarization direction. The component of the light with its polarization parallel to this axis will be transmitted and that perpendicular to it will be absorbed. The transmitted wave, just behind the Polaroid now contains two components, with polarizations parallel and perpendicular to that of the incident wave. A new state, degenerate in energy to the initial one, has been "diffracted into existence" by the piece of Polaroid.

Good and Walker developed an optical model for diffraction in the high energy scattering of hadrons as early as 1960 [1]. They consider the possibility that a beam of hadrons, of type A and mass M, incident on a heavy nucleus at rest, produces two new particles, B and C, of combined rest mass M^*, leaving the nucleus in its ground state. The longitudinal kick given to the nucleus needs to be fairly small to avoid breaking it up : $q_{||} \leq m_\pi / A_N^{1/3}$, where m_π is the mass of the pion and A_N is the nucleon number of the nucleus. As long as hadron A has a three momentum, P, of at least

$$P_{th} = \frac{(M^2 - M^{*2}) A_N^{1/3}}{m_\pi} \tag{1}$$

then the process is energetically possible. According to quantum mechanics the dressed nucleon $|\tilde{A}\rangle$ is a complicated superposition of 'bare' states, $|B_i\rangle$, which have the same quantum numbers as A. These, in turn, may be re-expanded in terms of a set of 'physical' dressed states $|\tilde{D}_j\rangle$,

$$|\tilde{A}\rangle = \sum_i b_{Ai}|B_i\rangle = \sum_{ij} b_{Ai}d_{ij}|\tilde{D}_j\rangle . \tag{2}$$

For very high beam energies $P \gg P_{th}$, the system A spends a very short time in the heavy nucleus on the timescale of its quantum fluctuations, we may consider these fluctuations as essentially frozen during the interaction.

In the nuclear medium A will have a different expansion, $|\tilde{A}\rangle = c_{Ak}|\tilde{C}_k\rangle$. Each component $|\tilde{C}_k\rangle$ constitutes an eigenstate of the interaction and may be absorbed differently by the nucleus. For the scattering matrix just behind the target we have

$$|S\rangle = |I\rangle - |T\rangle = (1-\bar{\eta})|\tilde{A}\rangle + \sum_i(\bar{\eta}-\eta_i)c_{Ai}|C_i\rangle. \tag{3}$$

Provided the eigenstates have different absorption coefficients ($\bar{\eta} \neq \eta_i$) then new physical states $|C_i\rangle$, "degenerate in mass" with A, may be diffracted into existence by the interaction with the nuclear target.

1.2. High energy hadron-hadron diffractive scattering, rapidity gaps, light cone co-ordinates and variables for diffractive DIS

Consider very high energy diffractive scattering of two hadrons: hadron A of momentum P_A, travelling in the $+z$ direction, loses only a very small amount of its momentum and remains intact ($P'_A \gtrsim 0.99P_A$); hadron B, travelling in the opposite direction, breaks up. It is very useful to introduce *light cone co-ordinates* for the patrons in these hadrons:

$$\begin{aligned} q_i^+ &= q_i^0 + q_i^z = m_i^\perp(\cosh y_i + \sinh y_i) = m_i^\perp e^{y_i}\,, & (4)\\ q_i^- &= q_i^0 - q_i^z = m_i^\perp(\cosh y_i - \sinh y_i) = m_i^\perp e^{-y_i}\,, & (5)\\ q_i^\perp &= (q_i^x, q_i^y)\,, & (6)\\ m_i^\perp &= \sqrt{q_i^{\perp 2} + m_i^2}\,, & (7)\end{aligned}$$

where y_i is the rapidity of parton i and is approximately equal to its pseudorapidity η_i provided $m_i^2 \ll q_i^{\perp 2}$. A typical parton in hadron B has a large q^- component and a small q^+ component and the converse is true for a parton in hadron A. Momentum conservation

$$P_A + P_B = P'_A + \sum_{i\epsilon B'} q_i \tag{8}$$

then yields

$$\xi = \frac{P_A^+ - P_A'^+}{P_A^+} = \frac{\sum_{i\epsilon B'} q_i^+ - P_B^+}{P_A^+} \approx \frac{\sum_i m_i^\perp e^{-(y_A - y_i)}}{m_A} \tag{9}$$

for the momentum fraction, ξ, lost by hadron A. We can immediately see that for small $\xi \lesssim 0.01$ we will necessarily have a *rapidity gap*, $\eta_{\rm gap} \approx y_A - y_{i,max}$, in the final state and that we can get a reasonable experimental measure of ξ, even without measuring the final state hadron, by considering hadrons close to the gap and those with high $q_i^{\perp 2}$ which contribute most in the above sum (for a more detailed discussion see Appendix A of Collins *et al* [2]).

In diffractive DIS one has to integrate over phase space, d^3P', of the scattered proton. Integrating over the angle, one has two additional scattering variables t and ξ. The scattering variables usually used to describe the events are

$$x = \frac{Q^2}{Q^2 + W^2} \quad , \quad M^2 = (P - P' + q)^2 , \tag{10}$$

$$t = (P - P')^2 \quad , \quad \xi \approx \frac{M^2 + Q^2}{W^2 + Q^2} , \tag{11}$$

$$\beta = \frac{Q^2}{2q.(P - P')} = \frac{Q^2}{M^2 + Q^2 - t} \approx \frac{Q^2}{M^2 + Q^2} \approx \frac{x}{\xi} , \tag{12}$$

where W^2 is the centre of mass energy of the $\gamma^* P$-subprocess, q and $q^2 = -Q^2$ are the four momentum and virtuality of the photon, and M is the mass of the diffractive final state. In order that the proton does not break up, the momentum transfer in the interaction should not be too large: $t \lesssim 1\text{GeV}^2$. We will be primarily concerned with events for which $M^2 \approx Q^2 \ll W^2$, i.e. with the diffractive production of mass states approximately degenerate with the virtuality of the photon.

1.3. Eigenstates of diffraction in DIS

An analogous process to the optical model discussed above is the production of an e^+e^- pair from a highly energetic virtual photon in a Coulomb field, such as that surrounding a heavy nucleus. The pair is a virtual fluctuation of the virtual photon and, provided the individual virtualities of the fermions are small, will require only a small 'kick' from the field to push them onto mass shell.

This notion forms the basis of the semiclassical approach to diffraction in DIS. Viewed from the proton's rest frame, DIS at very high energies (small x) corresponds to the scattering of 'frozen' parton configurations in the virtual photon from the colour field of the proton. Those configurations in which one of the partons carries only a small fraction of the photon's longitudinal momentum and has a low transverse momentum have a small 'off-shellness' and may be diffracted into existence by receiving a small momentum transfer from the proton. The presence of this 'wee' parton in the photon implies that the fluctuation develops a large transverse size by the time it reaches the proton.

To see this, consider the lowest order fluctuation of the photon into a $q\bar{q}$ -pair, of transverse momentum $p_\perp^2$. Momentum conservation at the photon vertex gives

$$\frac{\Delta_p^2}{1-\alpha}+\frac{\Delta_l^2}{\alpha}=-\left(Q^2+\frac{p_\perp^2+m_q^2}{\alpha(1-\alpha)}\right), \tag{13}$$

where l, Δ_l^2 and m_q are the four-momentum, virtuality and mass of the anti-quark; $\alpha = l_z/q_z \approx l_0/q_0$ is longitudinal momentum fraction of the photon carried by the anti-quark; Δ_p^2 is the quark virtuality. Hence, in order to produce partons that are close to mass shell we need the asymmetric configurations in which α or $1-\alpha \ll 1$, $p_\perp^2 \ll Q^2, m_q^2 = 0$. The uncertainty principle then gives the following estimate for the lifetime of these fluctuations

$$\Delta\tau \sim \frac{1}{\Delta E} \sim \frac{1}{m_p x_{bj}}\left(\frac{Q^2}{Q^2+p_\perp^2/\alpha(1-\alpha)}\right), \tag{14}$$

which is much longer, at small x_{bj}, than the typical timescales involved in the interaction of the partons with the proton, which are of the order of the inverse of the mass, m_p, of the proton.

Miettinen and Pumplin [3] were the first to recognize that frozen parton fluctuations, in general, correspond to the eigenstates of diffraction. A very nice discussion of many of the points made here can be found in Chapter 7 of the recent text book by Forshaw and Ross [4].

2. Two-gluon exchange models

The simplest QCD model one can think of for diffraction is the exchange of two gluons in a colour singlet in the t-channel. Several recent papers present predictions for diffractive production of charm [5] and high $p_\perp$-jets [6], for the pure $q\bar{q}$ final state, which have this mechanism of diffractive exchange in common. Unfortunately a complete $\mathcal{O}(\alpha_s)$ calculation, which would also include a gluon in the final state is not yet available. In terms of the diffractive structure function, these corrections are expected to be important when the diffractive mass is large compared to Q^2. The differential cross sections for transversely and longitudinally polarized photons can be written, in the double leading logarithmic approximation, in terms of the square of the gluon density, $G(\xi)$, as follows

$$\frac{d^2\sigma_L}{d\alpha dp_\perp^2} = \frac{2\sum_q e_q^2\alpha_{em}\alpha_s^2\pi^2[\xi G(\xi,(p_\perp^2+m_q^2)/(1-\beta))]^2 C}{3(a^2+p_\perp^2)^6}[\alpha(1-\alpha)]^2Q^2(a^2-p_\perp^2)^2, \tag{15}$$

$$\frac{d^2\sigma_T}{d\alpha dp_\perp^2} = \frac{\sum_q e_q^2\alpha_{em}\alpha_s^2\pi^2[\xi G(\xi,(p_\perp^2+m_q^2)/(1-\beta))]^2 C}{6(a^2+p_\perp^2)^6} \times \left[4(\alpha^2+(1-\alpha)^2)p_\perp^2a^4+m_q^2(a^2-p_\perp^2)^2\right], \tag{16}$$

$$a^2 = \alpha(1-\alpha)Q^2+m_q^2. \tag{17}$$

The factor C parameterizes the required extrapolation from $t \approx 0$ to the integrated cross section, $C \approx \Lambda^2$ (where Λ is a typical hadronic scale).

3. The semiclassical approach

In this section predictions for the diffractive production of heavy flavours and high-$p_\perp$ jets are presented for $q\bar{q}$ and $q\bar{q}g$ final states, in the semiclassical approach [7]. Phenomenological implications of this approach have been investigated in recent preprints [8 9].

3.1. $q\bar{q}$ final states

The longitudinal and transverse differential cross sections for the leading order $q\bar{q}$ fluctuation are as follows

$$\frac{d\sigma_L}{d\alpha dp_\perp'^2} = \frac{4\Sigma_q e_q^2 \alpha_{em}}{3(2\pi)}[\alpha(1-\alpha)]^2 \int_{x_\perp} \left| \int \frac{d^2p_\perp}{(2\pi)^2} \frac{\mathrm{tr}\tilde{W}^{\mathcal{F}}_{x_\perp}(p_\perp' - p_\perp)}{a^2 + p_\perp^2} \right|^2 , \tag{18}$$

$$\frac{d\sigma_T}{d\alpha dp_\perp'^2} = \frac{\Sigma_q e_q^2 \alpha_{em}}{3(2\pi)}(\alpha^2 + (1-\alpha)^2) f^{\mathcal{F}}(a^2, p_\perp') , \tag{19}$$

$$f^{\mathcal{F}}(a^2, p_\perp') = \int_{x_\perp} \left| \int \frac{d^2p_\perp}{(2\pi)^2} \frac{p_\perp \mathrm{tr}\tilde{W}^{\mathcal{F}}_{x_\perp}(p_\perp' - p_\perp)}{a^2 + p_\perp^2} \right|^2 , \tag{20}$$

$$a^2 = \alpha(1-\alpha)Q^2 + m_q^2 . \tag{21}$$

The eikonal factor

$$W^{\mathcal{F}}_{x_\perp}(y_\perp) = U^\dagger(x_\perp + y_\perp)U(x_\perp) - 1 \tag{22}$$

is built from the non-Abelian eikonal factors U and $U^\dagger$ of the quark and antiquark whose light-like paths penetrate the colour field of the proton at transverse positions $x_\perp$ and $x_\perp + y_\perp$, respectively. The superscript $\mathcal{F}$ is used because the quarks are in the fundamental representation of the gauge group. The function $f^{\mathcal{F}}$ depends on the Fourier transform of the colour trace of the eikonal factor which depends in general on the transverse momentum lost by parton. Taking its trace projects onto the colour singlet configurations relevant for diffraction. For a soft colour field, large transverse momentum transfers are exponentially suppressed. The factors in the denominator and numerator in the integrand of Eq.(20) reflect the propagator and spin of this parton, respectively.

In the region $\alpha(1-\alpha)Q^2 > \Lambda^2$, we can expand the denominator of Eq.(20) in the momentum transfer lost by the quark and, from the properties of the eikonal function, show that there is no leading twist diffraction in this region. For $\alpha(1-\alpha)Q^2 < \Lambda^2$, for low $p_\perp$ quarks, i.e. the asymmetric configurations, this expansion is no longer possible, at least for massless quarks, since the denominator becomes small. As a

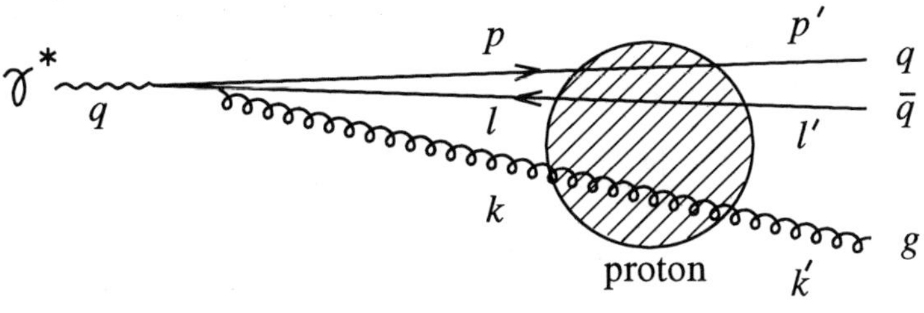

Fig. 1. Dominant partonic process for diffractive charm and high-$p_\perp$ jet production in the semiclassical approach.

result one gets a leading twist contribution to the transverse cross section in this region. The addition powers of $\alpha(1-\alpha)$ in Eq.(18) ensure that the contribution to the longitudinal cross section is higher twist.

In the case of a massive fermion, such as the charm quark, the expansion is always possible since the denominator is limited by the fermion mass and the cross section for diffractive charm production is suppressed by Λ^2/m_c^2. Similarly high-$p_\perp$ configurations are suppressed by $\Lambda^2/p_\perp^2$. All of this can be seen indirectly from size of the virtualities in Eq.(13).

The integration over t is implicit in Eq.(20), as reflected in the single integration over $x_\perp$. Reinstating the t-integration and expanding in $y_\perp$ (for high-$p_\perp$ configurations) we can relate the resulting $x_\perp$-integral to the gluon density and the cross section, $\sigma(y_\perp)$, for scattering a small dipole of size $y_\perp$ from the proton,

$$\int_{x_\perp} \mathrm{tr} W_{x_\perp}(y_\perp) = \frac{3\sigma(y_\perp^2)}{2} = \frac{\pi^2}{2}\alpha_s y_\perp^2 \xi G(\xi)\,. \tag{23}$$

The differential cross sections are then exactly those given in Eqs.(15,16) for the two-gluon exchange models. We see that the mechanism for the diffractive production of jets or open charm is essentially 'hard' and corresponds to the exchange of two gluons in the t-channel.

3.2. $q\bar{q}g$ final states

The $\mathcal{O}(\alpha_s)$ calculation with an additional gluon in the photon fluctuation proceeds along similar conceptual lines. Momentum conservation, for the fluctuation with a wee gluon and two high $p_\perp$-jets, now gives

$$\frac{\Delta_p^2}{1-\alpha} + \frac{\Delta_l^2}{\alpha} + \frac{\Delta_k^2}{\alpha'} = -\left(Q^2 + \frac{p_\perp^2 + m_q^2}{\alpha(1-\alpha)} + \frac{k_\perp^2}{\alpha'}\right), \tag{24}$$

for a gluon of transverse momentum $k_\perp$, virtuality Δ_k^2 and momentum fraction α'. Leading twist diffraction can now result in two cases: either the gluon is wee ($\alpha' \ll$

$1, k_\perp^2 \ll Q^2$) or the fermion is wee and massless (to avoid the mass suppression). The other two partons may emerge in a high transverse momentum configuration to produce jets. For the case with a wee gluon, shown in Fig.(1), we have

$$\frac{d\sigma_L}{d\alpha dp_\perp^2 d\alpha' dk_\perp'^2} = \frac{\sum_q e_q^2 \alpha_{em}\alpha_s}{16\pi^2} \frac{\alpha' Q^2 p_\perp^2}{[\alpha(1-\alpha)]^2 N^4} f^{\mathcal{A}}(\alpha' N^2, k_\perp'), \tag{25}$$

$$\frac{d\sigma_T}{d\alpha dp_\perp^2 d\alpha' dk_\perp'^2} = \frac{\sum_q e_q^2 \alpha_{em}\alpha_s}{128\pi^2} \frac{\alpha' \left\{[\alpha^2 + (1-\alpha)^2]\,[p_\perp^4 + a^4] + 2p_\perp^2 m_q^2\right\}}{[\alpha(1-\alpha)]^4 N^4} f^{\mathcal{A}}(\alpha' N^2, k_\perp'), \tag{26}$$

$$f^{\mathcal{A}}(\alpha' N^2, k_\perp') = \int_{x_\perp} \left| \int \frac{d^2 k_\perp}{(2\pi)^2} \left(\delta^{ij} + \frac{2k_\perp^i k_\perp^j}{\alpha' N^2}\right) \frac{\mathrm{tr}\tilde{W}_{x_\perp}^{\mathcal{A}}(k_\perp' - k_\perp)}{\alpha' N^2 + k_\perp^2} \right|^2, \tag{27}$$

$$N^2 = Q^2 + \frac{p_\perp^2 + m_q^2}{\alpha(1-\alpha)}. \tag{28}$$

A heavy quark mass or high-$p_\perp$ ensures that the $q\bar{q}$ pair stays small in transverse space and behaves like a gluon terms of colour. The superscript $\mathcal{A}$ is used since we are effectively testing the proton's field with two "gluons". The factor in round brackets and the denominator factor in Eq.(27) reflect the spin and propagator of the wee gluon, respectively. Integration over the final state variables of the wee gluon in the leading $\ln(1/x)$ approximation gives

$$\frac{d\sigma_L}{d\alpha dp_\perp^2} = \frac{\sum_q e_q^2 \alpha_{em}\alpha_s \ln(1/x) h_{\mathcal{A}}}{2\pi^3 (a^2 + p_\perp^2)^4} [\alpha(1-\alpha)]^2 Q^2 p_\perp^2, \tag{29}$$

$$\frac{d\sigma_T}{d\alpha dp_\perp^2} = \frac{\sum_q e_q^2 \alpha_{em}\alpha_s \ln(1/x) h_{\mathcal{A}}}{16\pi^3 (a^2 + p_\perp^2)^4} \left[(\alpha^2 + (1-\alpha)^2)\,(p_\perp^4 + a^4) + 2p_\perp^2 m_q^2\right], \tag{30}$$

$$h_{\mathcal{A}} = \int_{y_\perp} \int_{x_\perp} \frac{\left|\mathrm{tr} W_{x_\perp}^{\mathcal{A}}(y_\perp)\right|^2}{y_\perp^4}. \tag{31}$$

In the wee massless fermion case we have

$$\frac{d\sigma_L}{d\alpha dp_\perp^2 d\alpha' dk_\perp'^2} = \frac{4\Sigma_q e_q^2 \alpha_{em}\alpha_s}{9\pi^2} \frac{Q^2}{[\alpha(1-\alpha)]^2 N^4} f^{\mathcal{F}}(\alpha' N^2, k_\perp'), \tag{32}$$

$$\frac{d\sigma_T}{d\alpha dp_\perp^2 d\alpha' dk_\perp'^2} = \frac{\Sigma_q e_q^2 \alpha_{em}\alpha_s}{9\pi^2 p_\perp^2 N^4} \left[N^4 - 2Q^2(N^2 + Q^2) + \frac{N^4 + Q^4}{\alpha(1-\alpha)}\right] f^{\mathcal{F}}(\alpha' N^2, k_\perp'), \tag{33}$$

where the functional $f^{\mathcal{F}}$ is given in Eq.(20). Now $\alpha', k_\perp'$ refer to the final state variables of the slow quark (anti-quark) and $\alpha, p_\perp$ to the fast anti-quark (quark). Since the fast anti-quark and gluon stay very close together in transverse space, in terms of colour they behave like an anti-quark. Integration gives

$$\frac{d\sigma_L}{d\alpha dp_\perp^2} = \frac{16\Sigma_q e_q^2 \alpha_{em}\alpha_s}{27\pi^3}\frac{Q^2}{[\alpha(1-\alpha)]N^6}h_{\mathcal{F}}, \tag{34}$$

$$\frac{d\sigma_T}{d\alpha dp_\perp^2} = \frac{4\Sigma_q e_q^2 \alpha_{em}\alpha_s}{27\pi^3 N^6 p_\perp^2}\left[N^4 - 2Q^2(N^2+Q^2) + \frac{(N^4+Q^4)}{\alpha(1-\alpha)}\right]h_{\mathcal{F}}, \tag{35}$$

with $h_{\mathcal{F}}$ defined like $h_{\mathcal{A}}$.

Comparing Eqs.(29,30) with Eqs.(34,35) we see that the configurations with a wee gluon are enhanced by $\ln(1/x)$ at small x relative to those with a wee fermion. In addition there may be an additional large suppression of the slow fermion contributions due to colour factors ($h_{\mathcal{A}} \approx 16 h_{\mathcal{F}}$ [8]). As a result we claim that the configurations with a wee gluon dominate over those with a wee fermion in the small-x region relevant to diffraction and we shall ignore the latter from now on. In contrast to the $q\bar{q}$ case the diffractive production of high-$p_\perp$ jets and charm is now 'soft' and relies on the non-perturbative interaction of the wee gluon with the proton.

3.3. Boost to the Breit frame

We now consider what happens when one boosts from the Proton's rest frame to the Breit frame (in which q = (0,0,0,Q) and the proton is fast). As an example we will consider the $q\bar{q}g$ fluctuation with a wee gluon. The general relations for the boost along the z-axis, in light cone co-ordinates of Eqs.(4-7), are

$$a'^{+} = \gamma(1-\beta)a^{+} \quad ; \quad a'^{-} = \gamma(1+\beta)a^{-}. \tag{36}$$

For this particular boost

$$\gamma(1-\beta) = \frac{m_p x}{|Q|} = \frac{1}{\gamma(1+\beta)}. \tag{37}$$

The typical four momentum of the slow gluon before interaction is

$$k = (\frac{\Lambda^2}{m_p x}, -\Lambda x, \vec{k}^\perp), \tag{38}$$

where, crucially, its '-' component is *negative*. The boost merely rescales the components to give

$$k' = (\frac{\Lambda^2}{|Q|}, -\frac{\Lambda|Q|}{m_p}, \vec{k}^\perp). \tag{39}$$

This negative energy outgoing gluon may then be reinterpreted as a positive energy *incoming* parton in the proton [10]. In contrast, the final state wee gluon is forced to

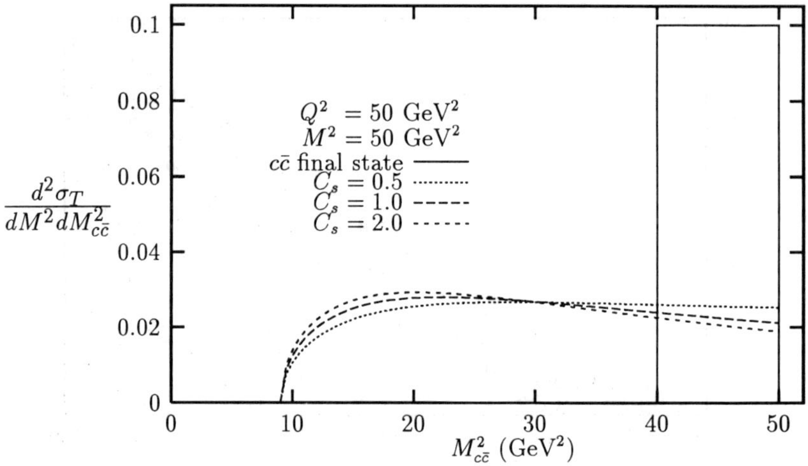

Fig. 4. Normalized mass spectra for the $c\bar{c}g$ and $c\bar{c}$ final states calculated in the semiclassical approach

First results on open charm production in diffraction for the 1994 running period were presented by the H1 collab., at Warsaw [11]. Both H1 and ZEUS have recently presented results on open charm and high-$p_\perp$ jets in diffractive DIS [12]. Given the increase in statistics of HERA for the 1996 running period, it is hoped that these phenomenological tests may be performed very soon.

An additional means of distinguishing the underlying diffractive mechanism is the energy dependence of the two processes. In the two gluon model a steeply rising gluon density, taken from a fit to F_2 for example, produces a rise in diffractive charm events that is twice as steep with energy. In contrast the semiclassical approach the energy dependence is flat, at least at this order, corresponding to a classical bremstrahl spectrum of gluons.

5. References

1. M. L. Good and W. D. Walker *Phys. Rev.* **120** (1960) 1857, 1860.
2. J. Collins *et al*, *Phys. Rev.* **D51** (1995) 3182.
3. H. I. Miettinen and J. Pumplin *Phys. Rev.* **D18** (1978) 1696.
4. J. R. Forshaw and D. A. Ross, *Quantum Chromodynamics and the Pomeron*, (CUP, Cambridge, 1997).
5. E. M. Levin, A. D. Martin, M. G. Ryskin, and T. Teubner, hep-ph/9606443;
   M. Genovese, N. N. Nikolaev, B. G. Zakharov, *Phys. Lett.* **B378** (1996) 347;
   H. Lotter, preprint, DESY 96-260, hep-ph/9612415;
   M. Diehl, preprint, CPTH-S492-0197, hep-ph/9701252.

6. N. N. Nikolaev and B. G. Zakharov, *J. Exp. Theor. Phys.* **78** (1994) 598; M. Diehl, *Z. Phys.* **C66** (1995) 181; J. Bartels, H. Lotter and M. Wüsthoff, *Phys. Lett.* **B379** (1996) 239; J. Bartels and M. Wüsthoff, *J. Phys.* **G22** (1996) 929; M. Wüsthoff, preprint, ANL-HEP-PR-97-03, hep-ph/9702201.
7. W. Buchmüller and A. Hebecker, *Nucl. Phys.* **B476** (1996) 203; W. Buchmüller, M. F. McDermott and A. Hebecker, *Nucl. Phys.* **B487** (1997) 283; *erratum ibid.*
8. W. Buchmüller, M. F. McDermott, and A. Hebecker, *Phys. Lett.* **B404** (1997) 353.
9. W. Buchmüller, M.F. McDermott, and A. Hebecker, hep-ph/9706354.
10. A. Hebecker, preprint DAMTP-97-10, hep-ph/9702373.
11. J Phillips, H1. Collab., *Proc. of XXVIII ICHEP* (Warsaw, 1996), Vol 1, 623.
12. C. Cormack, H1. Collab., T. Terron, ZEUS Collab., Talks at DIS 97, Chicago, Il., April 1997.

List of participants

Silvia Arcelli	U Bologna	arcelli@vxcern.cern.ch
William Bardeen	FNAL Batavia	bardeen@rdiv.fnal.gov
Jochen Bartels	U Hamburg	bartels@x4u2.desy.de
Riccardo Brugnera	U Padua	brugnera@vxdesy.desy.de
Jonathan Butterworth	UC London	jmb@hep.ucl.ac.uk
Matteo Cacciari	DESY Hamburg	cacciari@x4u2.desy.de
Tancredi Carli	MPI Munich	h01rtc@rec06.desy.de
Marcello Ciafaloni	U Florence	ciafaloni@fi.infn.it
Carsten Coldewey	Weizmann Rehovot	coldewey@mail.desy.de
Geoffrey Court	U Liverpool	court@hermes.desy.de
Jaroslav Cvach	ASCR Prague	cvach@mail.desy.de
Karin Daum	U Wuppertal	daum@mail.desy.de
Markus Diehl	CPT Palaiseau	diehl@orphee.polytechnique.fr
Stephan Egli	U Zurich	egli@physik.unizh.ch
Laurent Favart	LAL Orsay	favart@lal.in2p3.fr
Michel Fontannaz	LPTHE Orsay	fontanna@qcd.th.u-psud.fr
Dirk Graudenz	PSI Villigen	dirk.graudenz@cern.ch
Günter Grindhammer	MPI Munich	guenterg@desy.de
Jutta Hartmann	McGill Montreal	eisenmann@vxdesy.desy.de
Robert Klanner	U Hamburg	klanner@zow.desy.de
Michael Klasen	DESY Hamburg	klasen@mail.desy.de
Bernd Kniehl	MPI Munich	kniehl@vms.mppmu.mpg.de
Michael Krämer	RAL Chilton	michael.kraemer@rl.ac.uk
Gustav Kramer	U Hamburg	kramer@mail.desy.de
Michael Kuhlen	MPI Munich	kuhlen@desy.de
Zoltan Kunszt	ETH Zurich	kunszt@itp.phys.ethz.ch
Bodo Lampe	MPI Munich	bol@mppmu.mpg.de
Alan Martin	U Durham	a.d.martin@durham.ac.uk
Martin McDermott	DESY Hamburg	mcdermot@mail.desy.de
Beate Naroska	U Hamburg	naroska@mail.desy.de
Wolfgang Ochs	MPI Munich	wwo@mppmu.mpg.de
Emmanuel Paschos	U Dortmund	paschos@hal1.physik.uni-dortmund.de
Robert Roosen	U Brussels	roosen@mail.desy.de
Reinhold Rückl	U Würzburg	rueckl@physik.uni-wuerzburg.de
Renate Saffert	MPI Munich	rel@mppmu.mpg.de
Gavin Salam	INFN Milan	salam@mi.infn.it
Yves Sirois	LPNHE Palaiseau	sirois@polhp2.in2p3.fr
Jack Smith	Stony Brook/DESY	jacksmit@mail.desy.de
Stefan Söldner-Rembold	U Freiburg	soldner@ruhpa.physik.uni-freiburg.de
Werner Vogelsang	CERN Geneva	werner.vogelsang@cern.ch
Andreas Vogt	U Würzburg	avogt@physik.uni-wuerzburg.de
Albrecht Wagner	DESY Hamburg	wagnera@desy.de